IUTAM Symposium on Nonlinearity and Stochastic Structural Dynamics

SOLID MECHANICS AND ITS APPLICATIONS
Volume 85

Series Editor: G.M.L. GLADWELL
Department of Civil Engineering
University of Waterloo
Waterloo, Ontario, Canada N2L 3GI

Aims and Scope of the Series

The fundamental questions arising in mechanics are: *Why?*, *How?*, and *How much?* The aim of this series is to provide lucid accounts written by authoritative researchers giving vision and insight in answering these questions on the subject of mechanics as it relates to solids.

The scope of the series covers the entire spectrum of solid mechanics. Thus it includes the foundation of mechanics; variational formulations; computational mechanics; statics, kinematics and dynamics of rigid and elastic bodies: vibrations of solids and structures; dynamical systems and chaos; the theories of elasticity, plasticity and viscoelasticity; composite materials; rods, beams, shells and membranes; structural control and stability; soils, rocks and geomechanics; fracture; tribology; experimental mechanics; biomechanics and machine design.

The median level of presentation is the first year graduate student. Some texts are monographs defining the current state of the field; others are accessible to final year undergraduates; but essentially the emphasis is on readability and clarity.

For a list of related mechanics titles, see final pages.

IUTAM Symposium on

Nonlinearity and Stochastic Structural Dynamics

Proceedings of the IUTAM Symposium
held in Madras, Chennai, India
4–8 January 1999

Edited by

S. NARAYANAN
Department of Applied Mechanics,
Indian Institute of Technology,
Madras, India

and

R.N. IYENGAR
Department of Civil Engineering,
Indian Institute of Science, Bangalore, India
(Formerly Director, CBRI, Roorkee)

SPRINGER SCIENCE+BUSINESS MEDIA, B.V.

A C.I.P. Catalogue record for this book is available from the Library of Congress.

ISBN 978-94-010-3808-9 ISBN 978-94-010-0886-0 (eBook)
DOI 10.1007/978-94-010-0886-0

Printed on acid-free paper

Scientific Committee

Prof. S.T. Ariaratnam, University of Waterloo, Waterloo, Canada
Prof. V.V. Bolotin, Moscow Energetic Institute, Moscow, Russia
Prof. R.N. Iyengar, (Chairman), Central Building Research Institute, Roorkee, India
Prof. Y.K. Lin, Florida Atlantic University, Boca Raton, USA
Prof. A. Naess, Norwegian Institute of Technology, Trondheim, Norway
Prof. S. Narayanan, (Chairman), Indian Institute of Technology, Madras, Chennai, India
Prof. J.B. Roberts, University of Sussex, Brighton, UK
Prof. W. Schiehlen, University of Stuttgart, Stuttgart, FRO
Prof. F. Ziegler, University of Vienna, Vienna, Austria

National Advisory Committee

Dr. R.A. Mashelkar, Director general, CSIR, New Delhi
Prof. R. Narasimha, Director, Centre for Advanced Studies, Bangalore
Prof. R. Natarajan, Director, Indian Institute of Technology, Madras, Chennai
Prof. N.C. Nigain, Vice Chancellor, Roorkee University, Roorkee
Porf. R.N. Iyengar, Director, Central Building Research Insititute, Roorkee
Prof. S. Narayanan, Indian Institute of Technology, Madras, Chennai

Local Organising Committee

Dr. P. Chandraamouli, IIT Madras
Prof. Megha Singh, IIT Madras
Dr. A. Meherprasad, IIT Madras
Prof. S. Narayanan, IIT Madras
Prof. B.S. Prabhu, IIT Madras
Prof. M.R. Pranesh, IIT Madras
Prof. V. Ramamurti, IIT Madras
Dr. C. Sujatha, IIT Madras
Prof. C.P. Vendhan, IIT Madras

SYMPOSIUM SPONSORS

- International Union of Theoretical and Applied Mechanics (IUTAM)
- Council of Scientific and Industrial Research (CSIR), India
- Indian National Science Academy (INSA), India
- Council on Science and Technology in Developing Countries (COSTED)
- Indian Institute of Technology (IIT), Madras
- Central Building Research Institute (CBRI), Roorkee
- Indian Space Research Organisation (ISRO), Bangalore

CONTENTS

Preface

The IUTAM Symposium on Nonlinearity and Stochastic Structural Dynamics held in IIT, Madras, Chennai from January 4-8, 1999 is the sixth in the series of IUTAM symposia devoted to the topic. The earlier IUTAM Symposia on nonlinear problems in stochastic structural dynamics were organized in Southampton, UK (1976), Frankfurt / Oder, Germany (1982), Innsbruck (1987), Torino (1991) and Trondheim (1995).

The Symposium in Madras provided a forum and opportunity for international scientists and scholars to assemble, present and discuss important and recent developments in this field and exchange ideas between them.

A scientific committee was appointed by the bureau of IUTAM, which identified a number of active research workers in this area all over the world and invitations were sent to them for participation, and presenting papers at the Symposium and 34 of them accepted the invitation and sent abstracts of the papers to be presented. In the end 24 papers were presented at the Symposium in the lecture sessions. These papers are collected in this volume. There were in all 42 participants from 10 countries, namely Canada, China, Germany, India, Italy, Norway, Netherlands, UK, USA and Vietnam.

The lecture sessions were arranged according to the following topics

Bifurcation and Stochastic stability
Nonlinear system response
System identification
Maps and Hamiltonian systems
Solution of Fokker-Plank equation
Linearization and averaging
Simulation and stochastic nonlinear systems
Chaos
Reliability and stochastic parameter variation
Flow induced vibrations and other applications

Since there were overlaps in topics between the presentations, the papers in this volume are arranged in alphabetical order of the first author of the papers. The papers covered a wide spectrum of expertise of high analytical rigour and sound practical application. Not only did the presentations focus on nonlinearity in stochastic systems but also towards understanding intrinsic nonlinear phenomena like bifurcation and chaos in deterministic systems. In tune with the philosophy of IUTAM symposia enough scope and time were given for all paper presentations and discussions on them. We hope that the presentations and discussions during the Symposium will most definitely stimulate further research in the challenging area of nonlinearity and stochastic structural dynamics and that the proceedings of the Symposium will help in this regard.

There was a formal inaugural function at the start of the Symposium with an inaugural address by Prof. N.C. Nigam and welcome address by Prof. W. Schiehlen, the president of IUTAM, both of which are included in this volume. On the first day of the Symposium there was a cultural evening with a classical dance performance by Ms. C. Sailaja followed by the conference dinner. The Symposium banquet was held on the second day of the conference in the Trident with Prof. Y.K. Lin giving the banquet speech. A one day excursion on the third day to the neighbouring historic port town of Mahabalipuram and the temple town of Kancheepuram was organized which was enjoyed by all the participants. A group photo of the conference participants taken on the occasion with the backdrop of the motif of the Five Rathas is also printed in this volume. There was a concluding session in which Prof. W. Wedig summarized the contributions made during the Symposium and alluded to the success of the Symposium.

Apart from the Scientific Committee, to assist in the organizing of the Symposium a National Advisory Committee and a Local Organizing Committee were constituted. While the Scientific Committee members helped in identifying potential participants for the Symposium and reviewing the abstracts and full papers, the National Advisory Committee provided valuable advice from time to time in the conduct of the Symposium and the local organizing committee bore the brunt of the whole organization. The success of the Symposium would not have been possible without the excellent and able support of all these committees and especially the Local Organizing Committee. While expressing our sincere thanks to these committees for their support on behalf of these committees we would like to thank the IUTAM Bureau for promoting and sponsoring the Symposium. We would also like to thank the IUTAM for their travel support to a number of foreign participants as well as to the other sponsors of the Symposium for their financial support which made this Symposium possible. The contributions of the authors and participants in the Symposium provided the most important inputs and were instrumental in making the Symposium a success and produced this volume. We deeply appreciate their efforts.

In editing this volume the help received from Dr. M. Ramesh and Mr. V. Balamurugan is gratefully acknowledged. We would also like to thank Kluwer for their cooperation in bringing out this volume.

S. Narayanan **R.N. Iyengar**

Welcome Address of the President of IUTAM

Professor Nigam,
Professor Natarajan,
Dear Professors Iyengar and Narayanan,
Dear Colleagues from all over the world,
Ladies and Gentlemen,

It is my honor and pleasure to welcome all of you on behalf of the International Union of Theoretical and Applied Mechanics. We are very happy to be invited to the Indian Institute of Technology Madras here in Chennai, the capital of the State of Tamil Nadu. Let me use this opportunity for a short look on the past and present activities of IUTAM.

In the field of mechanics organized meetings between scientists were initiated 77 years ago, namely in 1922, when Prof. Theodore von Karman and Prof. Tullio Levi-Civita organized the world's first conference, in hydro- and aeromechanics. Two years later, in 1924, the First International Congress encompassing all fields of mechanics, that means analytical, solid and fluid mechanics, including their applications, was held in Delft, The Netherlands. From then on, with exception of the year 1942, International Congresses in Mechanics have been held every four years.

The disruption of international scientific cooperation caused by the Second World War was deeper than that caused by the First World War, and the need for reknotting ties seemed stronger than ever before when the mechanics community reassembled in Paris for the Sixth Congress in 1946. Under these circumstances, at the Sixth Congress in Paris, it seemed an obvious step to strengthen bonds by forming an international union, and as a result IUTAM was created and statutes were adopted. Then, the next year, in 1947, the Union was admitted to ICSU, the International Council for Science. This Council coordinates activities among various other scientific unions to form a tie between them and the United Nations educational, Scientific and Cultural Organization, well known as UNESCO.

Today, IUTAM forms the international umbrella organization of nearly 50 national Adhering Organizations of mechanics from nations all over the world. Furthermore, a large number of inter- national scientific organizations of general or more specialized branches of mechanics are connected with IUTAM as Affiliated Organizations. As a few examples, let me mention: the European Mechanics Society (EUROMECH), the International Association of Computational Mechanics (IACM) which held its Fourth World Congress on Computational Mechanics last year in Buenos Aires, Argentina and the International Association for Vehicle System Dynamics (IAVSD).

Within IUTAM the only division used so far is related to solid and fluid mechanics as indicated by our two Symposia Panels. But recently six Working Groups have been established by the General Assembly devoted to specific areas of mechanics. These areas are:

"Mechanics of Non-Newtonian Fluids",

"Dynamical Systems",

"Fracture Mechanics and Damage",

"Mechanics of Materials",

"Electromagnetic Processing",

"Computational Mechanics"

It is expected that in the long run the Working Groups will be developed into Standing Committees. In addition, other specific areas of mechanics may be identified to support the international cooperation in more branches of mechanics.

IUTAM carries out an exceptionally important task of scientific co-operation on mechanics on the international scene. Each national Adhering organisation of IUTAM, like the Indian National Science Academy (INSA), is represented by a number of scientists in IUTAM's General Assembly. In particular, the Indian delegates with IUTAM are

Professor A.K.,Mallik from IIT Kanpur,
Professor M.L. Munjal froiu IIS Bangalore,
Professor S.P. Sukhatme from IIT Mumbai.

I would like to mention that INSA held its 64th Annual General Meeting here in Chennai last week.

Mechanics is a well-developed science in India represented at most universities and national laboratories. Since 1949 there have been held more than 200 IUTAM symposia worldwide. Out of them 2 IUTAM symposia where organized in India, and both Symposia were dedicated to fluid dynamics. As early as 1977 the IUTAM Symposium on Monsoon Dynamics took place at the IIT Delhi in cooperation with the International union of Geodesy and Geophysics (IUGG). And today we are opening the first IUTAM Symposium in India related to solid mechanics.

As I mentioned before, IUTAM organizes international Congress and Symposia all over the world. Thus, the Twentieth International Congress of Theoretical and Applied Mechanics will be held in Chicago, USA, from 27 August - 2 September 2000, what means in one years from now. First Announcements of this forthcoming congress are already widely distributed and published in many scientific journals, and in due time, a final announcement of the congress will be available to all who have responded to the first announcement.

The present Symposium is exceptionally interesting because it deals with an important topic of mechanics. The Symposium includes Nonlinear System Responses and Simulation, Bifurcation, Chaos and Stochastic Stability, Identification and Reliability. IUTAM found that the proposal of Professors Iyengar and Narayanan for such a symposium was not only very timely, but also very well founded in the research carried out in India, and the proposal for the Symposium was readily accepted and granted by the General Assembly of IUTAM. This is the sixth IUTAM Symposium devoted to Nonlinear Dynamics after the Symposium held 1995 in Trondheim, Norway and earlier Symposia organized in Southampton (76), Frankfurt/Oder (82), Innsbruck (87) and Torino (91) . Thus, there is no doubt that IUTAM considers Stochastic Dynamics as an essential field of mechanics.

On behalf of IUTAM, I wish to express my sincere thanks to the Indian Institute of Technology, in particular to its Director, Prof. Natarajan, for the invitation to host this significant scientific event, and I thank all the invited participants for their readiness to come and to contribute to the success of the Symposium by very active participation in the lectures and the scientific discussions, as well as in the social program.

Finally, I would like to mention that to sponsor a scientific meeting is one thing, to organize one is another. A heavy burden is placed on the shoulders of the Chairmen who are in charge of the scientific and the practical local arrangements, and of the associates who are assisting him. All who have tried this before know perfectly well how much work has to be done in organizing a meeting like this one.

Thus, we should be thankful, not only to the International Scientific Committee, but also very much to the Local Organizing Committee, and in particular to the Chairman of both Committees. Professor S. Narayanan, who carried the heaviest load and responsibility. And even the date of our Symposium is chosen very well in the most pleasant period of the year here in Chennai.

It is up to us now, Ladies and Gentlemen, to harvest the fruits of the organizers' work. Let us contribute our share to make this IUTAM Symposium a meeting that will be long remembered as a very successful one!

On behalf of IUTAM, I greet you all and wish you great success!

Werner Schiehlen

Inaugural Address

I deem it an honour to be invited to inaugurate the IUTAM Symposium and to address the distinguished participants. The group of people, across the world, who work in the field of random vibration, constitute a small and intimate community. It has been my pleasure to be a part of this community for nearly three decades. We are delighted to host this symposium, and I join Prof. Iyengar in welcoming you to the Symposium and to India.

It is appropriate that this symposium is being held at IIT, Madras. Teaching of random vibration was started in India at IIT Madras, in mid 60's, by Prof. Hans Wagner while he was a Visiting Professor from Germany. The inauguration of the Symposium, this morning, is of special significance for me. I joined IIT, Kanpur in November 1968, on return from Caltech, and introduced the graduate course on random vibration in the Department of Aeronautical Engineering. It is exactly three decades to the date since the first lecture was delivered in this course. Professors S. Narayanan and C.P. Vendhan, now at IIT, Madras were both in this class. In 1988, I had the opportunity to teach a course on random vibration at IIT, Madras. I am happy that two students, among those who attended the course, are present at this symposium.

Uncertainty is inherent in almost all natural and manmade phenomena. For a long time, uncertainty was treated in engineering practice through concepts of factor of safety or load factor, which are essentially a cover-up of our ignorance based on engineering judgement. Advances in probability theory and random processes made it possible to treat uncertainty in a rational and consistent manner. The first applications of probability theory and random processes were made in Physics and Astronomy, and later in the Communication Theory. A robust mathematical framework, and applicable body of knowledge, became available by the middle of this century. Around this time, the Aerospace Engineering was in the midst of a bold initiatives in which failure, or malfunctioning, had serious consequences. A high level of reliability and minimisation of weight, therefore, became a major engineering challenge. Random vibration came to be recognised as a distinct specialisation around this time.

In 1958, Prof S.H. Crandall organised a Summer School at MIT, The School covered the state-of-art at that time in the area of random vibration and included: response of discrete linear systems subjected stationary random excitation; and design and testing of systems for random vibration environment. Five years later, in 1963, Prof. Crandall organised the second Summer School, which covered the response of linear and non-linear systems to random excitation, and application of random vibration theory to the design and testing of aerospace systems. In the preface to the second volume Prof. Crandall wrote:

> "The five years that have elapsed between the volumes have been active ones in the development of random vibrations. There have been significant theoretical advances, and there has been an accumulation of considerable practical experience. This progress is reflected in the contents of this second volume as compared with its predecessor".

The lecture notes of the two Summer Schools were published by The MIT Press as **Random Vibration, Volumes I** and **II**. The two volumes cover the state of knowledge in the theory and application of the random vibration in early 60's.

The decade of 60's was a period of rapid advances in random vibration. Three persons : Professors S.H. Crandall, T.K. Caughey, and V.V. Bolotin played a seminal role in this progress. Significant advances were made in determining closed form and approximate response of non-linear systems to random excitation, analytical treatment of fatigue, creep and plastic drift of systems, testing and reliability based design of systems operating in random vibration environment and measurement and analysis of random data.

Three text books : **Random Vibration in Mechanical Systems** by Crandall and Mark (1963); **An Introduction to Random Vibration** by Robson (1963); and **Statistical Method in Structural Mechanics** Bolotin (1965) contain the state-of-the-art in the field of of random vibration in early 60's. In 1967, Prof. Y.K. Lin, who is in the audience, authored a treatise on random vibration, entitled : **Probabilistic Theory of Structural Dynamics** which made it possible for a generation of students and research workers to understand and apply random vibration to a wide range of engineering problems. Several books on random vibration have been published in 80's and 90's covering the recent advances in theory and applications.

The advances in random vibration since the decade of 60's have been incremental covering the non-gaussian, non-stationary response, multi-dimensional random processes, non-linear systems, stochastic stability analysis, and stochastic differential equations. Advances have also been made in the areas of statistical energy analysis, parametric excitations, evolutionary random processes and modelling and simulation through Monte Carlo methods. Applications of random vibration have been extended to earthquake and wind engineering, ocean waves, and vehicle motion on tracks.

Recent work in the field of random vibrations covers response of non-linear systems determined through Fokker Planck equation, closure techniques, stochastic averaging methods, system identification, and simulation through Monte Carlo Methods and chaos. This symposium covers all these aspects. I am sure the deliberations of the symposium would stimulate many new ideas for the future work in the field of random vibration.

With these remarks, I am happy to inaugurate the symposium.

N.C. Nigam

Participants of the IUTAM Symposium on Nonlinearity and Stochastic Structural Dynamics,
held in Madras, Chennai, India, 4-8- January 1999.

Erratum

IUTAM Symposium on Nonlinearity and Stochastic Structural Dynamics
held in Madras, Chennai, India, 4–8 January 1999

S. Narayanan, R.N. Iyengar

ISBN 0-7923-6733-2

On page V, Prof. N.C. Nigain should be Prof. N.C. Nigam

P.T.O.

HIGHER ORDER AVERAGING METHOD OF COEFFICIENTS IN FOKKER - PLANCK EQUATION

N.D. ANH, N.D. TINH
Institute of Mechanics
224 Doi Can, Hanoi Vietnam

Abstract. Since the effect of some nonlinear terms is lost during the first order averaging procedure, the higher order stochastic averaging method is developed to predict approximately the response of linear and lightly nonlinear systems subject to weakly external excitation of wideband or narrow-band coloured noise random processes. The excitation of second order coloured noise is considered in more detail. It is shown that in the case of wideband random excitation the result obtained is different from the conventional replacement of wideband random excitation by an white noise excitation.

1. Introduction

For many years the well-known averaging method, originally given by Krylov and Bogoliubov and then developed by Mitropolskii (Bogoliubov and Mitropolskii, 1961) has proved to be a very powerful approximate tool for investigating deterministic weakly nonlinear vibration problems. The advantage of this method is that it reduces the dimension of the response coordinates. In the field of random vibration the averaging method was extended by Stratonovich (1963) and has a mathematically rigorous proof by Khasminskii (1963). At present, the stochastic averaging method (SAM) is widely used in different problems of stochastic mechanics such as vibration, stability and reliability problems (see e.g. Ariaratnam & Tam,1979; Bolotin, 1984; Ibrahim, 1985; Lin & Cai, 1995; Roberts & Spanos, 1986; Zhu, 1988).

It should be noted that principally only first order SAM has been applied in practice and usually to systems subject to white noise or wideband random processes. It is well-known, however, the effect of some non-linear terms is lost during the first order averaging procedure. In order to over come this insufficiency, different averaging procedures for obtaining approximate solutions have been developed (see e.g. Mitropolskii et al, 1992; Red-Horse & Spanos, 1992; Sri Namachchivaya & Lin, 1988; Zhu & Lin, 1994; Zhu et al, 1997). Recently, a higher order averaging procedure using Fokker-Planck (FP) equation was developed in (Anh, 1993, 1995) and then applied to Van der Pol oscillator under white noise excitation (Anh & Tinh, 1995). In the paper this procedure is further developed to linear and lightly nonlinear systems subject to weakly external excitation of wideband or narrow-band colored noise random processes.

S. Narayanan and R.N. Iyengar (eds.),
IUTAM Symposium on Nonlinearity and Stochastic Structural Dynamics, 1–14.

2. Coloured noise random processes

We consider a stationary coloured noise random process c(t) which is the result of the passage of a white noise though the linear forming filter L of order 2n :

$$L(c(t)) \equiv \frac{d^{2n}c(t)}{dt^{2n}} + \sum_{j=0}^{2n-1} \alpha_j \frac{d^j c(t)}{dt^j} = \sigma\dot{\xi}(t) \tag{1}$$

where $\alpha_{j,}$, σ are constants, $\dot{\xi}(t)$ is a the zero-mean white noise process with unit intensity

$$E(\dot{\xi}(t)\,\dot{\xi}(t+\tau)) = \delta(\tau) \tag{2}$$

where E(.) is the operator of expectation. It is supposed that all roots of the characteristic equation for the filter (1)

$$l(\mu) \equiv \mu^{2n} + \sum_{j=0}^{2n-1} \alpha_j \mu^j = 0 \tag{3}$$

have negative real parts. The behaviour of c(t) essentially depends on the roots of the characteristic equation (3) and one can get from (1) wideband or narrow-band coloured noise processes, respectively.

2.1. WIDEBAND COLOURED NOISE

Let the filter L in (1) can be expressed in the form

$$L(.) = \prod_{k=1}^{n} \left[\frac{d^2(.)}{dt^2} - 2\lambda_k \frac{d(.)}{dt} + \omega_k^2(.) \right] \tag{4}$$

where

$$\lambda_k < 0;\ \omega_k,\ |\lambda_k| >> \varepsilon \tag{5}$$

ε is a small positive parameter. The spectral density function of the process c(t) is to be found

$$S_c(\omega) = \frac{\sigma^2}{2\pi \prod_{k=1}^{n} \left[\left(\omega^2 - \omega_k^2\right)^2 + 4\lambda_k^2 \omega^2 \right]} \tag{6}$$

By investigating the spectral density function of the process c(t) it is supposed that, for the present analysis, c(t) is considered to be a wideband coloured noise process if the condition (5) is satisfied.

2.2. NARROW- BAND COLOURED NOISE

Let the filter L can be expressed in the form

$$L(.) \equiv L_0(.) + \varepsilon L_1(.) \equiv \prod_{k=1}^{m}\left[\frac{d^2(.)}{dt^2} + \omega_k^2(.)\right] + \varepsilon L_1(.) \tag{7}$$

where L_1 is also a linear filter of order q; m, q $\leq$ n; ω_k, k =1, 2,....., m, are distinct positive values and $\omega_k >> \varepsilon$. In this case it is supposed that one can get from (1), (7) a narrow- band coloured noise process.

3. Excitation of wideband coloured noise

Consider a single - degree - of - freedom lightly damping system subject to weakly external random excitation

$$\ddot{x} + \omega_0^2 x = \varepsilon f_1(x,\dot{x}) + \varepsilon^2 f_2(x,\dot{x}) + \sqrt{\varepsilon} c(t) \tag{8}$$

where the symbols have their customary meanings, f_1, f_2 are nonlinear functions of $x, \dot{x}$. The wideband coloured noise process c(t) is defined by filter (1), (4). Let a stationary solution of (8) exist. Eliminating c(t) from (8), (1) and using (4) gives

$$\prod_{k=1}^{n}\left[\frac{d^2(.)}{dt^2} - 2\lambda_k\frac{d(.)}{dt} + \omega_k^2(.)\right](\ddot{x} + \omega_0^2 x) = \varepsilon L f_1 + \varepsilon^2 L f_2 + \sqrt{\varepsilon}\sigma\xi(t) \tag{9}$$

The corresponding linear equation, (ε=0), has following solution

$$x(t) = \sum_{k=1}^{n} e^{\lambda_k t}\left(c_q \cos\sqrt{\omega_k^2 - \lambda_k^2}\, t + s_q \sin\sqrt{\omega_k^2 - \lambda_k^2}\, t\right) + a_0 \cos(\omega_0 t + \theta_0) \tag{10}$$

where c_q, s_q, a_0, θ_0 are constants. It is supposed that in the case of weakly nonlinear systems, ($\varepsilon \neq 0$), the summation in (10) will tend to zero for $t \to \infty$. Thus, an approximate expression for the stationary solution of (9) can be found in the form

$$\frac{d^j x(t)}{dt^j} \equiv x^{(j)}(t) = a_0 \omega_0^j \cos(\varphi_0 + j\frac{\pi}{2}), \; j = 0,1,..,2n+1 \tag{11}$$

where a_0, φ_0 are new slowly varying variables - amplitude and phase - and determined from the system of Ito stochastic differential equations

$$\dot{a}_0 = -\frac{\sqrt{2\pi S_c(\omega_0)}}{\sigma\omega_0}(\varepsilon\tilde{L}f_1 + \varepsilon^2\tilde{L}f_2)\sin(\varphi_0+\beta) + \frac{\varepsilon 2\pi S_c(\omega_0)}{2a_0\omega_0^2\sigma^2}\cos^2(\varphi_0+\beta) - \\ -\sqrt{\varepsilon}\frac{\sqrt{2\pi S_c(\omega_0)}}{\omega_0}\sin(\varphi_0+\beta)\dot{\xi}(t), \tag{12}$$

$$\dot{\varphi}_0 = \omega_0 - \frac{\sqrt{2\pi S_c(\omega_0)}}{a_0\omega_0\sigma}(\varepsilon\tilde{L}f_1 + \varepsilon^2\tilde{L}f_2)\cos(\varphi_0+\beta) - \frac{\varepsilon 2\pi S_c(\omega_0)}{a_0^2\omega_0^2\sigma^2}\sin 2(\varphi_0+\beta) - \\ -\sqrt{\varepsilon}\frac{\sqrt{2\pi S_c(\omega_0)}}{a_0\omega_0}\cos(\varphi_0+\beta)\dot{\xi}(t),$$

wherein

$$\tilde{L}f_v = L(f_v) \text{ with } x^{(j)}(t) \text{ replaced by (11), v=1, 2,}$$

$$\beta = \sum_{q=1}^{n}\beta_q\,, \quad \sin\beta_q = \frac{2\lambda_q\omega_0}{\sqrt{4\lambda^2{}_q\omega_0^2 + (\omega_q^2-\omega_0^2)^2}}$$

$$\cos\beta_q = \frac{\omega_q^2-\omega_0^2}{\sqrt{4\lambda_q^2\omega_0^2 + (\omega_q^2-\omega_0^2)}} \tag{13}$$

The well - known system of equations for amplitude and phase in the case of white noise excitation can be obtained from (12) by putting $\beta_q = 0$ and replacing $S_c(\omega_0)$ by the intensity of white noise. Hence, for the mechanical system subject to additive wideband coloured noise excitation, amplitude and phase equations are similar to the ones for the mechanical system subject to additive white noise excitation. Thus, a higher order averaging procedure as shown in (Anh, 1995) can be applied to (12). It is noted, however, the equations (12) are different from the ones obtained by the conventional replacement of wideband excitation by an white noise one.

4. Excitation of narrow-band coloured noise

Consider a single - degree - of - freedom lightly damping system subject to external random excitation

$$\ddot{x} + \omega_0^2 x = \varepsilon f_1(x.\dot{x}) + \varepsilon^2 f_2(x,\dot{x}) + c(t) \tag{14}$$

where the narrow-band coloured noise random process c(t) is obtained by the filter (1), (7) and $\sigma = \sqrt{\varepsilon}\sigma_1$. The excitation of narrow-band processes is considered in many papers (see e.g. Iyengar, 1991). Eliminating c(t) from (14), (1) and using (7) gives

$$\prod_{k=1}^{m}\left[\frac{d^2(.)}{dt^2} + \omega_k^2(.)\right](\ddot{x} + \omega_0^2 x) = \varepsilon F_1 + \varepsilon^2 F_2 + \varepsilon^3 .. + \sqrt{\varepsilon}\sigma_1\dot{\xi}(t) \tag{15}$$

where it is denoted

$$F_1 = L_0 f_1 - L_1(\ddot{x} + \omega_0^2 x) \ , \ F_2 = L_0 f_2 + L_1 f_1 \tag{16}$$

Suppose

$$\|\omega_s - \omega_\rho\| >> \varepsilon \ , \ s, \rho = 0,1,..,m. \tag{17}$$

According to the averaging concept, a stationary solution of (15) is found in the form

$$\frac{d^i x(t)}{dt^i} = a_0\omega_0^i \cos\left(\varphi_0 + i\frac{\pi}{2}\right) + \sum_{k=1}^{m} a_k\omega_k^i \cos\left(\varphi_k + i\frac{\pi}{2}\right), i = 0,1,..,2m+1. \tag{18}$$

where a_s, φ_s, s = 0, 1, .., m are new variables. By using Ito differentiation formula the equation (15) is transformed into the following system of equations:

$$\begin{aligned} \dot{a}_s &= \varepsilon A_{1s}(a,\varphi) + \varepsilon^2 A_{2s}(a,\varphi) + \varepsilon^3 + ... - \sqrt{\varepsilon}\,\sigma_1\frac{\sin\varphi_s}{\omega_s\Omega_s}\dot{\xi}(t) \\ \dot{\varphi}_s &= \omega_s + \varepsilon B_{1s}(a,\varphi) + \varepsilon^2 B_{2s}(a,\varphi) + \varepsilon^3 ... - \sqrt{\varepsilon}\,\sigma_1\frac{\cos\varphi_s}{a_s\omega_s\Omega_s}\dot{\xi}(t) \end{aligned} \tag{19}$$

where it is denoted

$$A_{1s}(a,\varphi) = \left\{-\frac{\tilde{F}_1(a,\varphi)}{\omega_s\Omega_s}\sin\varphi_s + \frac{\cos^2\varphi_s}{2a_s\omega_s\Omega_s^2}\right\}$$

$$B_{1s}(a,\varphi) = \left\{-\frac{\tilde{F}_1(a,\varphi)}{\omega_s\Omega_s a_s}\cos\varphi_s - \frac{\sin\varphi_s\cos\varphi_s}{\omega_s^2\Omega_s^2 a_s^2}\right\}$$

$$A_{2s}(a,\varphi) = -\frac{\tilde{F}_2(a,\varphi)}{\omega_s\Omega_s}\sin\varphi_s$$

$$B_{2s}(a,\varphi) = -\frac{\tilde{F}_2(a,\varphi)}{\omega_s\Omega_s a_s}\cos\varphi_s$$

$$\Omega_s = \prod_{\rho=0,\rho\neq s}^{m} (\omega_\rho^2 - \omega_s^2)$$

$$a = \{a_0, a_1, .., a_m\}, \; \varphi = \{\varphi_0, \varphi_1, .., \varphi_m\} \tag{20}$$

$$\widetilde{F}_v(a,\varphi) = F_v \text{ with } x^{(i)}(t) = \sum_{\rho=0}^{m} a_\rho \omega_\rho^i \cos\left(\varphi_\rho + i\frac{\pi}{2}\right), v = 1,2.$$

and F_v are defined in (16). The Fokker-Planck equation for the stationary probability density function W(a,φ) corresponding to the system (19) takes form

$$\sum_{s=0}^{m} \omega_s \frac{\partial W}{\partial \varphi_s} = -\varepsilon[A_1, B_1]L(W) - \varepsilon^2[A_2, B_2]L(W) + \varepsilon^3 ... \tag{21}$$

where the operators $[A_v, B_v]$ L(.), v = 1, 2, are defined as follows

$$[A_1, B_1]L(W) = \sum_{s=0}^{m} \left[\frac{\partial}{\partial a_s}(A_{1s}W) + \frac{\partial}{\partial \varphi_s}(B_{1s}W) \right] -$$
$$-\frac{\sigma_1^2}{2} \sum_{s=0}^{m} \sum_{\rho=0}^{m} \left\{ \frac{\sin\varphi_s \sin\varphi_\rho}{\omega_s \omega_\rho \Omega_s \Omega_\rho} \frac{\partial^2 W}{\partial a_s \partial a_\rho} + \frac{\sin\varphi_s \cos\varphi_\rho}{a_\rho \omega_s \omega_\rho \Omega_s \Omega_\rho} \frac{\partial^2 W}{\partial a_s \partial \varphi_\rho} + \right.$$
$$\left. + \frac{\cos\varphi_s \cos\varphi_\rho}{a_s a_\rho \omega_s \omega_\rho \Omega_s \Omega_\rho} \frac{\partial^2 W}{\partial \varphi_s \partial \varphi_\rho} \right\} \tag{22}$$
$$[A_2, B_2]L(W) = \sum_{s=0}^{m} \left[\frac{\partial}{\partial a_s}(A_{2s}W) + \frac{\partial}{\partial \varphi_s}(B_{2s}W) \right]$$

It can be shown that the operators $[A_v, B_v]$ L(.) are linear ones. We seek then the solution of (21) in the form

$$W(a,\varphi) = W_0(a,\varphi) + \varepsilon W_1(a,\varphi) + \varepsilon^2 W_2(a,\varphi) + ... \tag{23}$$

Substituting (23) into (21) yields

$$\sum_{s=0}^{m} \omega_s \left[\frac{\partial W_0}{\partial \varphi_s} + \varepsilon \frac{\partial W_1}{\partial \varphi_s} + \varepsilon^2 \frac{\partial W_2}{\partial \varphi_s} + ... \right] = -\varepsilon[A_1, B_1]L(W_0) -$$
$$- \varepsilon^2 \{[A_2, B_2]L(W_0) + [A_1, B_1]L(W_1)\} + ... \tag{24}$$

Comparing the coefficients of like powers of ε one obtains

$$\varepsilon^0: \qquad \sum_{s=0}^{m} \omega_s \frac{\partial W_0}{\partial \varphi_s} = 0, \tag{25}$$

$$\varepsilon^1: \qquad \sum_{s=0}^{m} \omega_s \frac{\partial W_1}{\partial \varphi_s} = -[A_1, B_1]L(W_0), \tag{26}$$

$$\varepsilon^2: \qquad .\sum_{s=0}^{m} \omega_s \frac{\partial W_2}{\partial \varphi_s} = -\{[A_2, B_2]L(W_0) + [A_1, B_1]L(W_1)\}. \tag{27}$$

From (25) it gives a periodic solution with respect to φ as follows

$$W_0 = W_0(a) \tag{28}$$

Substituting (28) into (26) yields

$$\sum_{s=0}^{m} \omega_s \frac{\partial W_1}{\partial \varphi_s} = -[A_1, B_1]L\big(W_0(a)\big) \tag{29}$$

The arbitrary integration function $W_0(a)$ must be chosen form the condition for the function $W_1(a, \varphi)$ to be periodic with respect to φ. Thus, one gets from (29)

$$\big\langle [A_1, B_1]L(W_0(a))\big] \big\rangle = 0 \tag{30}$$

where <.> is the averaging operator with respect to phase φ

$$<.>= \frac{1}{(2\pi)^{m+1}} \int_0^{2\pi}\int_0^{2\pi} \dots \int_0^{2\pi} (.)\, d\varphi_0 d\varphi_1 \dots d\varphi_m \tag{31}$$

Substituting (22) into (30) yields

$$\sum_{s=0}^{m} \left\{ \frac{\partial}{\partial a}\Big(< A_{1s} > W_0(a)\Big) - \frac{\sigma_1^2}{4\omega_s^2 \Omega_s^2} \frac{\partial^2 W_0(a)}{\partial a_s^2} \right\} = 0 \tag{32}$$

The averaged Fokker - Planck equation (32) can be considered as an extension into the case of narrow - band excitation of the well - known first order averaged FP equation obtained for the white noise excitation. Further, it is seen that the averaged FP equation (32) is obtained by equating zero the averaged value of the coefficient of power ε in the original FP equation (21). It should be noted that although the original FP equation (21) is reduced to the averaged one (30) the integration of (30) remains a difficult problem.

The second term $W_1(a, \varphi)$ in (23) is determined from (26) using Fourier expansion

$$[A_1, B_1]L(W_0(a)) = W_0(a)\sum_{k_0}\sum_{k_1}\ldots\sum_{k_m} C_{k_0k_1\ldots k_m}(a)\, e^{i(k_0\varphi_0+k_1\varphi_1+\ldots+k_m\varphi_m)} \tag{33}$$

where

$$C_{k_0k_1\ldots k_m}(a) = \frac{1}{W_0(a)(2\pi)^{m+1}}\int_0^{2\pi}\int_0^{2\pi}\ldots\int_0^{2\pi}[A_1, B_1]L(W_0(a)). \\ .e^{-i(k_0\varphi_0+k_1\varphi_1+\ldots k_m\varphi_m)}d\varphi_0 d\varphi_1\ldots d\varphi_m \tag{34}$$

Substituting (33) into (26) yields

$$W_1(a,\varphi) = W_0(a)\sum_{k_0}\sum_{k_1}\ldots\sum_{k_m}\frac{C_{k_0k_1\ldots k_m}(a)}{k_0\omega_0 + k_1\omega_1 + \ldots + k_m\omega_m}e^{k_0\varphi_0+k_1\varphi_1+\ldots+k_m\varphi_m} + \\ + W_{10}(a) \tag{35}$$

provided that

$$k_0\omega_0 + k_1\omega_1 + \ldots + k_m\omega_m \neq 0 \tag{36}$$

The arbitrary integration function $W_{10}(a)$ must be chosen from the condition for the function $W_2(a, \varphi)$ to be periodic. Analogously, one can find third term $W_2(a, \varphi)$ in (23).

5. Excitation of second order coloured noise

Coloured noise of second order p(t) can be obtained by the filter P of the form

$$P(p(t)) \equiv \ddot{p}(t) + 2\alpha\dot{p}(t) + \omega_1^2 p(t) = 2\sigma_2\sqrt{\alpha}\,\omega_1\,\dot{\xi}(t) \tag{37}$$

where $\alpha > 0$, ω_1, $\sigma_2 >> \varepsilon$. The auto-correlation function and the spectral density function of p(t) are, respectively,

$$R_p(\tau) = \sigma_2^2 e^{-\alpha\tau}\left(\cos\sqrt{\omega_1^2-\alpha^2}\,\tau + \alpha(\sqrt{\omega_1^2-\alpha^2})^{-1}\left|\sin\sqrt{\omega_1^2-\alpha^2}\,\tau\right|\right) \\ S_p(\omega) = \frac{2\sigma_2^2\alpha\omega_1^2}{\pi[(\omega^2-\omega_1^2)^2 + 4\alpha^2\omega^2]} \tag{38}$$

The bandwidth of the process p(t) is controlled by the value of the parameter α. For the present analysis the coloured noise p(t) is considered to be a wideband random process if $\alpha >> \varepsilon$,. If $\alpha \sim \varepsilon$, i.e.

$$\alpha = \varepsilon\alpha_1, \tag{39}$$

p(t) is a narrow - band process and the filter (37) has the form (7), where

$$P_0(.) = \frac{d^2(.)}{dt^2} + \omega_1^2(.) \ , \ P_1(.) = 2\alpha_1 \frac{d(.)}{dt} \tag{40}$$

Consider a linear system subject to external second order coloured noise excitation (37)

$$\ddot{x} + 2h\dot{x} + \omega_0^2 x = p(t) \tag{41}$$

The exact mean square displacement is to be found

$$E(x^2) = \frac{\sigma_2^2\left[\alpha\omega_1^2 + h\omega_0^2 + 4\alpha h(\alpha + h)\right]}{h\omega_0^2\left[(\omega_0^2 - \omega_1^2)^2 + 4(\alpha + h)(\alpha\omega_0^2 + h\omega_1^2)\right]} \tag{42}$$

and will be compared late with the corresponding approximate solutions.

5.1. WIDEBAND SECOND ORDER COLOURED NOISE

Consider a single - degree - of - freedom lightly damping system subject to weakly external wideband second order coloured noise excitation

$$\ddot{x} + \omega_0^2 x = \varepsilon f_1(x,\dot{x}) + \varepsilon^2 f_2(x,\dot{x}) + \sqrt{\varepsilon}p(t) \tag{43}$$

where p(t) is determined by (37) with $\alpha >> \varepsilon$. In the conventional manner one replaces the wideband random process p(t) by an white noise process using (38)

$$p(t) = \sqrt{2\pi S_p(\omega_0)}\dot{\xi}(t) = \frac{2\sigma_2\sqrt{\alpha}\omega_1\dot{\xi}(t)}{\sqrt{(\omega_0^2 - \omega_1^2)^2 + 4\alpha^2\omega_0^2}} \tag{44}$$

Thus, the conventional "equivalent" system subject to white noise corresponding to (43) is as follows

$$\ddot{x} + \omega_0^2 x = \varepsilon f_1(x,\dot{x}) + \varepsilon^2 f_2(x,\dot{x}) + \sqrt{\varepsilon}\sqrt{2\pi S_p(\omega_0)}\dot{\xi}(t) \tag{45}$$

In the case of linear system, $f_1 = -2h\dot{x}, \ f_2 = 0$, the equation (45) yields

$$E(x^2) = \frac{\sigma_2^2\alpha\omega_1^2}{h\omega_0^2[(\omega_0^2 - \omega_1^2)^2 + 4\alpha^2\omega_0^2]} \tag{46}$$

Comparing the approximate solution (46) with the exact one (42) shows that the conventional replacement (45), in the linear case, gives only zero power of ε of (42). Using now the proposed averaging procedure an approximate solution of (43), (37) is found from (12) for n = 1 where a_0, φ_0 are determined from the system of Ito differential equations:

$$\begin{aligned}
\dot{a}_0 &= -\frac{(\varepsilon\tilde{P}f_1+\varepsilon^2\tilde{P}f_2)}{\omega_0 R^2}[(\omega_1^2-\omega_0^2)\sin\varphi_0+2\alpha\omega_0\cos\varphi_0]+\\
&+\frac{2\varepsilon\sigma_2^2\alpha\omega_1^2}{a_0\omega_0^2R^4}[(\omega_1^2-\omega_0^2)\cos\varphi_0-2\alpha\omega_0\sin\varphi_0]^2-\\
&-\sqrt{\varepsilon}\frac{2\sigma_2\sqrt{\alpha}\omega_1}{\omega_0R^2}[(\omega_1^2-\omega_0^2)\sin\varphi_0+2\alpha\omega_0\cos\varphi_0]\dot{\xi}(t)\\
\dot{\varphi}_0 &= \omega_0-\frac{(\varepsilon\tilde{P}f_1+\varepsilon^2\tilde{P}f_2)}{a_0\omega_0R^2}[(\omega_1^2-\omega_0^2)\cos\varphi_0-2\alpha\omega_0\sin\varphi_0]-\\
&-\frac{4\varepsilon\sigma_2^2\alpha\omega_1^2}{a_0^2\omega_0^2R^4}[(\omega_1^2-\omega_0^2)\sin\varphi_0+2\alpha\omega_0\cos\varphi_0]x\\
&x[(\omega_1^2-\omega_0^2)\cos\varphi_0-2\alpha\omega_0\sin\varphi_0]-\\
&-\sqrt{\varepsilon}\frac{2\sigma_2\sqrt{\alpha}\omega_1}{a_0\omega_0R^2}[(\omega_1^2-\omega_0^2)\cos\varphi_0-2\alpha\omega_0\sin\varphi_0]\dot{\xi}(t)
\end{aligned}\tag{47}$$

wherein it is denoted

$$\begin{aligned}
&R=\sqrt{4\alpha^2\omega_0^2+(\omega_1^2-\omega_0^2)^2}\\
&\tilde{P}f_v=P(f_v)\ ,\ v=1{,}2\ \text{ with } x^{(i)}(t) \text{ is replaced by } (11), n=1
\end{aligned}\tag{48}$$

It is seen that the equations (47) are different from the conventional "equivalent" system subject to white noise (46). A higher order averaging procedure (Anh, 1995) can be applied to (47). It can be shown, in particular for linear system, $f_1=-2h\dot{x}$, $f_2=0$, the equations (47) give not only zero power but also higher powers of ε for $E(x^2)$.

5.2. NARROW-BAND SECOND ORDER COLOURED NOISE

5.2.1. Lightly damping system

Consider a lightly damping system subject to external narrow-band second order coloured noise excitation

$$\ddot{x}+\omega_0^2x=\varepsilon f_1(x.\dot{x})+\varepsilon^2f_2(x,\dot{x})+p(t)\tag{49}$$

where p(t) is determined from (37) and (39). Using the proposed averaging procedure an approximate solution of (49), (37), (39) is found in the form (18) for m = 1 where a_s, φ_s, s = 0, 1 are determined by the Ito differential equations (19), (20), with

$$\begin{aligned} \sigma_1 &= 2\sigma_2\sqrt{\alpha_1\omega_1} \\ \Omega_0 &= -\Omega_1 = \omega_1^2 - \omega_0^2 \ , \ a = (a_0, a_1) \ , \ \varphi = (\varphi_0, \varphi_1) \\ \widetilde{F}_1 &= \ddot{f}_1 + \omega_1^2 f_1 - 2\alpha_1(\ddot{x} + \omega_0^2 \dot{x}) \\ \widetilde{F}_2 &= \ddot{f}_2 + \omega_1^2 f_2 + 2\alpha_1 \dot{f}_1 \end{aligned} \tag{50}$$

in which

$$x^{(i)}(t) = \sum_{\rho=0}^{1} a_\rho \omega_\rho^i \cos(\varphi_\rho + i\frac{\pi}{2}) \ , \ i = 0,1,2,3 \tag{51}$$

Example 1. Linear system. Let in (49)

$$f_1 = -2h\dot{x}, \ f_2 = 0 \tag{52}$$

Substituting (52) into (50) yields

$$F_1 = -2(\alpha_1 + h)\ddot{x} - 2(\alpha_1\omega_0^2 + h\omega_1^2)\dot{x} \tag{53}$$

In this case the corresponding averaged FP equation (21) gives the following solution

$$W(a_0, a_1) = C a_0 a_1 \exp\left\{ -\frac{(\omega_1^2 - \omega_0^2)^2}{2\sigma_2^2}(a_1^2 + \frac{h\omega_0^2}{\alpha_1\omega_1^2} a_0^2) \right\} \tag{54}$$

where C is a constant of normalization. Thus, the corresponding approximate mean square $E(x^2)$ is to be found

$$E(x^2) = \frac{1}{2}(E(a_0^2) + E(a_1^2)) = \frac{\sigma_2^2(h\omega_0^2 + \alpha_1\omega_1^2)}{h\omega_0^2(\omega_1^2 - \omega_0^2)^2} \tag{55}$$

which is zero power of ε of (42) noting (39). Higher order approximate solution can be found by application to corresponding FP equation (21) the proposed higher order averaging procedure.

5.2.2. High damping system

Consider a high damping system subject to external narrow-band second order coloured noise excitation

$$\ddot{x} + 2h\dot{x} + \omega_0^2 x = \varepsilon f_1(x.\dot{x}) + \varepsilon^2 f_2(x,\dot{x}) + p(t) \tag{56}$$

where p(t) is determined from (37) and (38). Eliminating p(t) from (56), (37), (39) yields

$$\left(\frac{d^2(.)}{dt^2} + 2h\frac{d(.)}{dt} + \omega_0^2(.)\right)(\ddot{x} + \omega_1^2 x) = \varepsilon F_1 + \varepsilon^2 F_2 + \varepsilon^3 .. + 2\sqrt{\varepsilon}\sigma_2\alpha_1\omega_1\dot{\xi}(t) \tag{57}$$

where

$$F_1 = \ddot{f}_1 + \omega_1^2 f_1 - 2\alpha_1(\dddot{x} + 2h\ddot{x} + \omega_0^2\dot{x}),\ F_2 = \ddot{f}_2 + \omega_1^2 f_2 + 2\alpha_1\dot{f}_1 \tag{58}$$

The characteristic equation corresponding to (57), for $\varepsilon = 0$, has following roots

$$\mu_{1,2} = -h \pm i\sqrt{\omega_0^2 - h^2},\ \mu_{3,4} = \pm i\omega_1 \tag{59}$$

Thus, using (9) an approximate solution of (57) can be found in the form

$$x^{(k)}(t) = a_1\omega_1^k \cos(\varphi_1 + k\frac{\pi}{2}),\ k = 0,1,2,3 \tag{60}$$

where a_1, φ_1 are determined from

$$\begin{aligned}
\dot{a}_1 = &-\frac{(\varepsilon\tilde{P}f_1 + \varepsilon^2\tilde{P}f_2 + \varepsilon^3 ...)}{\omega_1 R^2}[(\omega_0^2 - \omega_1^2)\sin\varphi_1 + 2\alpha_1\omega_1\cos\varphi_1] + \\
&+\frac{2\varepsilon\sigma_2^2\alpha_1}{a_1 R^4}[(\omega_0^2 - \omega_1^2)\cos\varphi_1 - 2\alpha_1\omega_1\sin\varphi_1]^2 - \\
&-\sqrt{\varepsilon}\frac{2\sigma_2\sqrt{\alpha_1}}{R^2}[(\omega_0^2 - \omega_1^2)\sin\varphi_1 + 2\alpha_1\omega_1\cos\varphi_1]\dot{\xi}(t) \\
\dot{\varphi}_1 = \omega_1 &- \frac{(\varepsilon\tilde{P}f_1 + \varepsilon^2\tilde{P}f_2 + \varepsilon^3 ..)}{a_1\omega_1 R^2}[(\omega_0^2 - \omega_1^2)\cos\varphi_1 - 2\alpha_1\omega_1\sin\varphi_1] - \\
&-\frac{4\varepsilon\sigma_2^2\alpha_1}{a_1^2 R^4}[(\omega_0^2 - \omega_1^2)\sin\varphi_1 + 2\alpha_1\omega_1\cos\varphi_1] x \\
&x[(\omega_0^2 - \omega_1^2)\cos\varphi_1 - 2\alpha_1\omega_1\sin\varphi_1] - \\
&-\sqrt{\varepsilon}\frac{2\sigma_2\sqrt{\alpha_1}}{a_1 R^2}[(\omega_0^2 - \omega_1^2)\cos\varphi_1 - 2\alpha_1\omega_1\sin\varphi_1]\dot{\xi}(t)
\end{aligned} \tag{61}$$

wherein it is denoted

$$R = \sqrt{4h^2\omega_1^2 + (\omega_0^2 - \omega_1^2)^2}$$
$$\widetilde{F}_v = F_v \;,\; v = 1,2 \;\text{ with } x^{(i)}(t) \text{ is replaced by (60)} \tag{62}$$

The equations (61) are similar to the ones in the case of white noise excitation. Thus, a higher order averaging procedure as shown in (Anh, 1995) can be applied here.

6. Conclusion

For many years the stochastic averaging method has been a very useful tool for investigating non-linear random vibration systems. However, the effect of some non-linear terms cannot be investigated by using the classical first order SAM. In the paper, the higher order stochastic averaging method is developed to predict approximately the response of linear and lightly nonlinear systems subject to weakly external excitation of wideband or narrow-band coloured noise random processes. The excitation of second order coloured noise is considered in more detail. It is shown that in the case of wideband random excitation the result obtained is different from the conventional replacement of wideband random excitation by an white noise excitation.

Acknowledgment. Support from the fundamental research project in natural sciences is gratefully acknowledged.

7. References

1. Anh, N. D., (1993) Higher order approximate solutions in stochastic averaging method. *In Proc. of NCSR of VN, 5, 19-26.*
2. Anh, N. D., (1995) Higher order averaging method of coefficients in Fokker-Planck equation. *In special volume: Advances in nonlinear structural dynamics of SADHANA, Indian Acad. of Sc., 373-388.*
3. Anh, N. D. and Tinh, N. D. (1995) Higher order averaging solutions for Van der Pol oscillator. *In Proc. Int. Conf. On Nonlinear Stochastic Dynamics, Hanoi, Viet nam, 27-38.*
4. Ariaratnam, S.T., Tam, D. S. F., (1979) Random vibration and stability of a linear parametrically excited oscillator. *Z. Angew. Math. Mech., **59**, no 2, 79-84.*
5. Bogoliubov, N. N. and Mitropolskii Iu. A., (1961). *Asymptotic methods in the theory of nonlinear oscillations*, New York: Gordon and Breach.
6. Bolotin, V. V. (1984) *Random vibration of elastic systems*. Hague: Martinus Nijhoff.
7. Ibrahim, R. A. (1985). *Parametric random vibration* (Hertfordshire New York: Research Studies Press/John Wiley and Sons).
8. Iyengar, R. N. (1992). Approximate analysis of nonlinear systems under narrow band random excitation, *In Proc. of IUTAM Sympo-sium on Nonlinear Stochastic Mechanics, Turin, Italy, 309-320.*
9. Khasminskii, R. (1963) Averaging principle for the parabolic and elliptic differential equations and for Markov processes with small diffusion. *Theor. Prob. Appl., 8, 1- 21 (in Russian).*
10. Lin, Y. K, Cai G. Q. (1995). *Probabilistic structural dynamics*, Mc Graw-Hill, Inc.

11. Mitropolskii, Iu. A., Dao, N. V, Anh, N. D., (1992), *Nonlinear oscillations in systems of arbitrary order*, Naukova - Dumka, Kiev, (in Russian).

12. Red-Horse, J. R. Spanos, P. D. (1992). A generalization to stochastic averaging in random vibration. *Int. J. Nonlinear Mech.* ***27****: 85-101.*

13. Roberts, J. B, Spanos, P.T. D. (1986). Stochastic averaging: An approximate method of solving random vibration problem. *Int. J. Nonlinear Mech.* ***21****: 111-134.*

14. Sri Namachchivaya, N., and Lin, Y. K., (1988). Application of stochastic averaging for nonlinear systems with high damping, *Probabilistic Engineering Mechanics,* ***3****, no 3, 159-167.*

15. Stratonovich, R.L. (1963) *Topics in the theory of a random noise* New York: Gordon and Breach Vol. 1.

16. Zhu, W. Q. (1988). Stochastic averaging method in random vibrations. *Appl. Mech. Rev.* ***41****: 189-199.*

17. Zhu, W. Q, Yu, M. Q, Lin Y. K. (1994). On improved stochastic averaging procedure. Probab. Eng. Mech. **9**: 203-212.

18 Zhu, W. Q. et al., (1997) Stochastic averaging of quasi-integrable Hamiltonian systems, *J. Applied Mech.,* ***64****, 975- 984.*

RECENT PROGRESS
IN STOCHASTIC BIFURCATION THEORY

LUDWIG ARNOLD
Institut für Dynamische Systeme
Fachbereich 3, Universität, Postfach 33 04 40
28334 Bremen, Germany
email: arnold@math.uni-bremen.de

Abstract. We report on new results in stochastic bifurcation theory obtained in 1997 and 1998. These include: (i) a rather complete classification of the one-dimensional case by Crauel, Imkeller and Steinkamp, (ii) new insight into the stochastic Hopf bifurcation (made possible by the random version of the subdivision algorithm of Dellnitz et al.) by Keller and Ochs, (iii) a study of the stochastic Brusselator by Arnold, Bleckert and Schenk-Hoppé, (iv) Baxendale's further studies of an SDE at a bifurcation point, (v) a new method of proving the existence of a random attractor for an SDE by transforming it into a random differential equation, by Imkeller and Schmalfuß.

Key words and phrases: Random dynamical system, stochastic bifurcation, invariant measure, Lyapunov exponent, random attractor, stochastic differential equation, Fokker-Planck-Kolmogorov equation

AMS 1991 subject classification: primary 34F05; secondary 60H10, 93E03.

1. Introduction. D- and P-bifurcation

In this paper we report on progress made in stochastic bifurcation theory between August 1997 (when the manuscript of the monograph [1] was shipped to the publisher) and December 1998 (the deadline of this paper).

We will assume familiarity of the reader with the basic concepts of the theory, as laid down in Chapter 9 (Bifurcation Theory) of [1]. We will only discuss bifurcation phenomena in families (φ_α) of random dynamical

S. Narayanan and R.N. Iyengar (eds.),
IUTAM Symposium on Nonlinearity and Stochastic Structural Dynamics, 15–27.

systems (RDS) which are related to sign changes of Lyapunov exponents $\lambda_i(\mu_\alpha)$ of φ_α-invariant measures μ_α, α a bifurcation parameter. Those bifurcation phenomena were named *dynamical*, or *D-bifurcations* in [1], as opposed to qualitative changes of stationary densities of the corresponding Markov process (denoted *P-bifurcations*), the latter being not related to the stability measured by Lyapunov exponents of the RDS.

In Section 2 we give a brief account of the rather complete and beautiful analysis of the case of stochastic differential equations (SDE) with state space $\mathbb{R}$, done by Crauel, Imkeller and Steinkamp [7], a study which was overdue. It shows that in dimension 1 the concepts developed in [1] suffice: Invariant measures bifurcate from invariant measures at parameter values where the Lyapunov exponent is equal to zero, and this is all which can happen!

The level of invariant measures does, however, not suffice in dimension 2 and higher to describe all bifurcation phenomena. This is the conclusion of an exciting numerical study of the stochastic Duffing-van der Pol oscillator in the Hopf regime by Keller and Ochs [12] (see Section 3). It was known before that noise splits the pair of deterministic complex-conjugate eigenvalues into two simple Lyapunov exponents which cross 0 one after another, giving rise to the three consecutive regimes $\lambda_2 < \lambda_1 < 0$, $\lambda_2 < 0 < \lambda_1$ and $0 < \lambda_2 < \lambda_1$.

It was not known that there is a "chaotic" attractor in the second and third regime. This lack of knowledge was mainly due to inappropriate numerical algorithms which provided us with the (rather small) support of a bifurcating invariant measure, hence with an *inner* approximation of the (rather large) random attractor. Random analogues of the subdivision algorithm of Dellnitz et al. enabled Keller and Ochs [12] to obtain *outer* approximations of the attractor, also revealing its Cantor-type structure. These findings call for an extension of the concept of stochastic bifurcation from the level of invariant measures to the level of attractors and their (Hausdorff) dimension.

In Section 4 we present yet another example (the Brusselator with parametric noise) of the general fact that "additive noise destroys bifurcation".

In a series of profound papers, Baxendale gave general conditions for an SDE in $\mathbb{R}^d$ with fixed point 0 to develop a new invariant measure in the punctured space $\mathbb{R}^d \setminus \{0\}$ at the value the top Lyapunov exponent of 0 becomes positive, see [1, Section 9.5]. It was, however, not clear whether the new measure is stable, i.e. has a negative top Lyapunov exponent. In [4] he comes to the surprising conclusion that the new measure can also be unstable, see Section 5.

How does one prove the existence of random attractors for RDS? Schenk-Hoppé [14] (see also [1, Section 9.4]) formulated a "recipe" for the real noise case based on finding a (deterministic) Lyapunov function for the RDS by which one can construct an absorbing set. In the white noise case this procedure fails, and the only technique available so far is to find a random coordinate transformation by which an SDE is transformed into a random differential equation (RDE). This technique was recently systematically developed by Imkeller and Schmalfuß [11], and its principle will be described in Section 6.

Warning: This paper is an overview, written in a semi-technical style. We refer to the original papers for complete statements listing all technical assumptions.

2. Dimension One: Classification of Possible Cases

2.1. THE STRUCTURE OF INVARIANT MEASURES

It is an elementary fact that any ergodic invariant measure of a dynamical system generated by a vector field $\dot{x} = f(x)$ on $\mathbb{R}$ is a Dirac measure at an equilibrium point. This basically carries over to the random case [1, Theorem 1.8.4].

2.1. Proposition. *Let φ be a local continuous RDS with time $\mathbb{R}$ and state space $\mathbb{R}$. Then every ergodic φ-invariant measure μ_ω is a random Dirac measure, i.e. there exists a random variable a such that $\mu_\omega = \delta_{a(\omega)}$.*

This explains why in all examples and case studies in dimension 1, beginning with Arnold and Boxler [3], only Dirac measures were found. It does not explain the observation that new measures only bifurcate from *deterministic* Dirac measures, and that all invariant measures for SDE could have been found by solving the Fokker-Planck-Kolmogorov equation (FPKE). Are there invariant measures other than Markov measures?

We first study a single SDE in dimension 1,

$$dx = f(x)dt + \sigma(x) \circ dW = (f(x) + \frac{1}{2}\sigma'(x)\sigma(x))dt + \sigma(x)\, dW, \qquad (2.1)$$

with two-sided time $\mathbb{R}$, $\circ$ denoting the Stratonovich stochastic integral. We assume throughout that f and σ are regular enough so that (2.1) generates a smooth local RDS φ over the shift θ in Wiener space, i.e. (locally) satisfies the cocycle property

$$\varphi(t+s,\omega) = \varphi(t,\theta_s\omega) \circ \varphi(s,\omega),$$

where $\circ$ denotes composition of mappings.

There are two Markov processes embedded in the RDS φ, namely one forward, the other one backward in time, with respective transition probabilities

$$P^{\pm}(t,x,B) = \mathbb{P}\{\omega : \varphi(\pm t,\omega,x) \in B\}, \quad t \geq 0, \qquad (2.2)$$

with generators

$$L^{\pm} = (\pm f + \frac{1}{2}\sigma'\sigma)\frac{d}{dx} + \frac{1}{2}\sigma^2\frac{d^2}{dx^2} \,. \tag{2.3}$$

A measure ρ is called *stationary* under $P^{\pm}$ if it solves the FPKE $(L^{\pm})^*\rho = 0$.

The basic relation between invariant measures of φ and stationary measures of the corresponding Markov processes was discovered by Crauel [5].

2.2. Proposition. *Let φ be the RDS generated by (2.1). Then there is a one-to-one relation between*

(i) invariant measures which are measurable with respect to the past $\mathfrak{F}^-$ of the Wiener process, and P^+-stationary measures ρ^+, given by

$$\rho^+ \mapsto \mu_\omega := \lim_{t\to\infty} \varphi(t, \theta_{-t}\omega)\rho^+, \quad \mu_\omega \mapsto \rho^+ := \mathbb{E}\,\mu_\cdot,$$

(ii) invariant measures which are measurable with respect to the future $\mathfrak{F}^+$ of the Wiener process, and P^--stationary measures ρ^-, given by

$$\rho^- \mapsto \mu_\omega := \lim_{t\to\infty} \varphi(-t, \theta_t\omega)\rho^-, \quad \mu_\omega \mapsto \rho^- := \mathbb{E}\,\mu_\cdot.$$

We call a measure μ_ω a *Markov measure*, if it is either $\mathfrak{F}^-$ or $\mathfrak{F}^+$ measurable. The first basic result of Crauel, Imkeller and Steinkamp [7] is the following.

2.3. Theorem. *Let φ be generated by the SDE (2.1). Then every ergodic φ-invariant measure μ is a Markov measure, i.e. there is a random variable a which satisfies either $a \in \mathfrak{F}^-$ or $a \in \mathfrak{F}^+$ such that $\mu_\omega = \delta_{a(\omega)}$.*

The proof of Theorem 2.3 is not trivial and relies on the relative entropy theory developed by Crauel in earlier papers.

Theorem 2.3 says that every invariant measure μ for an SDE on $\mathbb{R}$ comes (in the sense of Proposition 2.2) from one of the two FPKE $(L^+)^*\rho^+ = 0$ or $(L^-)^*\rho^- = 0$, which thus remain to be studied.

2.2. EXISTENCE, UNIQUENESS, AND STABILITY OF INVARIANT MEASURES

It will turn out that a D-bifurcation can only occur out of a *deterministic* Dirac measure as only these measures can have a vanishing Lyapunov exponent. This motivates the following assumptions which are assumed to hold for the rest of this subsection:

$$f(0) = \sigma(0) = 0, \quad \sigma(x) \neq 0 \quad \text{for all} \quad x \neq 0.$$

Hence δ_0 is always an invariant measure, and $P^{\pm}$-stationary measures on $I^+ := (0,\infty)$ and $I^- := (-\infty, 0)$ are unique.

Define the *speed measure* $m^\pm$ of $P^\pm$ on $I \in \{I^-, I^+\}$ by

$$m^\pm(dx) := \rho^\pm(x)dx, \quad \rho^\pm(x) := \frac{2}{|\sigma(x)|}\exp\left(2\int_c^x \frac{\pm f(y)}{\sigma^2(y)}dy\right),$$

where $c \in I$. Clearly $(L^\pm)^*\rho^\pm = 0$, so that $m^\pm$ is the unique $P^\pm$-stationary measure if normalizable (which we assume to be done if possible).

An elementary consideration shows that

- $m^+(I) < \infty$ and $m^-(I) < \infty$ are not simultaneously possible,
- under $m^+(I) = \infty$ and $m^-(I) = \infty$ there is no invariant measure.

For determining a formula for the *Lyapunov exponent* $\lambda(\mu)$ of φ under μ we linearize (2.1) yielding

$$dv = f'(x)vdt + \sigma'(x)v \circ dW,$$

the solution of which is

$$v_t = v_0 \exp\left(\int_0^t (f'(x_s) + \frac{1}{2}\sigma(x_s)\sigma''(x_s))ds + \int_0^t \sigma'(x_s)dW_s\right).$$

Using the ergodic theorem and the FPKE we arrive at

$$\lambda(\mu) = \lim_{t\to\infty}\frac{1}{t}\log|v_t| = \mp 2\int_I \left(\frac{f(x)}{\sigma(x)}\right)^2 \rho^\pm(x)dx, \tag{2.4}$$

provided a normalizable $\rho^\pm$ exists.

It immediately follows that

- $\lambda(\delta_0) = f'(0)$,
- if $\lambda(\mu) = 0$ for some μ then μ is a deterministic Dirac measure.

Here is the main result of [7].

2.4. Theorem. *Let φ be the RDS generated by (2.1), $I \in \{I^-, I^+\}$, and denote by $U(0)$ a bounded neighborhood of $0 \in \mathbb{R}$.*

(i) If $m^+(I) < \infty$, then $m^-(I \setminus U(0)) = \infty$ if and only if φ is forward complete, in which case there exists a unique $\mathfrak{F}^-$ measurable ergodic φ-invariant measure $\mu_\omega^+ = \delta_{a(\omega)}$ corresponding to $m^+ = \rho^+dx = \mathbb{E}\mu^+$. This measure is stable, more precisely

$$\lambda(\mu^+) = -2\int_I \left(\frac{f(x)}{\sigma(x)}\right)^2 \rho^+(x)dx < 0,$$

and $A(\omega) := \{a(\omega)\}$ is the global random attractor of φ in I.

(ii) If $m^-(I) < \infty$, then $m^+(I \setminus U(0)) = \infty$ if and only if φ is backward complete, in which case there exists a unique $\mathcal{F}^+$ measurable ergodic φ-invariant measure $\mu^-_\omega = \delta_{r(\omega)}$ corresponding to $m^- = \rho^- dx = \mathbb{E}\,\mu^-_\cdot$. This measure is unstable, more precisely

$$\lambda(\mu^-) = 2\int_I \left(\frac{f(x)}{\sigma(x)}\right)^2 \rho^-(x)dx > 0,$$

and $R(\omega) := \{r(\omega)\}$ is the global random repeller of φ in I.

According to Theorem 2.4, we have the following **9 possible cases** for invariant measures of the RDS φ generated by the SDE (2.1):

1. No invariant measures: no measure on I^-, no measure on I^+, and δ_0 can be stable or unstable;
2. Stable invariant measures:
   (a) one stable measure on I^-, no measure on I^+,
   (b) no measure on I^-, one stable measure on I^+,
   (c) one stable measure on I^-, one stable measure on I^+,
   hence $\lambda(\delta_0) \geq 0$;
3. Unstable invariant measures:
   (a) one unstable measure on I^-, no measure on I^+,
   (b) no measure on I^-, one unstable measure on I^+,
   (c) one unstable measure on I^-, one unstable measure on I^+,
   hence $\lambda(\delta_0) \leq 0$;
4. Mixed cases:
   (a) one stable measure on I^-, one unstable measure on I^+,
   (b) one unstable measure on I^-, one stable measure on I^+,
   hence $\lambda(\delta_0) = 0$.

2.3. BIFURCATION SCENARIOS

We now consider a family of SDE parametrized by a real parameter $\alpha \in \mathbb{R}$,

$$dx = f_\alpha(x)dt + \sigma_\alpha(x) \circ dW = (f_\alpha(x) + \frac{1}{2}\sigma'_\alpha(x)\sigma_\alpha(x))dt + \sigma_\alpha(x)\, dW, \quad (2.5)$$

where $(x, \alpha) \mapsto f_\alpha(x)$ is twice and $(x, \alpha) \mapsto \sigma_\alpha(x)$ three times continuously differentiable with respect to x and satisfy

$$f_\alpha(0) = \sigma_\alpha(0) = 0, \quad \sigma_\alpha(x) \neq 0 \quad \text{for all} \quad x \neq 0 \text{ and all } \alpha \in \mathbb{R}.$$

For each fixed α the corresponding local C^1 RDS φ_α is in one of the above 9 cases, and letting α move permits us to switch between the various cases.

To fix ideas, let us restrict ourselves to the stochastic pitchfork bifurcation, which we define as follows.

Definition of Stochastic Pitchfork Bifurcation. Consider the family $(\varphi_\alpha)_{\alpha\in\mathbb{R}}$ of RDS on $\mathbb{R}$. It is said to undergo a stochastic pitchfork bifurcation at $\alpha = 0$, if

(i) for $\alpha \leq 0$, δ_0 is the only invariant measure for φ_α, with $\lambda(\delta_0) = f'_\alpha(0) < 0$ for $\alpha < 0$ and $\lambda(\delta_0) = f'_0(0) = 0$ for $\alpha = 0$,

(ii) for $\alpha > 0$, $\lambda(\delta_0) = f'_\alpha(0) > 0$, and the RDS φ_α possesses in addition to δ_0 the two invariant measures $\mu^1_\omega = \delta_{a_1(\omega)}$ on I^- and $\mu^2_\omega = \delta_{a_2(\omega)}$ on I^+ which are both stable,

(iii) we have $a_i \to 0$ in probability as $\alpha \downarrow 0$, $i = 1, 2$. □

Hence the stochastic pitchfork bifurcation consists of switching from case (1) with stable δ_0 for $\alpha < 0$ to case (2)(c) with unstable δ_0 for $\alpha > 0$ – for which the characterization of the corresponding cases given in Subsection 2.2 in terms of the speed measures $m^\pm_\alpha$ yields necessary and sufficient conditions.

Note that these conditions are global conditions for f_α and σ_α (finiteness of integrals over I etc.), in contrast to the deterministic case, where the pitchfork bifurcation can be characterized by the (local) behavior of $(x, \alpha) \mapsto f_\alpha(x)$ in a small neighborhood of $(x, \alpha) = (0, 0)$.

For details (and sufficient conditions covering all known cases) see [7].

3. Stochastic Hopf Bifurcation in the Noisy Duffing-van der Pol Oscillator

Let us consider the Duffing-van der Pol oscillator

$$\ddot{y} = \alpha y + \beta \dot{y} - y^3 - y^2 \dot{y}, \quad \alpha, \beta \in \mathbb{R},$$

for which the parameter α is perturbed by white noise. Putting $x = (y, \dot{y})$, we arrive at the SDE in $\mathbb{R}^2$ given by

$$dx = \begin{pmatrix} 0 \\ -x_1^3 - x_1^2 x_2 \end{pmatrix} + \begin{pmatrix} 0 & 1 \\ \alpha & \beta \end{pmatrix} x\,dt + \sigma \begin{pmatrix} 0 & 0 \\ 1 & 0 \end{pmatrix} x \circ dW. \tag{3.1}$$

For studying the stochastic analogue of the Hopf bifurcation it suffices to put $\alpha = -1$ and let β, the bifurcation parameter, increase from negative to positive values.

The first basic question is whether and at which values of β the two Lyapunov exponents $\lambda_2(\beta) < \lambda_1(\beta)$ of the trivial solution $x = 0$ of (3.1) change signs, as these values are the possible D-bifurcation points.

The second question is then whether there is in fact a bifurcation, and what kind of objects bifurcate out of $x = 0$.

3.1. LYAPUNOV EXPONENTS

Imkeller and Lederer [10] took a fresh look at the frequently studied stochastic linear oscillator

$$dv = \begin{pmatrix} 0 & 1 \\ -1 & \beta \end{pmatrix} v dt + \sigma \begin{pmatrix} 0 & 0 \\ 1 & 0 \end{pmatrix} v \circ dW, \qquad (3.2)$$

which is the linearization of (3.1) at $x = 0$ for $\alpha = -1$. They found the following surprisingly simple explicit representation of its Lyapunov exponents from which they derived various asymptotic, local and global properties.

3.1. Theorem. *The Lyapunov exponents of (3.2) can be represented as*

$$\lambda_{1,2}(\beta, \sigma) = \frac{\beta}{2} \pm \frac{\sigma^2}{4} C(\beta^2, \sigma^2),$$

where $C(\beta^2, \sigma^2)$ is the second moment of the stationary measure of the one-dimensional diffusion

$$dz = -U'(z)dt + \sigma dW, \quad U(z) = \frac{z^6}{12} - (\frac{\beta^2}{2} - 1)z^2.$$

3.2. Corollary. *Let $\sigma \neq 0$. Then the function $\beta \mapsto \lambda_1(\beta, \sigma)$ is smooth and possesses exactly one zero at a value $\beta_1 < 0$ with positive derivative. Likewise, $\beta \mapsto \lambda_2(\beta, \sigma) = -\lambda_1(-\beta, \sigma)$ is smooth and possesses exactly one zero at the value $\beta_2 = -\beta_1 > 0$ with positive derivative.*

We hence have a mathematically rigorous proof of the fact that there are exactly two possible values of β at which we can expect a bifurcation of (3.1) out of $x = 0$: The value $\beta_1 < 0$ at which λ_1 changes from negative to positive values, and the value $\beta_2 = -\beta_1 > 0$ at which λ_2 changes from negative to positive values.

3.2. THE STOCHASTIC HOPF BIFURCATION SCENARIO

We now address the second problem: Is there indeed a bifurcation at $\beta_1 < 0$ and $\beta_2 > 0$, and if yes, what is the object which bifurcates from 0?

The extensive study in [1, Chapter 9] was based on the idea of finding D-bifurcations on the level of invariant measures. This point of view suffices in dimension 1, as proved in Section 2. It does, however, not suffice anymore in dimension 2 and higher, the reason being that the support of an invariant measure (typically a finite random set) is just a very small subset of the invariant set which actually bifurcates.

We briefly report on some surprising numerical findings by Keller and Ochs [12], made possible by a random analogue of the subdivision algorithm of Dellnitz et al. The principle of this algorithm is that we let φ act on sets rather than points and produce a sequence of sets decreasing to the attractor, hence furnishing at each step an outer approximation of the attractor.

For the regime $\beta < \beta_1$ (i.e. $\lambda_2 < \lambda_1 < 0$), the set $\{0\}$ is the global attractor.

For the intermediate regime $\beta_1 < \beta < \beta_2$ (i.e. $\lambda_2 < 0 < \lambda_1$), Keller and Ochs found a "chaotic" attractor which is the closure of the one-dimensional unstable invariant manifold of 0. This manifold is not embedded, but only immersed into $\mathbb{R}^2$, and the transversal sections of its closure are Cantor-like sets.

When crossing into the third regime where $\beta > \beta_2$ (i.e. $0 < \lambda_2 < \lambda_1$), the attractor of the intermediate regime develops a "hole" around 0, and the new attractor in the punctured plane $\mathbb{R}^2 \setminus \{0\}$ is now the boundary of the two-dimensional unstable manifold of 0. This boundary again has a Cantor-like transversal sections.

We can offer the following intuitive explanation for the appearance of "chaos" in the intermediate regime: As the one-dimensional stable and unstable manifolds of 0 are statistically independent, they will intersect (transversally) with positive probability, hence, by ergodicity, with probability one. We thus have transversal homoclinic points, entailing "chaos" by the random version of Smale's theorem proved by Gundlach [9].

We think that it is an urgent task of stochastic bifurcation theory to replace the numerical evidence for the above scenario by rigorous proofs.

4. The Stochastic Brusselator

The Brusselator, a model for an oscillatory chemical reaction (limit cycle), was much studied in the 1970's and 1980's. Several authors investigated the Brusselator in the presence of external (parametric) noise, i.e. they looked at the SDE

$$\begin{aligned} dx &= (\alpha - (\beta+1)x + x^2 y)dt - \sigma x \circ dW, \\ dy &= (\beta x - x^2 y)dt + \sigma x \circ dW. \end{aligned}$$

The deterministic Brusselator ($\sigma = 0$) has a unique fixed point $(\alpha, \frac{\beta}{\alpha})$ which is stable for $0 < \beta \leq \alpha^2 + 1$ and unstable for $\beta > \alpha^2 + 1$, and undergoes a Hopf bifurcation for the bifurcation parameter β at $\beta = \alpha^2 + 1$.

Arnold, Bleckert and Schenk-Hoppé [2] made a mainly numerical study of the above SDE in the context of the theory of RDS.

Note first that noise on the parameter β is "additive" with respect to both the deterministic fixed point and the limit cycle in the sense that these solutions do not survive the perturbation.

Previous studies by Fronzoni et al. [8] showed that the unique stationary density $p_\beta(x,y)$ undergoes two P-bifurcations: It is bell-shaped for $\beta < \alpha^2+1$, from where on it develops a "dent", but all cross sections parallel to the xy plane are still simply connected. From a larger critical value β_c on some cross sections are doubly connected, i.e. the dent can "hold water".

The Lyapunov exponents of $p_\beta(x,y)$ remain, however, negative for the whole range of parameters. Thus a D-bifurcation is impossible. It was shown numerically that the factorization of $p_\beta(x,y)$ in the sense of Proposition 2.2 is a random Dirac measure, $\mu_\omega = \delta_{(\xi(\omega),\eta(\omega))}$, and $A(\omega) = \{(\xi(\omega),\eta(\omega))\}$ is the global random attractor of the system. This is yet another case where additive noise "destroys" bifurcation. A rigorous proof of this fact in the case of the pitchfork bifurcation was given by Crauel and Flandoli [6].

5. General Dimension: Baxendale's Results

Let the top Lyapunov exponent λ_α of the trivial solution $x=0$ of the SDE in $\mathbb{R}^d$

$$dx = f_\alpha^0(x)dt + \sum_{j=1}^m f_\alpha^j(x)\circ dW_j, \tag{5.1}$$

where $f_\alpha^j(0) = 0$ for all $j = 0,\ldots,m$ and $\alpha \in \mathbb{R}$, change from negative to positive values at the parameter value $\alpha = \alpha_0$ (say), and assume that a new stationary measure ρ_α bifurcates from $x=0$ at α_0.

It is often believed (and in certain cases true) that ρ_α is stable, i.e. has top Lyapunov exponent $\tilde\lambda_\alpha < 0$, which would be in agreement with the familiar "exchange of stability" picture.

Baxendale [4] has proved by way of examples that this need not be the case. His main result is that (modulo technical assumptions)

$$\lim_{\alpha\downarrow\alpha_0} \frac{\tilde\lambda_\alpha}{\lambda_\alpha} = \Gamma, \tag{5.2}$$

where the finite constant Γ depends only on the RDS generated by (5.1) for $\alpha = \alpha_0$. He gives examples (which are too involved to be reproduced here) for the case $\Gamma < 0$ as well as for the case $\Gamma > 0$.

6. Existence of Random Attractors in the White Noise Case

We now address the problem of constructing random attractors in the white noise case, i.e. of finding a random compact set $A(\omega)$ for the RDS φ in $\mathbb{R}^d$

generated by the SDE

$$dx = f(x)dt + \sigma(x) \circ dW \tag{6.1}$$

satisfying

- $\varphi(t,\omega)A(\omega) = A(\theta_t\omega)$ (invariance),
- $d(\varphi(t,\theta_{-t}\omega)D(\theta_{-t}\omega)|A(\omega)) \to 0$ for $t \to \infty$ for "many" random sets $D(\omega)$ (attractivity), where $d(B|A) := \sup_{x\in B}\inf_{y\in A}\|x-y\|$ denotes the semi-Hausdorff metric.

Until recently, only the real noise case, i.e. the case of an RDS ψ generated by an RDE

$$\dot{y} = g(\xi_t, y), \quad \xi_t(\omega) = \xi(\theta_t\omega) \quad \text{a nice stationary process,} \tag{6.2}$$

could be treated, according to the following procedure laid-out by Schenk-Hoppé [14]:

- Find a deterministic "Lyapunov function" $V : \mathbb{R}^d \to \mathbb{R}^+$ with the property that

$$\dot{V}(\psi(t,\omega,x)) \leq -\eta_t(\omega)V(\psi(t,\omega,x)) + \zeta_t(\omega),$$

  where η_t and $\zeta_t > 0$ are stationary processes with $\mathbb{E}\,\eta_t > 0$.
- The scalar RDE $\dot{u} = -\eta_t u + \zeta_t$ has a unique stationary solution $r_t > 0$ which is stable.
- The random set $B(\omega) := V^{-1}([0, 2r_0(\omega)])$ is compact, forward invariant and absorbing under ψ.

Then the random set

$$A(\omega) := \cap_{t\geq 0}\psi(t,\theta_{-t}\omega)B(\theta_{-t}\omega)$$

is a random attractor for ψ.

With this procedure, the existence of a random attractor for the noisy Duffing-van der Pol equation could be established [14, 1].

Unfortunately, this simple procedure is not available in the white noise case.

Suppose, however, that the RDS φ generated by (6.1) and the RDS ψ generated by (6.2) are (topologically) *cohomologous*, i.e. related by

$$\varphi(t,\omega) = \Phi(\theta_t\omega) \circ \psi(t,\omega) \circ \Phi(\omega)^{-1}, \tag{6.3}$$

where Φ is a random coordinate transformation (homeomorphism) of $\mathbb{R}^d$.

Then clearly $\Phi(\omega)A(\omega)$ is a random attractor of φ if and only if $A(\omega)$ is a random attractor of ψ.

How can we find the corresponding Φ in (6.3)? This problem was addressed by Keller and Schmalfuß [13] and systematically investigated by Imkeller and Schmalfuß [11].

We explain the basic idea of relating an SDE with an RDE via (6.3) in the one-dimensional case.

1. Let φ be the RDS generated by the SDE (6.1) on $\mathbb{R}$. Consider the stationary Ornstein-Uhlenbeck process $\xi_t(\omega) = \xi(\theta_t\omega)$ solving the SDE $d\xi = -\gamma\xi dt + dW$ with initial value
$$\xi(\omega) := \int_{-\infty}^{0} \exp(-\gamma t) dW_t(\omega).$$
Use this process to eliminate the Wiener process in the SDE (6.1) which yields
$$dx = (f(x) + \gamma\sigma(x)\xi_t)dt + \sigma(x) \circ d\xi_t$$
and still generates φ.
2. Solve the ODE $\dot{z} = \sigma(z)$ and denote its flow by $(t, z) \mapsto \chi(t, z)$. Put
$$\Phi(\omega, z) := \chi(\xi(\omega), z).$$
3. Denote by ψ the RDS generated by the RDE (6.2), where
$$g(\xi, y) := \exp\left(-\int_0^{\xi} \sigma'(\chi(s, y))ds\right)(f(\chi(\xi, y)) + \gamma\sigma(\chi(\xi, y))\xi).$$
Then φ, ψ and Φ are related by (6.3).

This procedure can be generalized to higher dimensions and is used in [11] to prove the existence of a random attractor for the Duffing-van der Pol oscillator with white parametric and additive noise as well as for other relevant systems.

Example. Consider the RDS φ generated by
$$dx = (\alpha x - x^3)dt + \sigma x \circ dW = (\alpha x - x^3 + \gamma\sigma x\xi_t)dt + \sigma x \circ d\xi_t$$
(φ could be explicitly written down, see [1, 2.3.7]), then solve $\dot{z} = \sigma z$ which yields $\chi(t, z) = \exp(\sigma t)z$, hence
$$\Phi(\omega, z) = \exp(\sigma\xi(\omega))z, \quad \xi(\omega) = \int_{-\infty}^{0} \exp(-\gamma t) dW_t(\omega).$$
The corresponding RDE is
$$\dot{y} = (\alpha + \eta_t)y - \zeta_t y^3, \quad \eta_t = \gamma\sigma\xi_t, \quad \zeta_t = \exp(2\sigma\xi_t),$$

with $\mathbb{E}\,\eta_t = 0$ and $\mathbb{E}\,\zeta_t > 0$.

This RDE can be explicitly solved [1, 9.3.5] to yield the RDS ψ which is known to have a random interval $A(\omega) = [-a(\omega), a(\omega)]$ as its random attractor. Hence

$$\varphi(t, \omega, x) = e^{\sigma\xi(\theta_t\omega)}\psi(t, \omega, e^{-\sigma\xi(\omega)}x)$$

also has a random interval as its random attractor given by

$$B(\omega) = e^{\sigma\xi(\omega)}A(\omega) = [-e^{\sigma\xi(\omega)}a(\omega), e^{\sigma\xi(\omega)}a(\omega)].$$

References

1. L. Arnold. *Random dynamical systems.* Springer-Verlag, Berlin Heidelberg New York, 1998.
2. L. Arnold, G. Bleckert, and K.-R. Schenk-Hoppé. The stochastic Brusselator: parametric noise destroys Hopf bifurcation. In H. Crauel and M. Gundlach, editors, *Stochastic dynamics*, pages 71–92. Springer-Verlag, New York, 1999.
3. L. Arnold and P. Boxler. Stochastic bifurcation: instructive examples in dimension one. In M. Pinsky and V. Wihstutz, editors, *Diffusion processes and related problems in analysis, volume II: Stochastic flows*, volume 27 of *Progress in Probability*, pages 241–255. Birkhäuser, Boston Basel Stuttgart, 1992.
4. P. Baxendale. Stability along trajectories at a stochastic bifurcation point. In H. Crauel and M. Gundlach, editors, *Stochastic dynamics*, pages 1–26. Springer-Verlag, New York, 1999.
5. H. Crauel. Markov measures for random dynamical systems. *Stochastics and Stochastics Reports*, 37:153–173, 1991.
6. H. Crauel and F. Flandoli. Additive noise destroys a pitchfork bifurcation. *Journal of Dynamics and Differential Equations*, 10:259–274, 1998.
7. H. Crauel, P. Imkeller, and M. Steinkamp. Bifurcations of one-dimensional stochastic differential equations. In H. Crauel and M. Gundlach, editors, *Stochastic dynamics*, pages 27–47. Springer-Verlag, New York, 1999.
8. L. Fronzoni, R. Mannella, P. McClintock, and F. Moss. Postponement of Hopf bifurcation by multiplicative colored noise. *Phys. Rev. A*, 36:834–841, 1987.
9. V. M. Gundlach. Random homoclinic orbits. *Random & Computational Dynamics*, 3:1–33, 1995.
10. P. Imkeller and C. Lederer. An explicit description of the Lyapunov exponent of the noisy damped harmonic oscillator. Submitted, 1998.
11. P. Imkeller and B. Schmalfuß. The conjugacy of stochastic and random differential equations and the existence of global attractors. Manuscript, Humboldt-Universität Berlin, 1998.
12. H. Keller and G. Ochs. Numerical approximation of random attractors. In H. Crauel and M. Gundlach, editors, *Stochastic dynamics*, pages 93–116. Springer-Verlag, New York, 1999.
13. H. Keller and B. Schmalfuß. Attractors for stochastic differential equations with nontrivial noise. *Buletinul Acad. Stiinte Republ. Moldova, Matematica*, 1(26):43–54, 1998.
14. K. R. Schenk-Hoppé. Random attractors – general properties, existence, and applications to stochastic bifurcation theory. *Discrete and Continuous Dynamical Systems*, 4:99–130, 1998.

NON-RESONANT AND RESONANT CHAOTIC DYNAMICS IN EXTERNALLY EXCITED CYCLIC SYSTEMS

ANIL. K. BAJAJ
School of Mechanical Engineering
Purdue University, West Lafayette, IN 47907

SOBITHA SAMARANAYAKE
Department of Mathematics and Computer Science
University of Wisconsin-Whitewater, Whitewater, WI 53190

Abstract The present work studies the vibratory response of geometrically nonlinear cyclic structures to harmonic excitations. These cyclic structure, in their linear approximations, possess pairwise double degenerate natural frequencies with orthogonal normal modes. The dynamic response of such systems, when excited near primary resonance, is studied using the method of averaging. The averaged equations, representing the evolution of the amplitudes and phases of the interacting normal modes, are known to exhibit complex dynamics including period-doubling bifurcations and Silnikov type chaos. The higher-dimensional Melnikov method and its extensions are utilized to detect the transversal intersection of stable and unstable manifolds of periodic orbits. Such intersections imply the existence of Smale horseshoes and, hence, chaotic dynamics for the averaged system.

1 Introduction

A vast literature exists on the nonlinear vibrations of multi degree-of-freedom dynamical systems that may exhibit internal resonances. The type of internal resonances a structure exhibits is determined by the linear natural frequencies and the nature of the nonlinearity inherent in the structure. When such structures are resonantly excited, energy can transfer between the modes in internal resonance. The resulting modal interactions are known to lead to periodic, quasi-periodic and chaotic motions (Nayfeh and Balachandran, 1989).

One-to-one internal resonances arise in many physical systems including the spherical pendulum (Miles, 1984) and the stretched string (Bajaj and Johnson, 1992). In many cases, these one-to-one resonances are a consequence of the geometric symmetry present in the idealized system, and the equations of motion reflect this symmetry. An important class of structures leading to one-to-one

S. Narayanan and R.N. Iyengar (eds.),
IUTAM Symposium on Nonlinearity and Stochastic Structural Dynamics, 29–41.

internal resonances are periodic structural systems with cyclic symmetry. Examples of these systems include bladed-disk assemblies, circular space antennas, and magnetic storage devices.

Motivated by the problem of mode localization in periodic and mistuned structures, Vakakis (1992) conducted an extensive study of the free and forced response of a nonlinear periodic structure. For the case of a harmonic excitation, he found that a stable branch of "traveling wave" type solutions bifurcates from a branch of "standing wave" type motions. In their studies on a similar model, Samaranayake *et al.*(1995) performed a careful computer-assisted local bifurcation analysis of the averaged equations. They showed that over some frequency interval the "traveling wave" type solutions can undergo a Hopf bifurcation, thus giving rise to limit cycle solutions for the averaged equations. These limit cycles and their transition to chaos was studied through direct numerical simulation. The present work details some results of investigations into the possibility of analytically predicting the existence of this chaotic dynamics.

The analysis here of the averaged equations is based on the higher-dimensional Melnikov method which is a perturbation technique that can be utilized to detect the presence of orbits homoclinic or heteroclinic to various types of invariant sets. The detailed developments for this analysis are given in Wiggins (1988). Banerjee and Bajaj (1997) applied this technique to determine parameter conditions needed for the existence of chaotic dynamics in systems with quadratic nonlinearities. As it turns out, this approach works only in the so-called "non-resonant" cases. For the "resonant" cases, a completely different approach to detecting homoclinic and heteroclinic orbits has to be employed. Kovacic and Wiggins (1992) and Haller and Wiggins (1993) have developed the required techniques for resonant cases for the non-Hamiltonian and Hamiltonian disturbances, respectively. The application of their analysis techniques to cyclic systems can be found in Samaranayake (1996).

2 Equations of Motion and Averaging

Consider a cyclic system consisting of n identical particles of mass m each, arranged in a ring, and interconnected by nonlinear extensional springs (see Figure 1).

Let x_i be the tangential displacement of the ith particle. Then,

$$f_{ij} = k_1(x_i - x_j) + \epsilon k_2(x_i - x_j)^3 \tag{1}$$

is the force in the spring between the ith particle and its jth neighbor, and

$$\Gamma_i = T_1 \frac{x_i}{a^2} + \epsilon T_2 \frac{{x_i}^3}{a^4} \tag{2}$$

is the force induced on the ith mass particle due to the torsional spring attached to it. Here k_1, T_1 and $\epsilon k_2,\ \epsilon T_2$ are, respectively, the linear and nonlinear stiffness coefficients for the extensional and torsional springs, and a is the radius of the

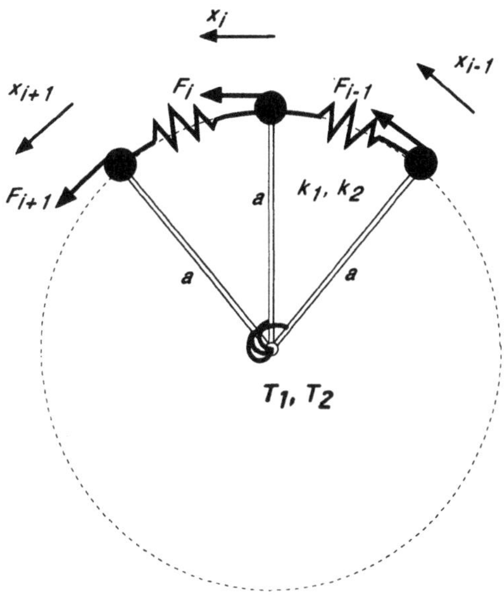

Figure 1: The cyclic system model.

ring. Each particle's motion is also assumed to be viscously damped with $\epsilon d\dot{x}_i$ as the damping force. The parameter ϵ is small, $0 \leq \epsilon \ll 1$, thus representing small nonlinearities and dissipation. Let $\tau = \bar{\omega}t$, where $\bar{\omega} = \left\{\frac{T_1}{ma^2}\right\}^{\frac{1}{2}}$. Here, $\bar{\omega}$ is the natural frequency of an individual substructure. Then, the equations of motion for the n degree-of-freedom cyclic structure can be written in a matrix form as

$$\underline{x}'' + \underline{K}\underline{x} + \frac{1}{m\bar{\omega}^2}(\epsilon\underline{F} - \underline{F}^{ex}) = 0, \tag{3}$$

where $\quad \underline{F} = [f_1 \quad f_2 \quad \dots \quad f_n]^T$, and where an external force given by

$$\underline{F}^{ex} = \mu\underline{\Gamma}\cos\bar{\Omega}t = \mu\underline{\Gamma}\cos\Omega\tau \qquad \text{and} \qquad \bar{\Omega} = \bar{\omega}\Omega, \tag{4}$$

has also been incorporated.

The matrix $\underline{K}$ is a real symmetric cyclic matrix. Thus, as is well known (Vakakis,1992), the linear system (when $\epsilon = 0$) has mostly pairwise degenerate natural frequencies with orthogonal normal modes. We wish to study nonlinear interactions between the pairs of modes with identical (or nearly identical) linear natural frequencies when the external harmonic excitation frequency is close to the linear natural frequency.

To study the dynamics of the system defined by equations (3), the method of averaging (Guckenheimer and Holmes, 1983) is utilized here. It transforms the equations of motion to an autonomous set of first-order ordinary differential equations. These averaged equations represent the slow-time evolution of the amplitudes and phases of harmonic motion of the interacting modes in resonance.

In the primary resonance case, we assume that the excitation frequency Ω is close to one of the degenerate natural frequencies, ω_j, and the excitation force is spatially distributed like one of the corresponding two modes in resonance, that is $\underline{F}^{ex} = \epsilon(\frac{\mu}{e})\underline{U}_j^c \cos\bar{\Omega}t$, where e is the magnitude of $\underline{U}_j^c$. Then, as shown in Samaranayake *et al.*(1995), the steady-state "mono-frequency" response of the system can be expressed in the form

$$\underline{x}(\tau) = \frac{1}{e}\{B_j(\tau)\underline{U}_j^c \cos(\Omega\tau + \beta_j(\tau)) + C_j(\tau)\underline{U}_j^s \cos(\Omega\tau + \gamma_j(\tau))\}, \tag{5}$$

where $e = \{\sum_{i=1}^n \cos^2 i\alpha_j\}^{\frac{1}{2}} = \{\sum_{i=1}^n \sin^2 i\alpha_j\}^{\frac{1}{2}} = \sqrt{\frac{n}{2}}$. Then, following the averaging procedure, we finally obtain the following averaged equations:

$$\begin{aligned} B_j' &= \epsilon\left\{pB_jC_j^2 sin2(\beta_j - \gamma_j) - dB_j - \sigma\sin\beta_j\right\}, \\ B_j\beta_j' &= \epsilon\left\{B_j\left\{3pB_j^2 + [2+\cos 2(\beta_j - \gamma_j)]pC_j^2 - \lambda_{jc}\right\} - \sigma\cos\beta_j\right\}, \\ C_j' &= \epsilon\left\{pB_j^2C_j\sin 2(\gamma_j - \beta_j) - dC_j\right\}, \\ C_j\gamma_j' &= \epsilon C_j\left\{[2+\cos 2(\gamma_j - \beta_j)]pB_j^2 + 3pC_j^2 - \lambda_{js}\right\}, \end{aligned} \tag{6}$$

where

$$p = \frac{1}{2m\bar{\omega}^2\Omega n^2}\left\{k_2L + \frac{T_2M}{a^4}\right\}, \quad d = \frac{d}{2m\bar{\omega}^2},$$
$$\sigma = \frac{\mu}{2m\bar{\omega}^2\Omega}, \quad \lambda_{jc} = \frac{\bar{\lambda}_{jc}}{2\Omega}, \quad \lambda_{js} = \frac{\bar{\lambda}_{js}}{2\Omega}. \tag{7}$$

These averaged equations depend on four parameters: the two frequency mistunings for the external excitation, the damping, and the amplitude of the external forcing. A detailed bifurcation analysis of these equations was performed by Samaranayake *et al.*(1995). In the following section, we give a sample of the numerical results clearly indicating the existence of chaotic behavior.

3 Constant, Periodic, and Chaotic Solutions of the Averaged Equations

A numerical study of constant solutions of the averaged equations was performed by using AUTO (a bifurcation analysis package) and by direct time integration. A representative set of response curves for excitation amplitude $\sigma = 0.1$ and damping parameter $d = 0.12$ is shown in Figure 2. These are given in terms of the four Cartesian components x_1, y_1, x_2 and y_2, where $x_1 = B_j\cos\beta_j$, $y_1 = B_j\sin\beta_j$, $x_2 = C_j\cos\gamma_j$, and $y_2 = C_j\sin\gamma_j$. For $\lambda_c < \lambda_{PF_1}$, the only solution is the single-mode solution directly excited by the external forcing in the form of a "cos $-j$" mode of the cyclic structure. The single-mode solution branches have turning points at λ_{SNS_1} and λ_{SNS_2} with the middle branch between the turning points being always unstable, and the lower branch being always stable. The upper

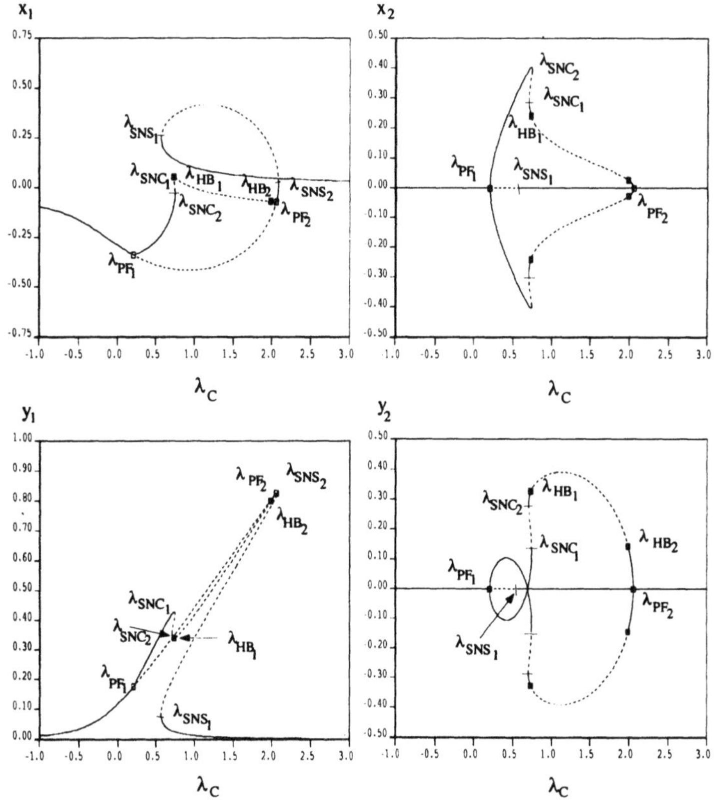

Figure 2: Steady-state response amplitudes as a function of the excitation frequency λ_c; $\sigma = 0.1$, $d = 0.12$, $\mu = 0$.

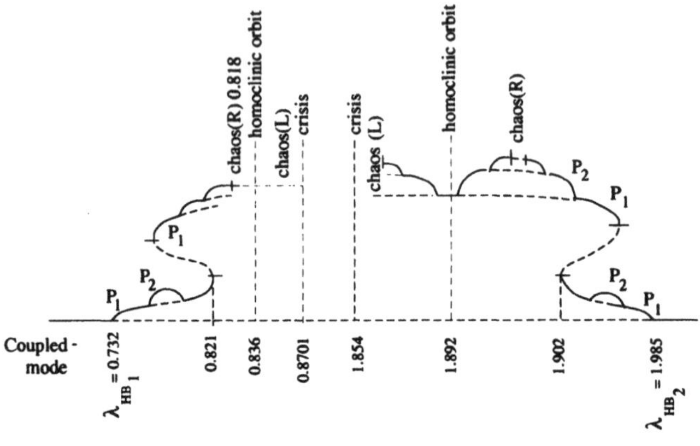

Figure 3: Qualitative bifurcation diagram for steady-state coupled-mode dynamic motions; $\sigma = 0.1$, $d = 0.12$, $\mu = 0$.

single-mode branch undergoes pitchfork bifurcations to coupled-mode solutions that exist between the frequency interval $(\lambda_{PF_1}, \lambda_{PF_2})$. There is a multi-valued region for coupled-mode constant solution in the frequency interval $(\lambda_{SNC_1}, \lambda_{SNC_2})$. There is also a wide frequency interval $\lambda_{HB_1} \leq \lambda_c < \lambda_{HB_2}$ where the coupled-mode constant solution is unstable by a Hopf bifurcation and we expect limit cycle solutions of the averaged equations, at least near the bifurcation frequencies λ_{HB_1} and λ_{HB_2}. At this level of damping, however, the dynamic steady-state solutions, or the solutions in the frequency interval $(\lambda_{HB_1}, \lambda_{HB_2})$ are very involved (Samaranayake *et al.*,1995).

Figure 3 shows a qualitative bifurcation diagram with approximate values of the excitation frequency λ_c at which changes in types of solutions take place. At $\lambda_{HB_1} = 0.732$, a limit cycle solution (denoted as P_1 solution) arises due to a Hopf bifurcation. It undergoes a period-doubling bifurcation to a solution with twice the period (P_2 solution), and then reverses to a P_1 solution. Around $\lambda_c = 0.821$, the P_1 solution seems to jump to a chaotic solution encircling only a single coupled-mode equilibrium point, and is thus identified as a Rössler type chaotic solution. This solution, when continued for lower frequencies, seems to arise due to period-doubling bifurcations from a limit cycle solution which also jumps back, for $\lambda_c < 0.811$, to the P_1 solution. This branch of solutions, most likely, arises at $\lambda_c = 0.811$ due to a saddle-node bifurcation, where a stable periodic (P_1) and an unstable periodic solution are created. This unstable periodic solution continues up to $\lambda_c = 0.821$ where it annihilates the stable P_1 solution, thus causing the jump from solutions branch arising at the Hopf bifurcation point. The Rössler type chaotic solution merges with its twin around $\lambda_c = 0.836$ via a homoclinic orbit and results in a symmetric Lorenz type chaotic attractor. The Lorenz type chaotic solution undergoes a "crisis" around $\lambda_c = 0.8701$, whereby, the chaotic attractor collides with the stable manifold of the saddle-type middle single-mode constant solution and after long "chaotic transients", settles down to the lower stable single-mode constant solution. A somewhat similar behavior is observed for frequencies beginning with the right Hopf frequency λ_{HB_2}, with the "crisis" arising near $\lambda_c = 1.854$. Thus, over the frequency interval $0.8701 < \lambda_c < 1.854$, almost all initial conditions, after some long transients, lead to the stable lower single-mode solution.

In the remainder of this work, we give some results of analysis of the averaged equations, required to predict the existence of chaotic behavior.

4 Standard Form: The Perturbed Hamiltonian System

In order to apply global perturbation techniques, we need to transform our averaged system to a form in which it can be considered as a perturbation of a Hamiltonian system. In addition, the reduced system, which is Hamiltonian, must be completely integrable. Thus, the averaged system (6) is transformed, using the transformations

$$x_1 = C\cos Q_1, x_2 = C\sin Q_1, Q_1 = \frac{\pi}{2} - (\gamma - \beta),$$
$$Q_2 = -\beta, P_2 = \frac{1}{2}(B^2 + C^2), \tag{8}$$

and substitutions $d = \delta d, \mu = \delta\mu$, to the following general form

$$\begin{aligned} \dot{X} &= JD_x H_0(X, P_2) + \delta g^X(X, P_2, Q_2), \\ \dot{P}_2 &= \delta g^{P_2}(X, P_2, Q_2), \\ \dot{Q}_2 &= D_{P_2} H_0(X, P_2) + \delta g^{Q_2}(X, P_2, Q_2), \end{aligned} \tag{9}$$

where $H_0 = -x_1^2x_2^2 + 2x_1^2P_2 - x_1^4 - \frac{\mu}{2}(x_1^2 + x_2^2) + \lambda_c P_2 - 3P_2^2$, $(X, P_2, Q_2) \in R^2 \bigotimes R \bigotimes T^1$, $g^X = (g^{x_1}, g^{x_2})$, $g^{x_1} = -dx_1 + D_{x_2}H_\delta$, $g^{x_2} = -dx_2 - D_{x_1}H_\delta$, $g^{P_2} = -2dP_2 - D_{Q_2}H_\delta$, $g^{Q_2} = D_{P_2}H_\delta$, $H_\delta = \sigma\sqrt{2(P_2 - P_1)}\cos Q_2$, and

$$J = \begin{bmatrix} 0 & 1 \\ -1 & 0 \end{bmatrix}. \tag{10}$$

In the above, H_0 represents the unperturbed Hamiltonian for the system (equations (9) with $\delta = 0$), H_δ represents the perturbation terms that are derivable from a Hamiltonian, and $D(.)$ is the partial derivative of the appropriate function with respect to the subscript variable, δ is an arbitrary small parameter, $0 < \delta << 1$, $\mu = \lambda_{jc} - \lambda_{js}$ is the internal mistuning, $\lambda_c = \lambda_{jc}$ is the external mistuning, and $P_1 = \frac{x_1^2 + x_2^2}{2}$. Note here that the external excitation and the damping are assumed to be of $O(\delta)$ compared to the nonlinearity, and therefore, the system for $\delta = 0$ represents the averaged two degree-of-freedom system in the absence of external forcing and dissipation.

This system in equations (9), which can be identified as system I in Wiggins (1988), is completely integrable when $\delta = 0$, and both the Hamiltonian function H_0 and the variable P_2 are constants of motion for the unperturbed system.

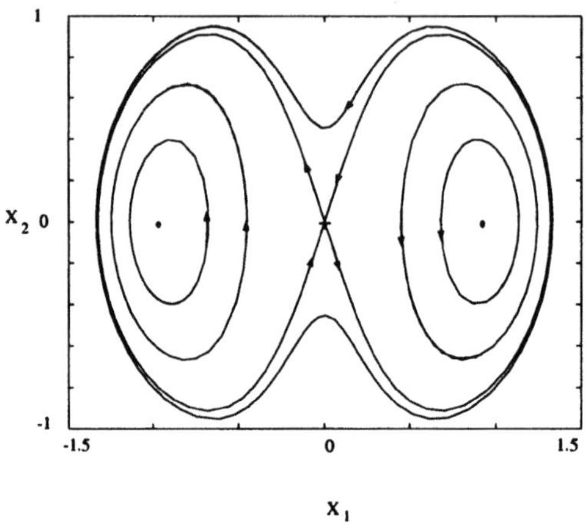

Figure 4: Phase portrait in the x_1-x_2 invariant plane for the unperturbed system; $P_{20} = 1$, $\mu = 0.5$.

5 Geometry of the Unperturbed System

The system obtained by setting $\delta = 0$ is referred to as the unperturbed system. Note that $P_2 = \text{constant} = P_{20}$, and Q_2 does not appear in the first three equations. Thus, the dynamics in the (x_1, x_2) plane is decoupled from that in the (P_2, Q_2) plane and is parameterized by P_{20}. The variable Q_2 is a periodic variable, $0 \leq Q_2 \leq 2\pi$, and $0 \leq P_1 \leq P_{20}$. The dynamics in the (x_1, x_2) plane is determined by the equations

$$\begin{aligned} \dot{x}_1 &= -x_2(2x_1^2 + \mu), \\ \dot{x}_2 &= -x_1[4P_2 - 2x_2^2 - 4x_1^2 - \mu]. \end{aligned} \tag{11}$$

Equations (11) are invariant under the transformation $(x_1, x_2) \to (-x_1, -x_2)$. Therefore, any homoclinic orbit that we find in the (x_1, x_2) plane has a symmetric twin. Also note that these equations have a fixed point at the origin for all values of P_{20}.

When $\mu = 0$, entire x_2-axis is an invariant set, thus leading to a degenerate system which is excluded from further consideration. The origin, $(x_1, x_2) = (0, 0)$, is an equilibrium point which exists for all μ and all $P_{20} \geq 0$. For $\mu > 0$, when $0 < P_{20} < \frac{\mu}{4}$, the origin is a center and for $P_{20} > \frac{\mu}{4}$, the origin is a saddle. The equilibrium point at the origin undergoes a pitchfork bifurcation at $P_{20} = \frac{\mu}{4}$, generating two additional equilibrium points $(x_1, x_2) = \pm(\sqrt{P_{20} - \frac{\mu}{4}}, 0)$ which are centers. Since the planar unperturbed system is Hamiltonian, the saddle point at the origin has a pair of homoclinic orbits, orbits bi-asymptotic to it as $t \to \pm\infty$. A typical phase portrait in the invariant (x_1, x_2) plane is shown in Figure 4 for $P_{20} > \frac{\mu}{4} > 0$.

It is easy to show that the homoclinic orbits satisfy

$$P_1 = P_{20} - \frac{\mu}{2(1+\cos 2Q_1)}, \tag{12}$$

where $Q_1 = \arctan(\frac{x_2}{x_1})$.

When $\mu < 0$, the origin is always stable and a center. There also arise other equilibrium solutions for which $P_1 = \frac{(x_1^2+x_2^2)}{2} > P_2$ and thus, they are not physically relevant. Thus, in the present analysis, only the case of $\mu > 0$ is considered. In particular, we restrict the parameter region where $P_{20} \geq \frac{\mu}{4}$ and the origin is a saddle point with homoclinic orbits connecting to it.

In the full four-dimensional phase space, the set

$$M_0 = \{(x_1, x_2, P_2, Q_2) \mid x_1 = x_2 = 0, 0 < \frac{\mu}{4} \leq P_2 < P_2^{max}, 0 \leq Q_2 < 2\pi\} \tag{13}$$

is a two-dimensional manifold invariant under the flow generated by the unperturbed vector field. Let $P_2 < P_2^{max}$ for some finite value P_2^{max}, so that M_0 is a bounded (compact) set. Since $(x_1, x_2) = (0, 0)$ is a saddle point, M_0 is a normally hyperbolic invariant manifold.

The manifold M_0 has a three-dimensional stable and a three-dimensional unstable manifold denoted by $W^s(M_0)$ and $W^u(M_0)$, respectively. The existence of the homoclinic orbits connecting the origin implies the non-transversal intersection of the stable manifold of M_0, $W^s(M_0)$, and the unstable manifold of M_0, $W^u(M_0)$, along a three-dimensional homoclinic manifold denoted by Γ.

This homoclinic manifold Γ can be expressed as

$$\Gamma = \{(X, P_2, Q_2) \mid X = X^h(t, P_2), 0 < \frac{\mu}{4} \leq P_2 < P_2^{max}, Q_2 = Q_2(t, P_2, Q_{20})\} \tag{14}$$

where $X^h = (x_1^h(t, P_2), x_2^h(t, P_2))$ gives the time parameterization of the homoclinic orbits, each of which is described by

$$H_0(x_1, x_2, P_2) - H_0(0, 0, P_2) = 0, \tag{15}$$

and $Q_2(t, P_2, Q_{20}) = \int_0^t D_{P_2} H_0(x_1^h, x_2^h, P_2) ds + Q_{20}$ which can be readily integrated to obtain the following:

$$Q_2(t) = (\lambda_c - 6P_2)t + \arctan\left(\sqrt{\frac{4P_2 - \mu}{\mu}} \coth \sqrt{\mu(4P_2 - \mu)}t\right) + Q_{20}. \tag{16}$$

Clearly, the trajectories in Γ approach a trajectory in M_0 asymptotically as $t \to \pm\infty$. Also, Q_{20} is a constant which is determined by the initial conditions.

6 Dynamics on M_0

The dynamics on the normally hyperbolic manifold M_0 is determined by the equations

$$\begin{aligned} \dot{P}_2 &= 0, \\ \dot{Q}_2 &= \lambda_c - 6P_2. \end{aligned} \tag{17}$$

Since P_2 is a constant on M_0, all orbits on the invariant manifold are concentric circles centered at the origin. Thus, when $6P_2 - \lambda_c \neq 0$, we have for each P_2, periodic orbits which are circles. Each orbit, defined by equation (12) is homoclinic to one of these concentric circles in M_0. When $P_2 = \frac{\lambda_c}{6}$, the corresponding circle is a circle of fixed points and the homoclinic orbits are indeed heteroclinic to fixed points on that circle. The case $P_2^r = P_2 = \frac{\lambda_c}{6}$ is called the resonance case.

When subjected to small perturbations, the dynamics on M_0 will be quite different in the resonance case as compared to the nonresonance case, and different techniques of analysis are needed to investigate resonance and nonresonance cases (Samaranayake,1996). In this work, results for only the non-resonance case are presented.

7 Non-resonance Case: $(P_2 \neq \frac{\lambda_c}{6})$

Since M_0 is normally hyperbolic, an invariant manifold M_δ (whose limit, as $\delta \to 0$, is the manifold M_0) can be obtained by a perturbation expansion of M_0 in powers of δ. The dynamics on M_δ is determined by the equations

$$\begin{aligned} \dot{P}_2 &= \delta[-2dP_2 + \sigma\sqrt{2P_2}\sin Q_2] + O(\delta^2), \\ \dot{Q}_2 &= \lambda_c - 6P_2 + \delta\frac{\sigma\cos Q_2}{\sqrt{2P_2}} + O(\delta^2). \end{aligned} \tag{18}$$

Since we have excluded the case satisfying the condition of $\lambda_c = 6P_2$, equations (18) can be studied by using the method of averaging for sufficiently small δ. Then, $\frac{dP_2}{dQ_2} = \delta[\frac{-2dP_2 + \sigma\sqrt{2P_2}\sin Q_2}{\lambda_c - 6P_2}] + O(\delta^2)$, which is 2π-periodic in Q_2. Averaging over the independent variable Q_2, we get the averaged equation $\frac{dP_2}{dQ_2} = -2\frac{dP_2}{\lambda_c - 6P_2}$, which, when combined with the fact that $\dot{Q}_2 = \lambda_c - 6P_2$ for small δ, gives $P_2 \to 0$ as $\tau \to \pm\infty$. Consequently orbits on M_δ leave M_δ, violating the persistence. Therefore, in the nonresonance case, the present analysis approach has to be restricted to the undamped system ($d = 0$). With $d = 0$, the entire vector field in equations (9) can be derived from the Hamiltonian $H = H_0 + \delta H_\delta$.

In the unperturbed system, the invariant manifold M_0 consists of concentric circles centered at the origin, and the stable and unstable manifolds of M_0 intersect along the homoclinic manifold Γ. When subjected to a Hamiltonian perturbation, KAM theory (Wiggins 1988) holds and most of the circles persist for sufficiently small δ. M_0 persists under small perturbations as a locally invariant manifold with boundary, which we have denoted above by M_δ. Furthermore, the manifolds $W^s(M_0)$ and $W^u(M_0)$ also persist under small perturbations as locally invariant manifolds $W^s_{loc}(M_\delta)$ and $W^u_{loc}(M_\delta)$, respectively. The global stable and unstable manifolds of M_δ can be determined by the time evolution of the manifolds $W^s_{loc}(M_\delta)$ and $W^u_{loc}(M_\delta)$, respectively. Transversel intersection of the stable and unstable manifolds of M_δ will lead to the existence of Smale horseshoes and, consequently, the chaotic dynamics.

The distance between the stable and the unstable manifolds of M_δ can be calculated using the Melnikov method. The main point behind the Melnikov method is to measure the distance between $W^s(M_\delta)$ and $W^u(M_\delta)$ along a trajectory of the unperturbed homoclinic manifold. This distance measure uses the global geometry of the unperturbed Hamiltonian structure.

Following Wiggins(1988), the Melnikov function given by

$$\begin{aligned} M^I &= \int_{-\infty}^{\infty} D_{Q_2} H_\delta(q_0(t))dt \\ &= \sigma \int_{-\infty}^{\infty} \sqrt{\frac{\mu}{2}} \frac{\sin Q_2}{\cos Q_1} dt. \end{aligned} \tag{19}$$

where, $q_0(t)$ refers to the trajectory on the unperturbed homoclinic manifold which contains the point p, and $Q_1 = \arctan \frac{x_2}{x_1}$. This integral converges conditionally. It has been shown in Wiggins (1988) that this improper integral makes sense if the limits $\pm\infty$ are approached along the sequence of times $\{T_j^s\}$ and $\{-T_j^u\}$, j = 1,2,....., where $T_j^s = T_j^u = \frac{2\pi j}{\lambda_c - 6P_2}$. Then, the integral can be easily evaluated to yield

$$M^I = 4\sigma\sqrt{\frac{\mu}{2}} b \sin\psi_0 \{\frac{\pi}{4c}\coth(\frac{\pi a}{2c}) - \frac{1}{2a}\}, \tag{20}$$

where $a = \lambda_c - 6P_2$, $b = \sqrt{\frac{4P_2 - \mu}{\mu}}$ and $c = \sqrt{\mu(4P_2 - \mu)}$.

If M^I has a simple zero with respect to ψ_0, which is the case when the coefficient of $\sin\psi_0$ is not zero, the stable and the unstable manifolds intersect transversely near the simple zero of M^I. A single intersection of the manifolds $W^u(\tau_\delta(\bar{P}_2))$ and $W^s(\tau_\delta(\bar{P}_2))$ implies an infinity of intersections, and the dynamical consequences associated with such an intersection are Smale horseshoes, and hence chaotic dynamics for the perturbed Hamiltonian system.

In terms of the system parameters, the Melnikov function in equation (20) indicates that any nonzero level of external excitation ($\sigma \neq 0$) will result in M^I having simple zeros so long as the other conditions of $P_2 > \frac{\mu}{4}$ and $\lambda_c \neq 6P_2$ are satisfied. Thus, given an internal mistuning μ, a minimum total energy level of initial conditions consistent with the nonresonance condition is required before the undamped system can exhibit chaotic behavior.

Figures 4 and 5 show some results of numerical simulations for equations (6), obtained by using DSTOOL. Figure 4 represents a phase portrait of the unperturbed system for $\mu = 0.5$ and $P_2 = 1$. The origin in the (P_1, Q_1) plane is a saddle point and has a pair of symmetric homoclinic orbits which encircle two center-type nontrivial equilibrium points. Figures 5a ($\sigma = 0.5$, $\delta = 0.01$) and 4b ($\sigma = 0.5$, $\delta = 0.1$) represent Poincare sections, at $Q_2 = 0$, of solutions of the perturbed system. The orbits for the unperturbed system, which are away from the homoclinic orbit, survive under small perturbations, as is expected from the KAM theory. Only the orbits which are sufficiently close to the homoclinic orbit of the unperturbed system lead to chaotic motions, that is, chaotic behavior is confined to a small region of the phase space. When the strength of the Hamiltonian perturbation is increased (Figure 5b), chaotic motions occupy a much larger region of the phase space.

We should note here that, even though the analysis is valid only for the undamped system,the chaotic behavior uncovered appears to persist under sufficiently small damping. Figure 5c represents a Poincare section at $Q_2 = 0$, of a chaotic solution for $\sigma = 0.5$, $\delta = 0.01$, $\mu = 0.5$ and $d = 0.0001$. Figure 5d is for damping $d = 0.001$. On increasing the damping, it is seen that the chaotic behavior persists for a while before it starts diffusing towards the zero solution; thus, suggesting that it is a "transient" chaotic motion.

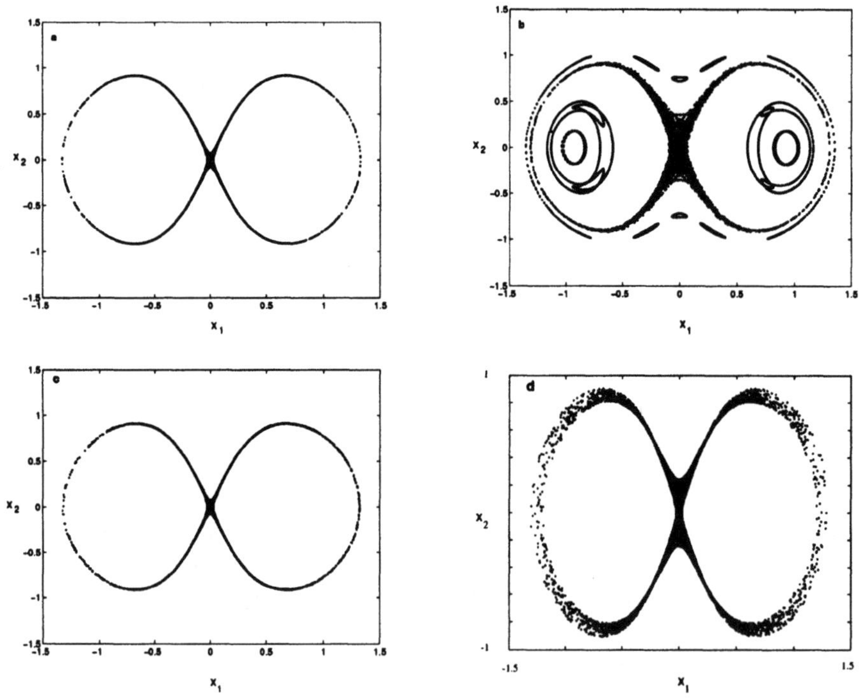

Figure 5: Poincare section of solutions with the surface of section defined by $Q_2 = 0$; $\lambda_c = 1$, $\mu = 0.5$, $\sigma = 0.5$. Undamped cases: **a** $\delta = 0.01$, **b** $\delta = 0.1$, Damped cases: **c** $\delta = 0.01$, $d = 0.0001$ **d** $\delta = 0.01$, $d = 0.001$.

8 Summary

In this work, global behavior of the averaged equations for the strongly coupled system in the primary resonance case is investigated. The analysis is based on methods which are applicable to a class of systems that are perturbations of completely integrable Hamiltonian systems. Both the cases of Hamiltonian and non-Hamiltonian perturbations are considered. In the non-resonance case, using higher-dimensional Melnikov method, a criterion to detect the transversal intersection of stable and unstable manifolds of periodic orbits, which are preserved under Hamiltonian perturbations, is derived. Dynamical consequences associated with such intersections are Smale horseshoes and, hence, chaotic dynamics for the perturbed Hamiltonian system. When subjected to non-Hamiltonian perturbations, analytical methods fail to detect chaotic dynamics in the system.

9 References

1. A. H. Nayfeh and B. Balachandran, "Modal interactions in dynamical and structural systems", *Applied Mechanics Reviews* **42 part 2** (11), S175-S201 (1989).

2. J. W. Miles, "Resonant motion of a spherical pendulum", *Physica D* **11**, 309-323 (1984).

3. A. K. Bajaj and J. M. Johnson, "On the amplitude dynamics and crisis in resonant motion of stretched strings", *Philosophical Transactions of the Royal Society London* **A338**,1-41 (1992).

4. A. F. Vakakis, "Dynamics of a nonlinear periodic structure with cyclic symmetry", *Acta Mechanica* **95**, 197-226 (1992).

5. S. Samaranayake, A. K. Bajaj and O. D. I. Nwokah, "Amplitude modulated dynamics and bifurcations in the resonant response of a structure with cyclic symmetry", *Acta Mechanica* **109**, 101-125 (1995).

6. S. Wiggins, *Global Bifurcations and Chaos*, Springer-Verlag, New York (1988).

7. B. Banerjee and A. K. Bajaj, "Amplitude modulated chaos in two degree-of-freedom systems with quadratic nonlinearities", *Acta Mechanica* **124**, 131-154 (1997).

8. G. Kovacic and S. Wiggins, "Orbits homoclinic to resonances, with an application to chaos in a model of the forced and damped sine-gordon equation", *Physica D* **57**, 185-225 (1992).

9. G. Haller and S. Wiggins, "Orbits homoclinic to resonances", *Physica D* **66**, 298-346 (1993).

10. S. W. Samaranayake, *Studies in Resonant Responses of Nonlinear Structures with Cyclic Symmetry*, Ph. D. Dissertation, Purdue University, West Lafayette, IN (1996).

SIMULATION OF NON-GAUSSIAN STOCHASTIC PROCESSES WITH NONLINEAR FILTERS

G. Q. CAI AND Y. K. LIN
Center for Applied Stochastics Research
Florida Atlantic University, Boca Raton, FL 33431, USA

Abstract

Nonlinear filters are designed to generate stationary stochastic processes with the knowledge of their spectral densities and first-order probability densities. Two types of spectral densities are considered: the low-pass type and the narrow-band type. In particular, the nonlinear filters are represented in the form of Itô stochastic differential equations, in which the drift coefficients are adjusted to match the spectral density, and the diffusion coefficients are determined according to the probability distribution. The scheme can be used to generate excitation random processes which are clearly non-Gaussian. Since such excitations are described by stochastic differential equations, they can be combined with the governing equations of the excited dynamical systems in analytical investigation, or as a basis for Monte Carlo simulation.

1. Introduction

When investigating a dynamical system under random excitations, it is important that the excitation processes should be modeled properly according to their estimated statistical and probabilistic properties. Emphasis in the past has been focused on the spectral densities, with little attention paid to the probability distribution. In fact, Gaussian distribution was often implied in earlier studies. The well-known simulation scheme of Fourier series expansion (e.g. Shinozuka, 1972; Shinozuka and Jan, 1972), which is capable of matching an arbitrary spectral density, leads asymptotically to a Gaussian distribution. The effect of inaccurate account for the probability distribution of the excitation process can be significant, especially when the behaviors of the system during the transient non-stationary state are of interest (Cai and Lin, 1997, 1998).

Several procedures have been proposed to generate non-Gaussian processes. For example, Yamazaki and Shinozuka (1988) incorporated a numerical iterative procedure in the Fourier expansion scheme. Liu and Munson (1982), Sondhi (1983), Nakayama (1994), and Gurley et al. (1996) proposed various memoryless nonlinear transformations to convert Gaussian processes to non-Gaussian ones. A randomized sinusoidal model was proposed independently by Dimentberg (1988) and Wedig (1989) with a constant

S. Narayanan and R.N. Iyengar (eds.),
IUTAM Symposium on Nonlinearity and Stochastic Structural Dynamics, 43–54.

amplitude and a random phase varying as a Wiener process. This model is capable of simulating both narrow-band and broad-band processes, and it has been used, for example, to investigate motion stability of long-span bridges in turbulent winds (Li and Lin, 1995). The use of nonlinear filters was considered by Kontorovich and Lyandres (1995). The simulated process was obtained as the output from a nonlinear system under a Gaussian white-noise excitation. The coefficients in the equation governing the filter were chosen to match simultaneously the given probability density and the spectral density. A simpler nonlinear filtering scheme was developed by Cai and Lin (1996), in which the nonlinear filter was described by an Itô stochastic differential equation, with the drift coefficient adjusted to match a low-pass spectral density and the diffusion coefficient adjusted to match the probability density. The latter procedure has been used to investigate the effects of excitation probability distribution on the response and reliability of linear and nonlinear oscillators (Cai and Lin, 1997, 1998).

The present paper is an extension of the nonlinear filter scheme to the case of narrow-band spectral density and arbitrary probability distribution. In this case, the filter is governed by two coupled Itô stochastic differential equations, again with the drift coefficients adjusted to match the spectral density and the diffusion coefficient adjusted to match the probability density. The procedure is considered to be of practical significance since some real excitation processes are narrow-banded.

2. Generation of Low-Pass Non-Gaussian Random Processes

It is illuminative to review briefly the nonlinear filter technique for generating non-Gaussian processes of low-pass spectral density, following our earlier paper of 1996.

Let the generated random process be a stationary diffusive Markov process $X(t)$ with zero mean, governed by the following Itô stochastic differential equation (Itô, 1951)

$$dX = -\lambda X\,dt + D(X)\,dB(t) \tag{1}$$

where λ is a positive constant, and $B(t)$ is a unit Wiener process. Multiplying (1) by $X(t_0)$ and taking the ensemble average, we obtain

$$\frac{dR_{XX}(\tau)}{d\tau} = -\lambda R_{XX}(\tau) \tag{2}$$

where $R_{XX}(\tau)$ is the correlation function of $X(t)$, namely,

$$R_{XX}(\tau) = \mathrm{E}[X(t_0)\,X(t)]\,; \qquad \tau = t - t_0 \tag{3}$$

Let the mean-square value of $X(t)$ be

$$R_{XX}(0) = \mathrm{E}[X^2(t)] = \sigma^2 \tag{4}$$

Solution for equation (2) subject to condition (4) is given by

$$R_{XX}(\tau) = \sigma^2 \exp(-\lambda\,|\tau|) \tag{5}$$

The corresponding spectral density of $X(t)$, the Fourier transform of $R_{XX}(\tau)$, is of the low-pass type,

$$\Phi_{XX}(\omega) = \frac{1}{2\pi}\int_{-\infty}^{\infty} R_{XX}(\tau)\, e^{-i\omega\tau}\, d\tau = \frac{\lambda\,\sigma^2}{\pi\,(\omega^2+\lambda^2)} \tag{6}$$

The Fokker-Planck equation, governing the probability density $p_X(x)$ of $X(t)$ at the stationary state, is given by

$$\frac{d}{dx}\left\{\lambda x p_X(x) + \frac{1}{2}\frac{d}{dx}[D^2(x)\, p_X(x)]\right\} = 0 \tag{7}$$

This stationary probability density exists if and only if

$$D^2(x) = -\frac{2\lambda}{p_X(x)}\int_{x_l}^{x} u\, p_X(u)\, du \tag{8}$$

where x_l is the lower bound for $X(t)$. The value of x_l may be finite or $-\infty$. Function $D^2(x)$, computed from equation (8), is non-negative, as it should, since $p_X(x) \geq 0$ and the mean value of $X(t)$ is zero. Thus, we have shown that the stochastic process $X(t)$ generated from the nonlinear filter, equation (1), with $D(X)$ given by (8), possesses a prescribed stationary probability density $p_X(x)$ and a low-pass spectral density (6). Note that a change of the λ value only changes the spectral density, not the probability density. On the other hand, the diffusion coefficient $D(X)$ is determined only from the probability density, not related to the spectral density.

Fir illustration, consider a family of probability densities

$$p_X(x) = C\exp(-\gamma x^2 - \delta x^4); \quad -\infty < x < \infty; \quad \delta > 0 \tag{9}$$

Substitution of (9) into (8) leads to

$$D^2(x) = \frac{\lambda}{2}\sqrt{\pi/\delta}\, \exp[\delta(x^2+\frac{\gamma}{2\delta})^2]\ \mathrm{erfc}[\sqrt{\delta}\,(x^2+\frac{\gamma}{2\delta})] \tag{10}$$

where erfc($\cdot$) is the complementary error function

$$\mathrm{erfc}(u) = \frac{2}{\sqrt{\pi}}\int_{u}^{\infty} e^{-t^2}\, dt \tag{11}$$

Three examples of $X(t)$ are generated from equations (1) and (10) with the same $\lambda = 1$

and $\sigma^2 = 1$, but with different sets of γ and δ values as follows:

Case 1: $\gamma = 0.4$, $\delta = 0.0185$

Case 2: $\gamma = -1$, $\delta = 0.4376$

Case 3: $\gamma = -5$, $\delta = 2.209$

The same λ and σ^2 values guarantee that the same spectral density is shared by the three generated processes, while different sets of γ and δ give rise to different probability densities. It should be noted that values for γ and δ are dependent of each other in order to meet the condition $\sigma^2 = 1$.

Figures 1 and 2 depict, respectively, the spectral densities in the positive ω-domain and the probability densities in the positive x-domain for the three cases. The plots in the negative ω-domain or the negative x-domain are mirror images. In these figures, the lines are calculated from (6) and (9), respectively, while the diamonds, squares and triangles represent the results obtained from simulation, using equations (1) and (10).

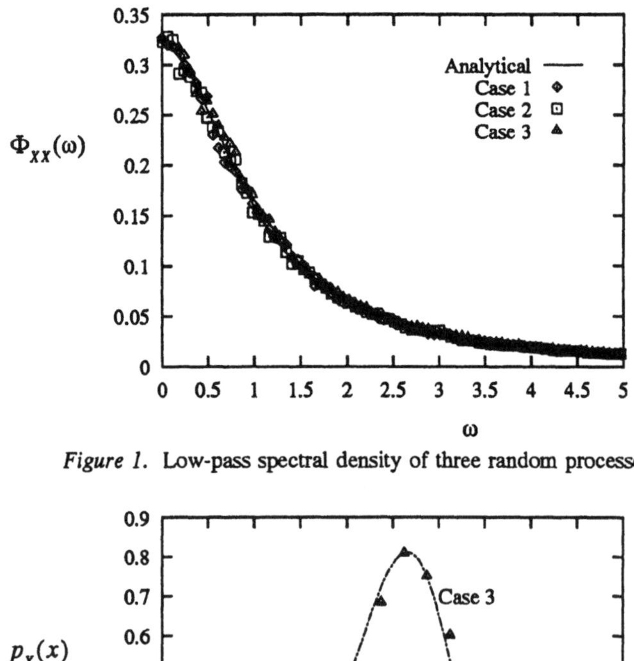

Figure 1. Low-pass spectral density of three random processes.

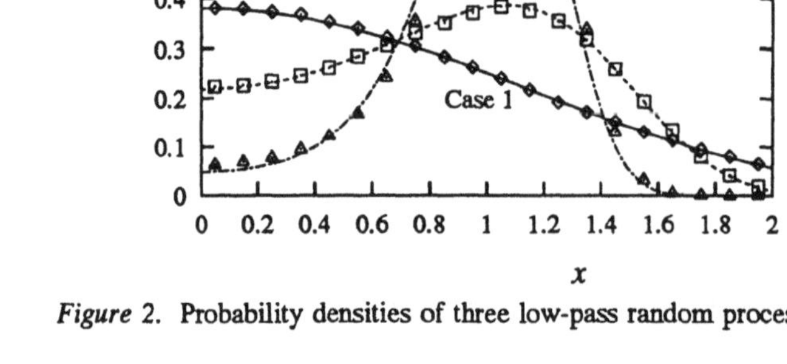

Figure 2. Probability densities of three low-pass random processes.

It is seen from figures 1 and 2 that the analytical and simulation results are in very good agreement. Indeed the three random processes have the same spectral density, but different probability densities. Case 1 is close to the Gaussian distribution, whereas case 3 corresponds to a bimodal distribution, far from being Gaussian.

3. Generation of Narrow-Band Non-Gaussian Random Processes

A narrow-band stationary process $X(t)$ may be expressed as

$$X(t) = A(t)\cos[\omega_d t + \Phi(t)] \tag{12}$$

where ω_d is a constant, $A(t)$ and $\Phi(t)$ are two stationary random processes, referred to as the envelope process and the phase process, respectively. It is assumed that process $\Phi(t)$ is uniformly distributed in $[0, 2\pi)$, independent of process $A(t)$. Equation (12) can be recast into

$$X(t) = Y(t)\cos\omega_d t - Z(t)\sin\omega_d t \tag{13}$$

where

$$Y(t) = A(t)\cos\Phi(t), \qquad Z(t) = A(t)\sin\Phi(t) \tag{14}$$

Process $A(t)$ is defined on $[0, a_u]$, while processes $X(t)$, $Y(t)$ and $Z(t)$ are defined on $[-a_u, a_u]$, where a_u can be finite or infinite.

3.1. PROBABILISTIC DESCRIPTIONS

The characteristic function of process $X(t)$ is given by

$$\begin{aligned} M_X(\theta) &= \int_{-a_u}^{a_u} \exp(i\theta x)\, p_X(x)\, dx \\ &= \int_0^{a_u} da \int_0^{2\pi} \exp\{i\theta a\cos[\omega_d t + \phi(t)]\}\, p_{A\Phi}(a,\phi)\, d\phi \end{aligned} \tag{15}$$

In the present case,

$$p_{A\Phi}(a,\phi) = \frac{1}{2\pi} p_A(a) \tag{16}$$

and the exponential term in (15) can be expanded as (Gradshteyn and Ryzhik, 1980)

$$\exp\{i\theta a\cos[\omega_d t + \phi(t)]\} = \sum_{n=-\infty}^{\infty} i^n \mathrm{J}_n(\theta a)\exp\{in[\omega_d t + \phi(t)]\} \tag{17}$$

where $\mathrm{J}_n(\cdot)$ is the nth-order Bessel function of the first-kind. Using (16) and (17), equation (15) is recast into

$$\begin{aligned} M_X(\theta) &= \frac{1}{2\pi}\sum_{n=-\infty}^{\infty} i^n \int_0^{a_*} \mathrm{J}_n(\theta a)\,p_A(a)\,da \int_0^{2\pi} \exp\{in[\omega_d t + \phi(t)]\}\,d\phi \\ &= \int_0^{a_*} \mathrm{J}_0(\theta a)\,p_A(a)\,da \end{aligned} \tag{18}$$

The probability density $p_X(x)$ can then be expressed as

$$p_X(x) = \frac{1}{2\pi}\int_{-\infty}^{\infty} \exp(-i\theta x)\,d\theta \int_0^{a_*} \mathrm{J}_0(\theta a)\,p_A(a)\,da \tag{19}$$

Using the following formula (Gradshteyn and Ryzhik, 1980)

$$\frac{1}{2\pi}\int_{-\infty}^{\infty} \mathrm{J}_0(\theta a)\exp(-i\theta x)\,d\theta = \begin{cases} \dfrac{1}{\pi\sqrt{a^2 - x^2}}, & a \ge |x| \\ 0, & a < |x| \end{cases} \tag{20}$$

(19) is recast into

$$p_X(x) = \frac{1}{\pi}\int_{|x|}^{a_*} \frac{p_A(a)\,da}{\sqrt{a^2 - x^2}} \tag{21}$$

which can also be obtained directly from (12) with a uniformly distributed $\Phi(t)$.

Conversely, the probability density of the envelope process $A(t)$ can also be determined from that of $X(t)$, as shown below. Making use of the relation (Arfken, 1985)

$$\delta(\alpha - a) = a\int_0^{\infty} \theta\,\mathrm{J}_0(\theta\alpha)\,\mathrm{J}_0(\theta a)\,d\theta \tag{22}$$

where $\delta(\cdot)$ is the Dirac-delta function, we have

$$p_A(a) = \int_0^{a_*} p_A(\alpha)\delta(\alpha - a)d\alpha = \int_0^{a_*} a p_A(\alpha)\, d\alpha \int_0^{\infty} \theta J_0(\theta\alpha) J_0(\theta a)\, d\theta$$

$$= a\int_0^{\infty} \theta J_0(\theta a)\left[\int_0^{a_*} p_A(\alpha) J_0(\theta\alpha)\, d\alpha\right] d\theta \tag{23}$$

Noticing that, according to (18), the inner integral is the characteristic function $M_X(\theta)$ of $X(t)$, we have

$$p_A(a) = a\int_0^{\infty} \theta J_0(\theta a) M_X(\theta)\, d\theta = a\int_0^{\infty} \theta J_0(\theta a)\, d\theta \int_{-a_*}^{a_*} p_X(x)\exp(i\theta x)\, dx \tag{24}$$

Equations (19), (21) and (24) show a one-to-one correspondence between the probability density functions of a stationary process $X(t)$ and its envelope process $A(t)$, under the assumption that the phase process $\Phi(t)$ is uniformly distributed, independent of $A(t)$.

With the knowledge of $p_A(a)$, the joint probability density of $Y(t)$ and $Z(t)$ can be obtained from transformation (14) to yield

$$p_{YZ}(y,z) = \frac{p_A(\sqrt{y^2+z^2}\,)}{2\pi\sqrt{y^2+z^2}} \tag{25}$$

It follows from (13), (14) and the uniform distribution of $\Phi(t)$ that

$$\mathrm{E}[X^2] = \mathrm{E}[Y^2] = \mathrm{E}[Z^2] = \frac{1}{2}\mathrm{E}[A^2], \qquad \mathrm{E}[YZ] = 0 \tag{26}$$

3.2. SPECTRAL DESCRIPTIONS

Now let processes $Y(t)$ and $Z(t)$ be governed by the following pair of Itô equations

$$\begin{aligned} dY &= -\lambda Y\,dt + D_1(Y,Z)\,dB_1(t) \\ dZ &= -\lambda Z\,dt + D_2(Y,Z)\,dB_2(t) \end{aligned} \tag{27}$$

where λ is a positive constant, and $B_1(t)$ and $B_2(t)$ are two independent unit Wiener processes. Multiplying the two equations in (27) by $Y(t_0)$ and taking the ensemble average, we obtain

$$\frac{d}{dt} R_{YY}(t_0, t) = -\lambda R_{YY}(t_0, t)$$
$$\frac{d}{dt} R_{YZ}(t_0, t) = -\lambda R_{YZ}(t_0, t) \tag{28}$$

where

$$R_{YY}(t_0, t) = \mathrm{E}[\,Y(t_0)\,Y(t)], \quad R_{YZ}(t_0, t) = \mathrm{E}[\,Y(t_0)\,Z(t)] \tag{29}$$

Let $Y(t)$ and $Z(t)$ be jointly stationary processes, $R_{YY}(t_0, t) = R_{YY}(t - t_0)$ and $R_{YZ}(t_0, t) = R_{YZ}(t - t_0)$. Letting $\tau = t - t_0$, (28) is recast into

$$\frac{d}{d\tau} R_{YY}(\tau) = -\lambda R_{YY}(\tau)$$
$$\frac{d}{d\tau} R_{YZ}(\tau) = -\lambda R_{YZ}(\tau) \tag{30}$$

Subject to the conditions

$$R_{YY}(0) = \mathrm{E}[Y^2] = \sigma^2, \quad R_{YZ}(0) = \mathrm{E}[YZ] = 0 \tag{31}$$

the two equations in (30) are solved to yield

$$R_{YY}(\tau) = \sigma^2 \exp(-\lambda\,|\tau|), \quad R_{YZ}(\tau) = 0 \tag{32}$$

Similarly, it can be shown that

$$R_{ZZ}(\tau) = \sigma^2 \exp(-\lambda\,|\tau|), \quad R_{ZY}(\tau) = 0 \tag{33}$$

The correlation function of process $X(t)$ can then obtained from (13) to yield

$$R_{XX}(\tau) = \sigma^2 \exp(-\lambda\,|\tau|) \cos(\omega_d \tau) \tag{34}$$

which corresponds to a spectral density

$$\Phi_{XX}(\omega) = \frac{\lambda \sigma^2 (\omega^2 + \omega_d^2 + \lambda^2)}{\pi\,[\,(\omega^2 - \omega_d^2)^2 + 2\lambda^2(\omega^2 + \omega_d^2) + \lambda^4\,]} \tag{35}$$

Letting $\omega_n^2 = \omega_d^2 + \lambda^2$, (35) is simplified to

$$\Phi_{XX}(\omega) = \frac{\lambda\sigma^2(\omega^2+\omega_n^2)}{\pi\,[\,(\omega^2-\omega_n^2)^2+4\lambda^2\omega^2\,]} \tag{36}$$

Equation (36) indicates that the stationary process $X(t)$ has a narrow-band spectral density if λ is small, in which case the peak of the spectral density is near ω_n, and the band-width depends on the magnitude of λ.

3.3. GENERATION OF PROCESS $X(t)$ WITH A GIVEN PROBABILITY DENSITY

It is noted that the spectral density (36) is unrelated to the diffusion coefficients $D_1(Y,Z)$ and $D_2(Y,Z)$ in equations (27); therefore, these coefficients may be adjusted to match a target probability density $p_X(x)$. Corresponding to (27), the Fokker-Planck equation for the joint probability density $p_{YZ}(y,z)$ of $Y(t)$ and $Z(t)$ is given by

$$\frac{\partial}{\partial y}(\lambda y p)+\frac{\partial}{\partial z}(\lambda z p)+\frac{1}{2}\frac{\partial^2}{\partial y^2}[D_1^2(y,z)\,p]+\frac{1}{2}\frac{\partial^2}{\partial z^2}[D_2^2(y,z)\,p] = 0 \tag{37}$$

Equation (37) is satisfied if D_1^2 and D_2^2 are given by

$$\begin{aligned} D_1^2(y,z) &= -\frac{2\lambda}{p_{YZ}(y,z)}\int_{-\sqrt{a_0^2-z^2}}^{y} u\,p_{YZ}(u,z)\,du \\ D_2^2(y,z) &= -\frac{2\lambda}{p_{YZ}(y,z)}\int_{-\sqrt{a_0^2-y^2}}^{z} v\,p_{YZ}(y,v)\,dv \end{aligned} \tag{38}$$

Since both $Y(t)$ and $Z(t)$ have zero means, the right-hand-sides of (38) are guaranteed to be non-negative. Furthermore, it can be shown that

$$D_1^2(y,z) = D_2^2(y,z) = \frac{2\lambda a}{p_A(a)}\int_a^{a_0} p_A(\alpha)\,d\alpha \tag{39}$$

where $a=\sqrt{y^2+z^2}$. Note that the right hand side of (39) is non-negative, as expected.

For a given probability density $p_X(x)$ and a spectral density in the form of (36), $Y(t)$ and $Z(t)$ can be generated from (27), where D_1 and D_2 are calculated from (39), and $X(t)$ is obtained from (13).

As an example, consider the following probability density for the envelope process $A(t)$

$$p_A(a) = C\,a\exp(-\alpha a^2-\beta a^4); \qquad 0\le a<\infty; \quad \beta>0 \tag{40}$$

Using (40), processes $Y(t)$ and $Z(t)$ can be generated from (27) and (39), and $X(t)$ from

(13). The following three sets of parameters are chosen for illustration

Case 4: $\lambda = 0.2$, $\sigma^2 = 1$, $\alpha = 0.4$, $\beta = 0.01351$

Case 5: $\lambda = 0.2$, $\sigma^2 = 1$, $\alpha = -2$, $\beta = 0.5249$

Case 6: $\lambda = 0.2$, $\sigma^2 = 1$, $\alpha = -8$, $\beta = 1.9952$

The narrow-band spectral densities in the positive ω domain are plotted for the three cases in figure 3, where the line are calculated from (36) and the diamonds, squares and triangles represent the results obtained from simulation, using equations (13), (27) and (39). It is seen that the analytical and simulation results are in very good agreement. Indeed the three random processes share the same narrow-band spectral density. The probability densities for the three processes are depicted in figure 4, with lines representing the analytical results calculated from (19), and triangles, squares and diamonds representing the simulation results. Case 6 corresponds to a bimodal distribution, far from being Gaussian.

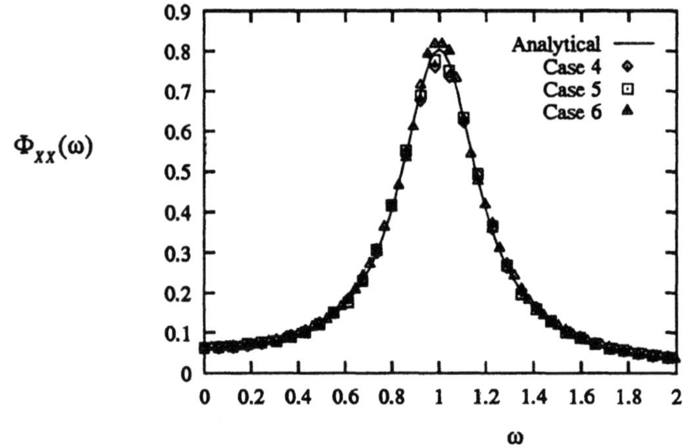

Figure 3. Narrow-band spectral density of three random processes.

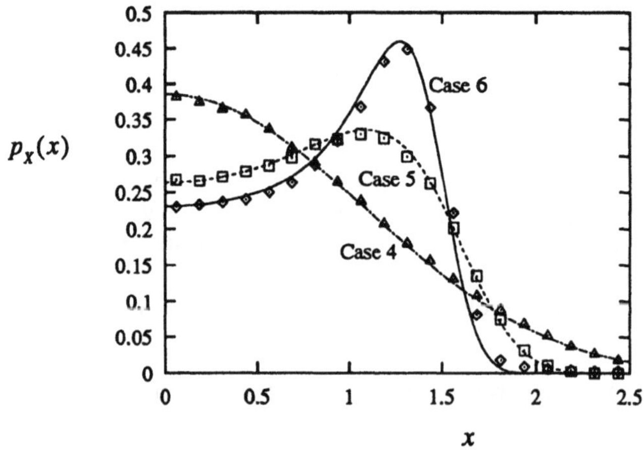

Figure 4. Probability densities of three narrow-band random processes.

4. Conclusions

Random processes with arbitrary probability densities and low-pass or narrow-band spectral densities can be generated from nonlinear filters, which are governed by a set of stochastic differential equations. The linear drift coefficients are chosen to approximate a target spectral density, while the nonlinear diffusion coefficients are determined to match the target probability distribution. Algorithm developed herein may be used to investigate the case of randomly excited dynamical systems, in which the probability distribution of the excitation process is strongly non-Gaussian. In principle, excitations of more complicated spectral densities can also be generated with nonlinear filters of higher dimensions, which will be the subject of a subsequent paper.

5. Acknowledgement

The work reported in this paper is supported by the National Science Foundation under Grant CMS-9531719. Opinions, findings, and conclusions or recommendations expressed are those of the writers, and do not necessarily reflect the view of the NSF.

6. References

Arfken, G. (1985) *Mathematical Methods for Physics*, 3rd Edition, Academic Press, Orlando, Florida.

Cai, G. Q. and Lin, Y. K. (1996) Generation of non-Gaussian stationary stochastic processes, *Physical Review E*, **54**, 299-303.

Cai, G. Q. and Lin, Y. K. (1997) Reliability of dynamical systems under non-Gaussian random excitations, *The 7th International Conference on Structural Safety and Reliability*, Paper No. WeC1-06, Kyoto, Japan.

Cai, G. Q., Lin, Y. K., and Xu, W. (1998) Response and reliability of nonlinear systems under stationary non-Gaussian excitations, *The 4th International Conference on Structural Stochastic Dynamics*, Notre Dame, Indiana.

Dimentberg, M. F, (1988) *Statistical Dynamics of Nonlinear and Time-Varying Systems*, Wiley, New York.

Gradshteyn, I. S. and Ryzhik, I. M., (1980) *Table of Integrals, Series, and Products*, Academic Press, New York.

Gurley, K. R., Kareem, A. and Tognarelli, M. A. (1996) Simulation of a class of non-normal random processes, *International Journal of Non-Linear Mechanics*, **31**, 601-617.

Itô, K. (1951) On stochastic differential equations, *Memoirs of the American Mathematical Society*, **4**, 289-302.

Kontorovich, V. Y. and Lyandres, V. Z. (1995) Stochastic differential equations: An approach to the generation of continuous non-Gaussian processes, *IEEE Transactions on Signal Processing*, **43**, 2372-2385.

Li, Q. C. and Lin, Y. K. (1995) New stochastic theory for bridge stability in turbulent flow, II, *Journal of Engineering Mechanics*, **121**, 102-116.

Liu, B. and Munson, D. C. (1982) Generation of a random sequence having a joint specified marginal distribution and autocovariance, *IEEE Transactions on Acoustics, Speech, and Signal Processing*, **ASSP-30**, 973-983.

Nakayama, J. (1994) Generation of stationary random signals with arbitrary probability distribution and exponential correlation, *IEICE Transactions on Fundamentals*, **E77-A(5)**, 917-922.

Shinozuka, M. (1972) Monte Carlo solution of structural dynamics, *Computers and Structures*, **2**, 855-874.

Shinozuka, M. and Jan, C-M. (1972) Digital simulation of random processes and its applications, *Journal of Sound and Vibration*, **10**, 111-128.

Sondhi, M. M. (1983) Random processes with specified spectral density and first-order probability density, *Bell System Technical Journal*, **62**, 679-701.

Wedig, W. V. (1989) Analysis and simulation of nonlinear stochastic systems, in W. Schiehlen (ed.), *Nonlinear Dynamics in Engineering Systems*, Springer-Verlag, Berlin, pp. 337-344.

Yamazaki, F. and Shinozuka, M. (1988) Digital generation of non-Gaussian stochastic field, *Journal of Engineering Mechanics*, **114**, 1183-1197.

STOCHASTIC NONLINEAR CONTROLLERS

F. CASCIATI AND L. FARAVELLI
Department of Structural Mechanics, University of Pavia
Via Ferrata 1 - 27100 Pavia - Italy

Abstract

Attention is focused on building a controller for driving a hybrid control scheme capable of working under stochastic excitation. The passive component of the controller, a base isolation bearing, is first described and the resulting structural system identified.
Then, among the available control algorithms, a fuzzy scheme is selected to drive the active component of the hybrid control scheme.

1. Introduction

There are many aspects which push toward the adoption of nonlinear active control solutions (Slotine and Li, 1991) for civil buildings and infrastructures (Yao, 1972; Housner and al., 1992; Housner and Masri, 1993; Soong, 1990).

- Nonlinear control may permit the use of less expensive components, as actuators and sensors, with nonlinear characteristics.
- Nonlinearities can be intentionally introduced into the controller so that model uncertainty can be tolerated.
- The availability of response deadbands and the adoption of limited values of the control forces are standard features of nonlinear control.
- Nonlinear control offers the possibility of a direct incorporation of the time delay.

This paper considers a further aspect. The adoption of base isolation (or passive damper) devices could dissipate most of the incoming (external) energy, but it does not solve any control problem. However, a coupling with active control devices (hybrid control) shows cost and performance optimality. The main inconveniences of passive control are avoided, while

S. Narayanan and R.N. Iyengar (eds.),
IUTAM Symposium on Nonlinearity and Stochastic Structural Dynamics, 55–69.

the energy amount and the force values required by the active component become reasonable.

2. The passive device

The base isolator considered in this paper belongs to the class of laminated-rubber bearings (LDRB = low damping rubber bearing) and is produced by the Italian Company ALGA spa. They are designed to support large weights while providing only small resistance to large horizontal displacements. The materials involved are steel and rubber. The latter one is characterized by its elastic shear modulus G (1 MPa being its reference value, at 100% of strain) and its bulk modulus κ (usually 2000 MPa). In (Skinner et al., 1993) as well as in (Kelly, 1993), one finds the main parameters and relations governing the rubber technology. One starts with a single disk of unstrained diameter D and thickness t. Let A denote the area ($= \pi D^2/4$) and S the ratio between the loaded area and the free area (it is a shape factor: for the circular disk $S = D/4t$). Indeed the bearing consists of n equal rubber layers for a total rubber length $t_r = nt$. The actual length of the bearing l also accounts for the thickness of the steel layers which alternate with the rubber layers and the top steel redistribution plate. It is shifted of Δl with respect to the bottom when horizontal actions act. This leaves an overlap area A' of bearing top and bottom which determines the maximum allowable compression weight as the product of A', G, S and the allowable shear strain due to the weight. This allowable strain is not usually exploited and a reserve shear strain capacity allows for the accommodation of a horizontal (seismic) displacement. Its value can be increased by increasing the rubber height: some algebra leads one to the allowable value $D/3$ for this horizontal displacement.

The problem, sometimes with no solution, of operating a scale reduction of the actual commercial product was avoided in the present study thanks to the chance offered by a special order the supplier company received: to produce a series of supports to preserve statues in a museum. The authors just obtained for their study a further item from that series.
The geometry of the device is given in Figure 1. The bottom steel disk has a companion (of diameter 60.3 mm) on the top: it shows the same thickness. The remaining length is occupied by $n = 20$ layers in rubber of thickness $t = 1.5$ mm separated by 19 steel disks of thickness 1 mm and diameter 55 mm. It follows that the rubber total length is $t_r = 30$ mm, the disk area $A = 2854$ mm^2 and the shape factor $S = 10$.
The response to normal load was tested on an universal testing machine. The range of testing loads was limited to (- 4.5 kN, 0) and the maximum

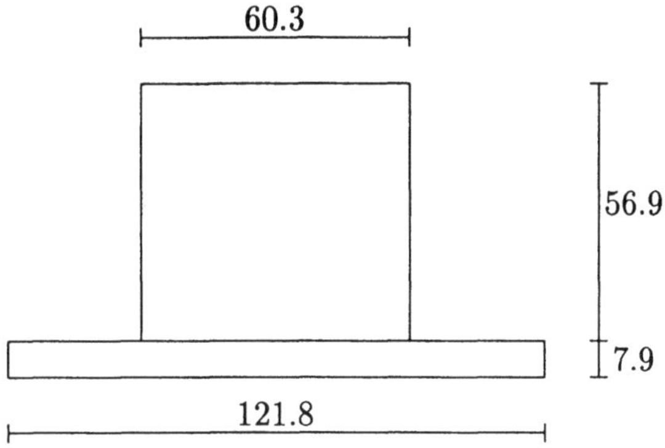

Figure 1. The base isolation device: lengths in millimeters.

contraction was 0.5 mm. Combining these values with the known geometry (Figure 1), the equivalent vertical stress, vertical strain and normal stiffness result to be 1.58 N/mm^2, 16.6 millistrain and 95 N/mm^2. The quantity of interest however is the normal stiffness, K_v, given by the ratio of load and contraction: $K_v = 9000$ N/mm.
The device was also tested to transversal actions, see Figure 2, in the range (0.0, 0.6 kN), with a maximum elongation of $\Delta l/l$=30% as summarized in Figure 3. Loading and unloading give rise to different behaviours depending on the maximum force achieved during the test. Therefore, the horizontal stiffness K_b, i.e. the ratio between transversal load and Δl depends on Δl: $K_b(\Delta l = 1) = 75$ N/mm; $K_b(\Delta l = 4) = 50$ N/mm; $K_b(\Delta l = 12.5) = 40$ N/mm. Further extended testing also provided $K_b(\Delta l = 17) = 35.2$ N/mm.
According to (Skinner et al., 1993), a rubber bearing may be approximated as a vertical shear beam. Indeed, the steel laminations inhibit any flexural deformation, but they do not provide impediment to shear deformation. Therefore the horizontal stiffness K_b can be written:

$$K_b = GA/t_r \tag{1}$$

t_r being the total rubber height, as above specified: it follows $G = 0.37$ N/mm^2 at strain 50% , a value not exactly in agreement with the one suggested by the supplier (0.8 N/mm^2 at 100% strain).

3. The experimental apparatus

The experimental apparatus models the foundation of a building. The need of scaling the structure, however, would be in conflict with the possibility of scaling the device. Therefore, the experiment was built around the smallest

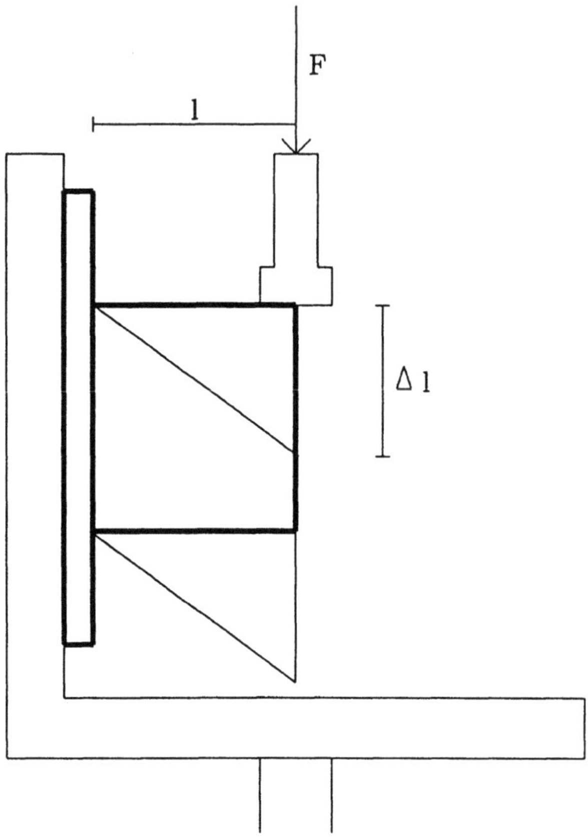

Figure 2. The base isolation device mounted for the shear static test.

base isolator available on the market as it was discussed in the previous section. Its vertical stiffness is too high in view of the testing environment and this prevents from using four bearings to support the superstructure. The problem was solved by using the bearing in conjunction with rubber teflon supports (Figures 4 and 5).

The steel plate which forms the superstructure has the mass M of 100 kg. The bearing period T_b one expects for the testing apparatus is:

$$T_b = 2\pi\sqrt{\frac{M}{K_b}} \tag{2}$$

The variability of K_b with the transversal displacement is the cause of the system nonlinearity, whose period varies with the load intensity from 0.33

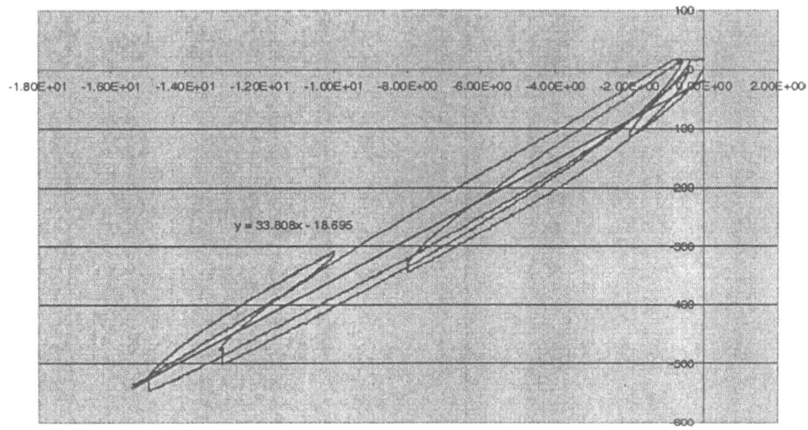

Figure 3. Shear force [N] versus displacement [mm] in the static shear test on the base isolation device

to 0.4 sec (from 3 to 2.5 Hz. if one thinks in terms of frequency). A correct design procedure would pursue $T_b = 2.5$ sec., but, as explained above, the authors built the test on the available device, rather than designing the device.

4. System identification

A first screening of the response was obtained by using sine excitations obtaining the dynamic magnification factors of Figure 6. The ordinate is the ratio between the output and the input acceleration.

A white noise was eventually adopted for the same response characterization, but low frequencies must be cut in order to achieve large intensities (Casciati et al., 1998). Frequencies over 27 Hz. are filtered because of no interest in earthquake engineering. The results are summarized in Figures 7, 8, 9, 10 and 11.

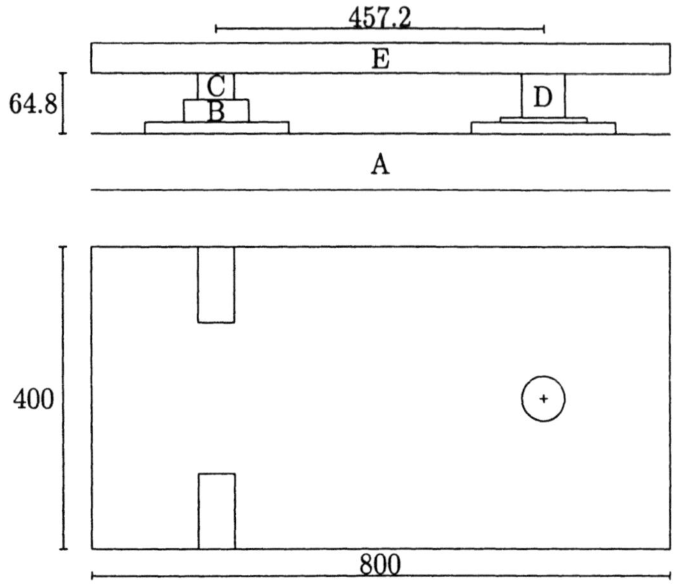

Figure 4. The experimental apparatus tested on the shaking table: A shaking table; B supporting plate; C teflon support; D base isolation device; E mass of 100 kg.

Figure 5. Photograph of the experimental apparatus.

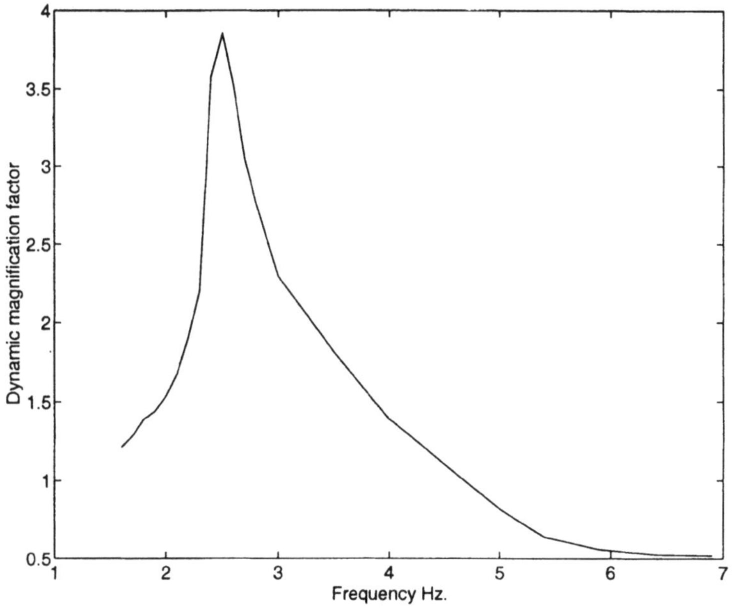

Figure 6. Dynamic magnification factors obtained by the shaking table test: sine waves of different frequencies were adopted to drive the table; 20 mm is the amplitude of the displacement wave at low frequency and it was decreased as the frequency increases in order to maintain the same intensity in terms of acceleration.

5. Application of hybrid control

As underlined in section 2, the basic isolator fails when large displacements are achieved. The idea is to add, to the experimental apparatus of Fig. 4, an electromagnetic brake which is only active when the displacement threshold is too close.

A control design requires to go through the following standard procedure, possibly with a few iterations (Slotine and Li, 1991):

1. specify the desired behaviour and select actuators and sensors: the accelerometer on the supported plate is the sensor. From this measure and the one of the table accelerometer, one derives the relative velocity and hence the isolation slide. An electromagnetic brake is the actuator;
2. model the structural system by a set of differential equations; by contrast in this paper an equivalent analogic circuit was derived from Figure 6 (damping 0.08) in order to incorporate digital/analog and analog/digital converters as well as any source of uncertainty in the model (Casciati et al., 1999);

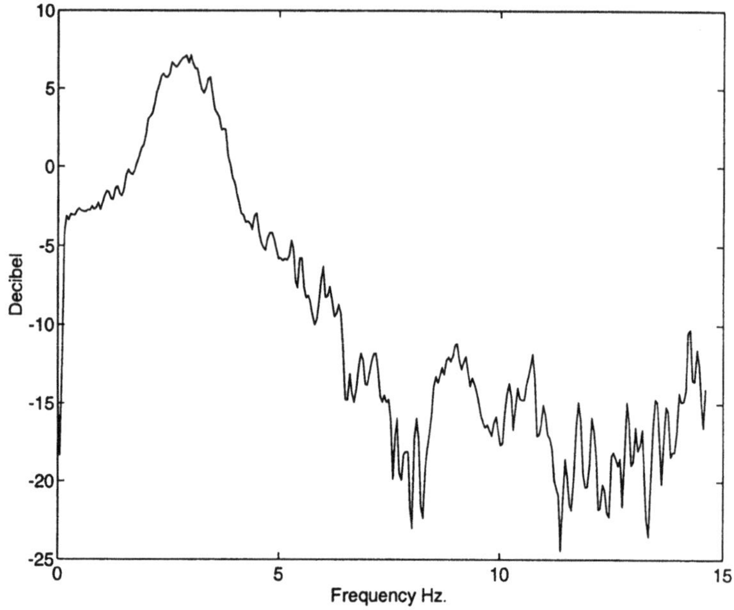

Figure 7. Transfer function derived by using a input signal of duration 20 minutes which in terms of acceleration is a white noise filtered below the range (0.1, 0.15 Hz.): the amplitude of the maximum displacement is 75 mm (the maximum allowable value).

3. design a control law for the system: a fuzzy chip controller was selected (Faravelli and Yao, 1995 ; Battaini et al., 1998);
4. analyse and simulate the resulting control system;
5. implement the control system in hardware.

Nevertheless, in the theory of nonlinear control systems there is no general method for designing nonlinear controllers; one has just a rich collection of alternative and complementary techniques. In particular expert control makes effective use of symbolic computation as classic application of artificial intelligence in the design process of any control algorithm. This methodology gives a new insight into heuristic, permitting the design of simple regulators, as well as multivariable controllers with sophisticated control laws (Pedrycz, 1989). A very special subset of expert controller consists of the fuzzy controllers based on fuzzy logic reasoning (Faravelli and Yao, 1995). The results are summarized in Figures 12, 13 and 14, which illustrate the goal of the process. Figure 12 compares the controlled and uncontrolled response, the latter one showing a reduction of 40%. Figure 13 compares the control force required for the control in Figure 12 with

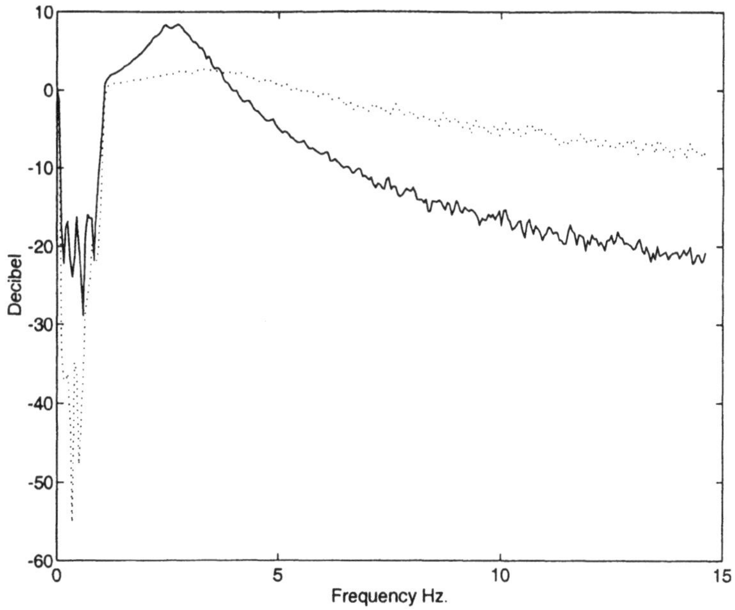

Figure 8. Transfer function derived by using a input signal of duration 20 minutes which in terms of acceleration is a white noise filtered below the range (1., 1.5 Hz.): the amplitude of the maximum displacement is 60 mm for the solid line and 20 mm for the dotted one.

the one resulting in the case the base excitation has a lower intensity. Figure 14 shows the situation when the problem excitation is the same as used in Figure 10. In order to achieve these results, rules were established on rebuilt displacements and velocities using five membership functions for each of them. The value of the force (described by five membership functions) was prescribed at label *zero* for all the situations where no large measures, of at least one of displacement and velocity, have been detected:

		DISPLACEMENT				
		NL	NE	ZE	PO	PL
	NL	PL	PO	PO	ZE	ZE
	NE	PO	ZE	ZE	ZE	ZE
VELOCITY	ZE	PO	ZE	ZE	ZE	NE
	PO	ZE	ZE	ZE	ZE	NE
	PL	ZE	ZE	NE	NE	NL

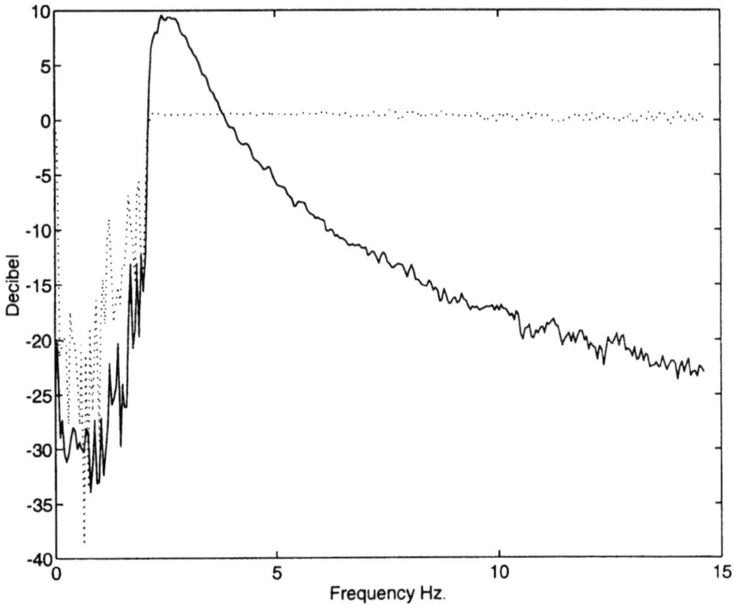

Figure 9. Transfer function derived by using a input signal of duration 20 minutes which in terms of acceleration is a white noise filtered below the range (2., 2.5 Hz.): the amplitude of the maximum displacement is 28 mm for the solid line and 2 mm for the dotted one.

where ZE = zero; PO = positive; PL = positive large and so on. The previous table shows the fuzzy rule in a matrix form: f.i. when the displacement is negative and the velocity is negative large, the resulting control force will be positive.

6. Conclusions

This paper emphasizes three main points:

- Nonlinearity is becoming a need for Civil Engineering structural control;
- The nature of the external disturbance is well modelled in a stochastic context;
- Robustness can easily achieved by special control schemes.

Within this framework a hybrid control scheme is pursued coupling a base isolation device (a Low Damping Rubber Bearing) with an actively controlled brake. A fuzzy controller is adopted and it is designed on the basis of an equivalent analogic circuit.

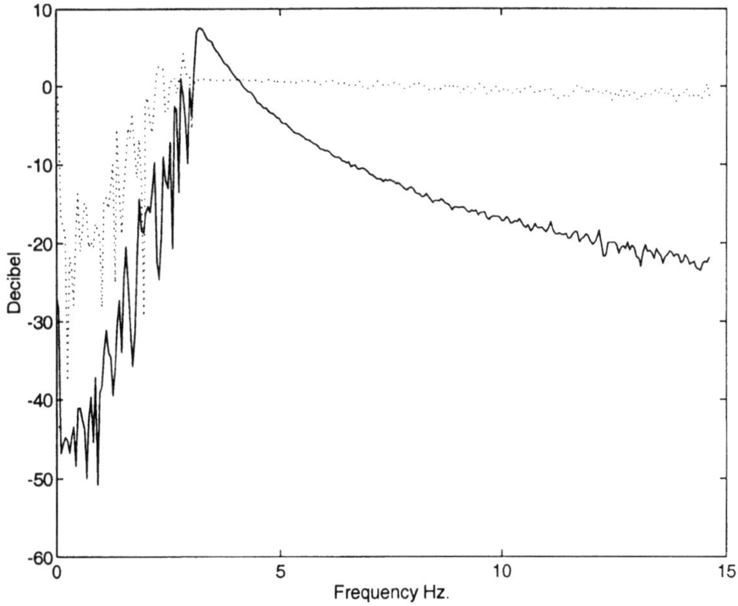

Figure 10. Transfer function derived by using a input signal of duration 20 minutes which in terms of acceleration is a white noise filtered below the range (3., 3.5 Hz.): the amplitude of the maximum displacement is 25 mm for the solid line and 2 mm for the dotted one.

Acknowledgement

This research was supported by grant ARS 98-207 from the Italian Space Agency (ASI).
The authors are indebted with Ing. Roberto Rossi and Mr. Alberto D'Andrea for managing the tests illustrated in the paper.

References

Battaini M, Casciati F. and Faravelli L. (1998), Fuzzy Control of Structural Vibration. An Active Mass System Driven by a Fuzzy Controller, *Earthquake Engineering and Structural Dynamics*, **27**, 1267-1276.

Casciati F., Faravelli L. and Petrini L. (1998), Input-Signal Characterization for Shaking-Tables, *Journal of Structural Control*, **5**, 2, 33-56.

Casciati F., Faravelli L. and Torelli G. (1999), A Fuzzy Chip Controller for Nonlinear Vibration, *Nonlinear Dynamics*, **20**, 85-98.

Faravelli L. and Yao T. (1995), Use of Adaptive Network in Fuzzy Control of Civil Structures, *Microcomputers in Civil Engineering*, **11**, 1, 67-76.

Housner G.W., Masri S.F., Casciati F. and Kameda H. (eds.) (1992), *Proceedings of the*

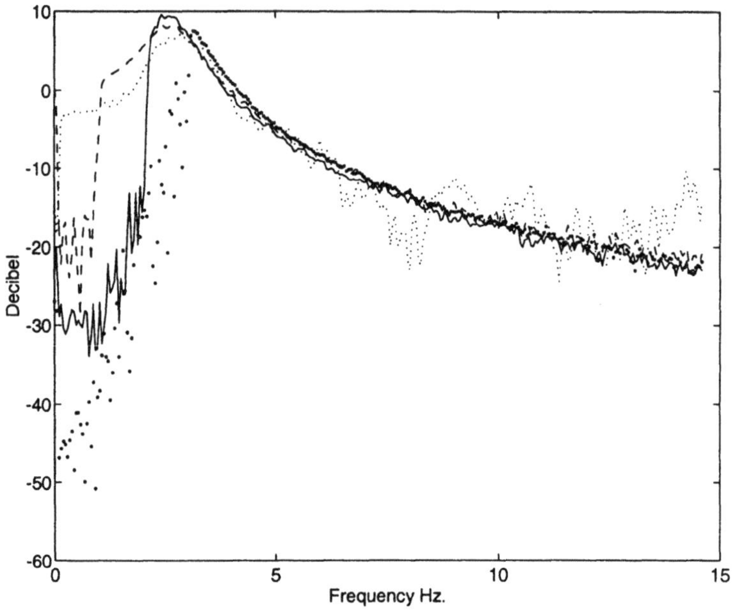

Figure 11. Transfers function derived by using a input signal of duration 20 minutes which in terms of acceleration is a white noise filtered below the range (0.1, 0.15 Hz.), (1., 1.5 Hz.), (2., 2.5 Hz.) and (3., 3.5 Hz.): the amplitudes of the maximum displacement are the largest in the previous figures.

U.S.-Italy-Japan Workshop/Symposium on Structural Control and Intelligent Systems, University of Southern California CE-9210.

Housner G.W. and Masri S.F. (eds.)(1993) *Proceedings of International Workshop on Structural Control - 1993*, University of Southern California CE-9311.

Kelly J.M. (1993), *Earthquake-Resistant Design with Rubber*, Springer-Verlag, London, UK.

Pedrycz W. (1989), *Fuzzy Control and Fuzzy Systems*, Research Studies Press, Taunton.

Skinner R.I., Robinson W.H. and McVerry G.H. (1993), *An Introduction to Seismic Isolation*, John Wiley and Soons, Chichester, UK.

Slotine J.E. and Li W. (1991), *Applied Nonlinear Control*, Prentice-Hall Inc., Englewood Cliffs.

Soong, T.T. (1990), *Active Structural Control: Theory and Practice*, Longman Scientific and Technical, Harlow.

Yao, J.T-P. (1972), Concept of structural control, *Journal of the Structural Division, ASCE*, 98 (7), 1567-1574.

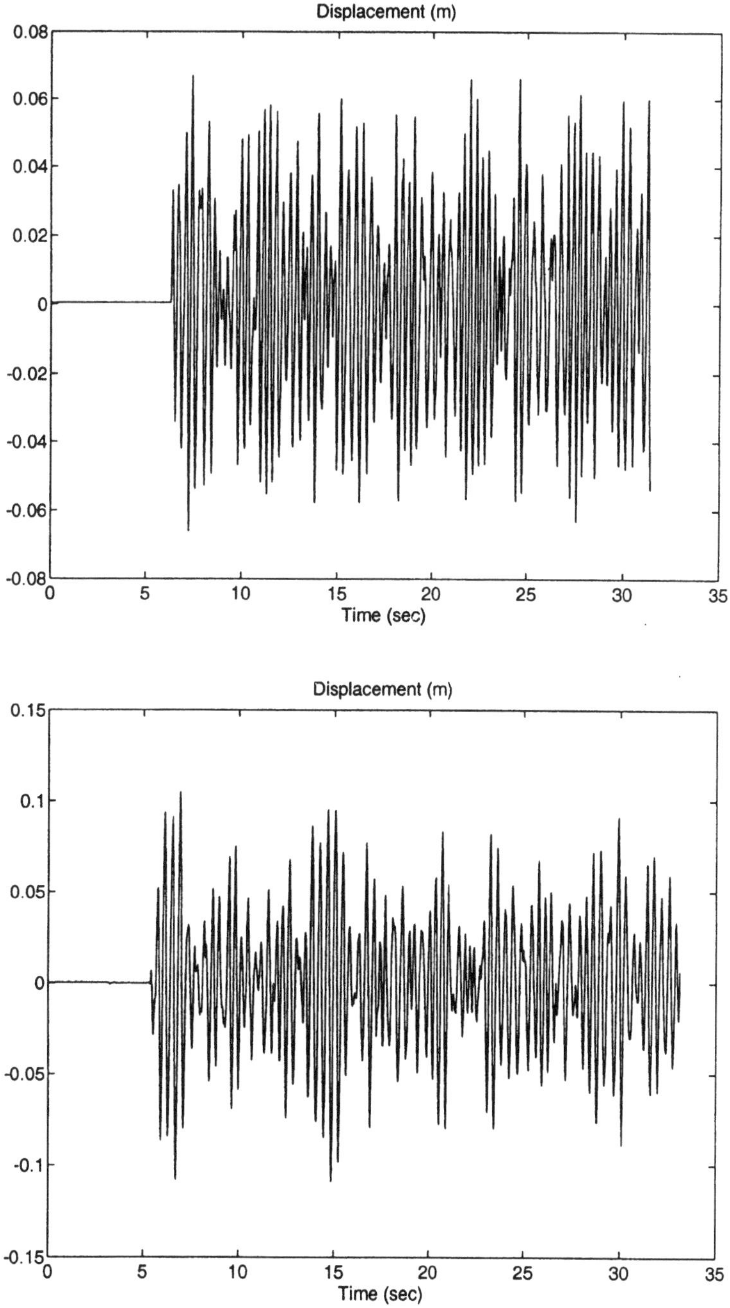

Figure 12. Controlled and uncontrolled displacement for the white noise input filtered below the range (1., 1.5 Hz) and high intensity.

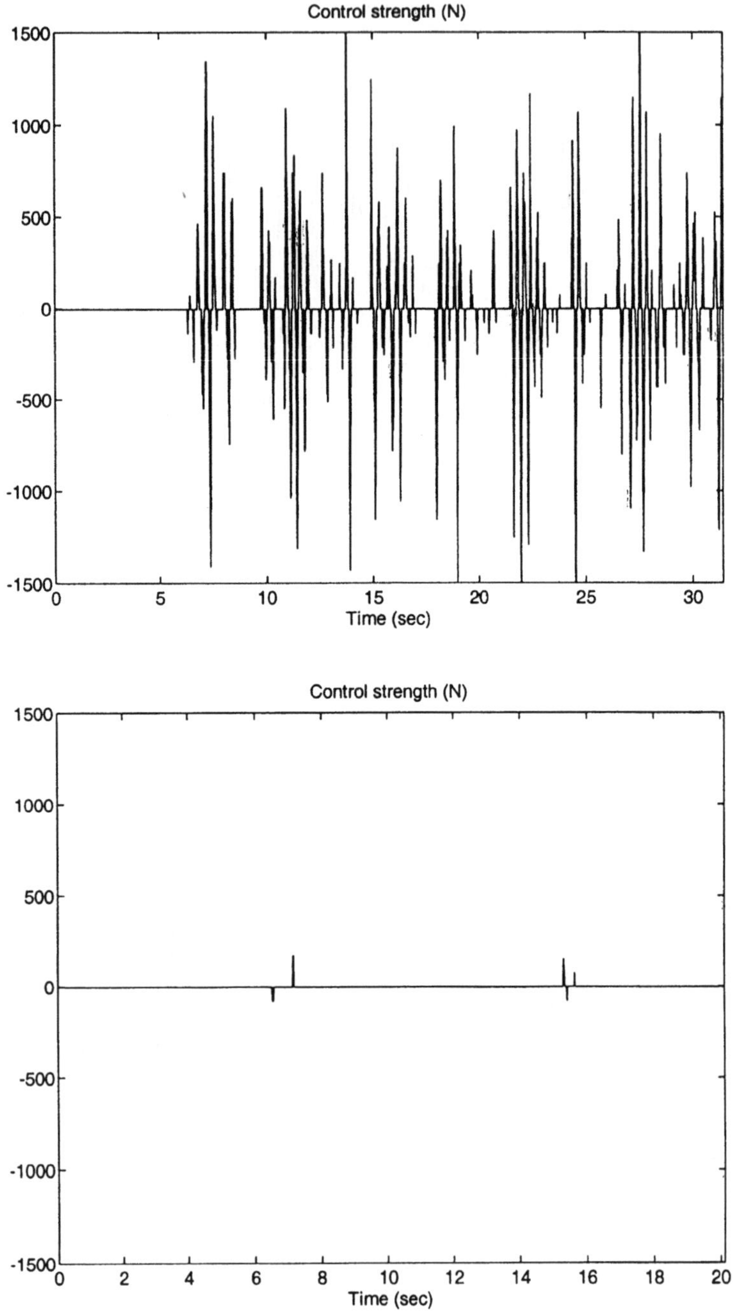

Figure 13. Control force for the white noise input filtered below the range (1., 1.5 Hz): high intensity (top) and low intensity (bottom).

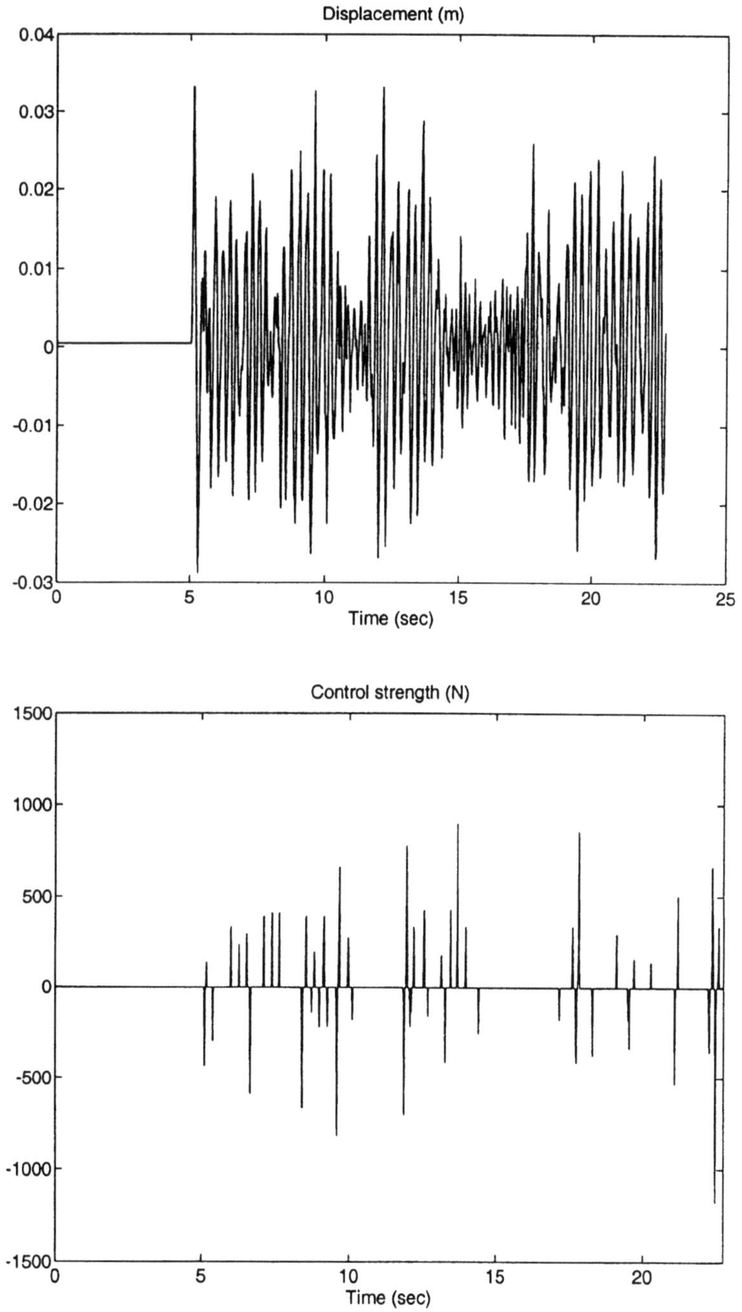

Figure 14. Controlled displacement and control force for the white noise input filtered below the range (3., 3.5 Hz) and high intensity.

VECTOR FIELDS AND MAPS

A Perturbation Approach

H. G. DAVIES and K. KARAGIOZIS
Department of Mechanical Engineering
University of New Brunswick
Fredericton, NB E3B 5A3 Canada

1. Introduction

This paper looks at some relationships between continuous vector field descriptions of nonlinear systems and associated return maps defined on suitable Poincaré sections. Limit cycles of the vector field become fixed points of the return map. We find descriptions of the return map in terms of perturbation expansions about the fixed points. The first order coefficients, of course, are just the Floquet multipliers; so our approach is an extension of Floquet ideas to higher order. A time correction is used to ensure that a perturbed limit cycle trajectory is evaluated after almost one period on the Poincaré section.

We use the Lorenz system as an example, looking in particular at the maps that arise as, with variation of a parameter, the response bifurcates from a symmetric limit cycle, through a symmetry breaking bifurcation to two anti-symmetric limit cycles, and then through a period-doubling bifurcation to two period-2 anti-symmetric limit cycles.

We consider the Lorenz system in the form

$$\begin{aligned} \dot{x} &= \sigma(x_2 - x_1) \\ \dot{x}_2 &= rx_1 - x_2 - x_1x_3 \\ \dot{x}_3 &= -bx_3 + x_1x_2 \end{aligned} \qquad (1)$$

with $\sigma = 10$, b = 8/3 and r varies [2]. The derivations of the map equations given below are fairly general, but numerical comparisons are made for this Lorenz system, with r in the range $220<r<350$.

Lorenz himself [1] used a return map based on successive maxima to characterise the response. This approach was developed by Sparrow [2], and Fang [3], and developed

S. Narayanan and R.N. Iyengar (eds.),
IUTAM Symposium on Nonlinearity and Stochastic Structural Dynamics, 71–76.

further by Michielin and Phillipson [4]. Symbolic dynamics can be used to characterise the types of trajectories that can be obtained. Henon and Pomeau [5] introduced a two-dimensional map that gives rise to a strange attractor similar to that of the Lorenz system, although no attempt was made to relate the map parameters to those of the vector field.

A common choice of Poincaré section for this system is $x_3 = (r-1)$. ($x_1 = x_2 = \pm (b(r-1))^{1/2}, x_3 = (r-1)$ are fixed points.) Typical maps obtained on this section are shown in Figures (1) and (2). The map shows $x_2(P)$ versus $x_2(0)$, where P indicates the return time to the Poincaré section. Corresponding values of $x_1(P)$ versus $x_2(0)$ show little variation, so the one-dimensional x_2 return map can be used effectively to characterise the system. (Note that in Figures (1) and (2) the map is defined for x_3 decreasing through the Poincaré section.)

Ding and Hao [6] have used a cubic map $\zeta^P = (A-1)\zeta^0 - A\zeta^{03}$ where ζ is a perturbation away from the fixed point, to approximate this map, and shown that such a map has a sequence of periodic windows that are similar to those of the original Lorenz system. We look for analytic approximations to the actual map. Averaging techniques may be used to find such maps (see, for example, the text by Wiggins [7]). Here, however, an extension of Floquet theory is used to develop a perturbation expansion for the map about a, possibly unstable, fixed point. The time correction used is based on the approach of Sparrow [2, Appendix E].

2. A Perturbation Expansion for the Map

We deal with systems, such as (1), in the form

$$\dot{x}_i(t) = f_i(x(t)) \qquad \begin{cases} x \in R^N \\ x \equiv (x_1, x_2, \dots x_N) \end{cases} \tag{2}$$

Suppose system (2) exhibits a limit cycle $x^*(t)$, with $x^*(T) = x^*(o)$. Consider perturbations away from this limit cycle, defining $x_i(t) = x_i^*(t) + \zeta_i(t)$. Then the perturbations ζ_i satisfy

$$\dot{\zeta}_i = \frac{\partial f_i}{\partial x_j}\zeta_j(t) + \frac{\partial^2 f_i}{\partial x_j \partial y_k}\zeta_j\zeta_k + \dots \tag{3}$$

(For the Lorenz system, only terms up to the quadratic terms are required.) In order to define a map, choose a Poincaré section, define perturbations ζ^o_i at t=0 on the section, and suppose that subsequent time histories may be approximated by a power series

$$\zeta_i(t) = a_{ij}(t)\zeta^o_j + b_{ijk}(t)\zeta^o_j\zeta^o_k + \ldots \tag{4}$$

Substitution into the perturbation expansion (3) yields

$$\begin{aligned} \dot{a}_{ij} &= \frac{\partial f_i}{\partial x_m} a_{mj} & a_{ij}(0) &= \delta_{ij} \\ b_{ijk} &= \frac{\partial f_i}{\partial x_m} b_{mjk} + \frac{\partial^2 f_i}{\partial x_m \partial x_n} a_{mj}\, a_{nk} & b_{ijk}(0) &= 0 \end{aligned} \tag{5}$$

The coefficients are periodic functions of the limit cycle $x^*(t)$. The terms a_{ij}, b_{ijk} etc are first evaluated at t = T. (The terms a_{ij} (T) are just the coefficients of the Floquet monodromy matrix.)

A perturbed trajectory $x(T) = x^*(T) + \zeta(T)$ does not necessarily define a point x(T) on the Poincaré section. Consider $x(T+\tau)=x^*(T+\tau)+\zeta(T+\tau)$, and choose τ so that $x(T+\tau)=x^*(T) + \zeta^P$ is on the section. Then we have,

$$x^*(T)+\zeta^P = x^*(T+\tau) + \zeta(T+\tau) \tag{6}$$

We can write

$$x^*(T+\tau) = x^*(T)+\dot{x}^*(T)\tau + \ddot{x}^*(T)\tau^2/2 + \ldots \tag{7}$$

and

$$\begin{aligned} \zeta(T+\tau) &= a(T+\tau)\zeta^\circ + b(T+\tau)\zeta^{\circ 2} + \ldots \\ &= (a(T) + \dot{a}(T)\tau + \ddot{a}(T)\, \tau^2/2)\zeta^\circ \\ &\quad + (b(T)+\dot{b}(T)\tau)\zeta^{\circ 2} + \ldots \end{aligned} \tag{8}$$

Equations (6), (7), and (8) after solving for τ define the mapping.

3. The Lorenz Equations

Figures (1) and (2) and the associated discussion suggest that we look for a one-dimensional return map on the Poincaré section $x_3 = (r-1)$ involving just $\zeta_{\cdot 2}$ We choose $\zeta_1^\circ = \zeta_3^\circ = 0$ on that section. Equations (6), (7) and (8) reduce to

$$\begin{pmatrix} \zeta_1^p \\ \zeta_2^p \\ \zeta_3^p \end{pmatrix} = \begin{pmatrix} \dot{x}_1 * \\ \dot{x}_2 * \\ \dot{x}_3 * \end{pmatrix} \tau + \begin{pmatrix} \ddot{x}_1 * \\ \ddot{x}_2 * \\ \ddot{x}_3 * \end{pmatrix} \tau^2/2 + \begin{pmatrix} a_{12} + \dot{a}_{12}\tau \\ a_{22} + \dot{a}_{22}\tau \\ a_{32} + \dot{a}_{32}\tau \end{pmatrix} \zeta_2^\circ + \begin{pmatrix} b_{122} \\ b_{222} \\ b_{322} \end{pmatrix} \zeta_2^{\circ 2} \quad (9)$$

where all coefficients are evaluated at T. Terms such as $\dot{x}_1^*$, $\ddot{x}_1^*$, $\dot{a}_{12}$, $\ddot{a}_{12}$ etc. can be found in terms of $x^*(T)$ and $a(T)$ from equations (1) and (5). We further assume a power series expansion for τ, so that

$$\tau = A\zeta_2^\circ + B\zeta_2^{\circ 2} + \ldots$$

finding the coefficients A,B, etc from the condition $\zeta_3^p = 0$ in (9). We finally obtain the map

$$\zeta_2^p = (\dot{x}_2^* A + a_{22})\zeta_2^\circ + (\dot{x}_2^* B + \ddot{x}_2^* A^2/2 + \dot{a}_{22} A + b_{222})\zeta_2^{\circ 2} \quad (10)$$

Although terms to $O(\zeta_2^{\circ 2})$ only are shown explicitly in (10), the numerical comparisons below involve series expansions to $O(\zeta_2^{\circ 5})$.

4. Some Numerical Comparisons

Figures (1) and (2) show comparisons of the map (10) with direct numerical computation of the first-return value of x_2, given that $x_1(0)$ and $x_3(0)$ have the fixed point values on the Poincaré section $x_3 = (r-1)$. The inserts are (x_1,x_2) projections of the limit cycles, showing a symmetric limit cycle at r = 350 and two anti-symmetric limit cycles at r = 280. The maps agree surprisingly well in the vicinity of the fixed points of the map. For r = 280, the map is derived from perturbations about the unstable symmetric limit cycle.

6. Conclusions

A perturbation expansion has been used to obtain return maps. The expansion is in terms of initial perturbations on a Poincaré section. A time correction, also in terms of these same perturbations is used to ensure that perturbed trajectories are sampled actually on the Poincaré section. Agreement with some numerical computations using the Lorenz system are good.

Work supported by NSERC Canada.

7. References

1. E.N. Lorenz (1963) *J. Atmos. Sci* 20 130-141.
2. C. Sparrow (1982) *The Lorenz Equations*, Springer-Verlag, New York.
3. H.P. Fang (1995), *Zeitschrift für Physik* B, 96, 567-552.
4. O. Michielin and P.E. Phillipson (1997), *Int. J. Bifur. And Chaos* 7, 373-382.
5. M. Hénon, Y. Pomeau, (1976), *Lect. Notes in Math* 565, 29-68.
6. M.Z. Ding and B.L. Hao, (1988) *Commun. Theor. Phys* 9 375.
7. S. Wiggins (1990) *Introduction to Applied Nonlinear Dynamical Systems and Chaos*, Springer Verlag, New York.

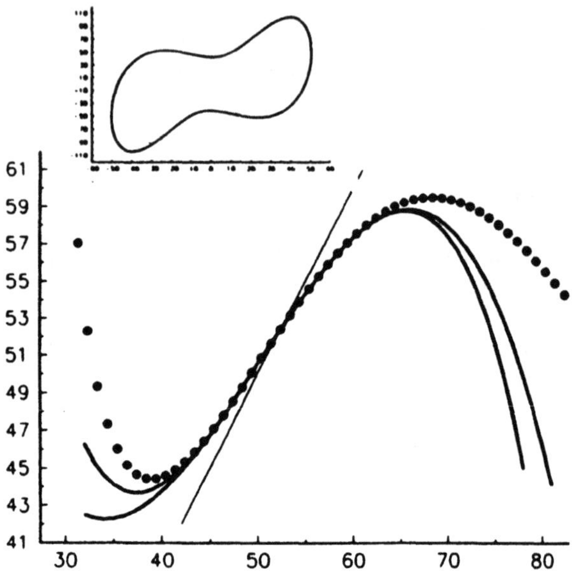

Figure 1. Lorenz oscillator at r=350. Return map of $x_2(P)$ vs. $x_2(0)$, showing cubic and quintic expansions. The insert shows the symmetric limit cycle.

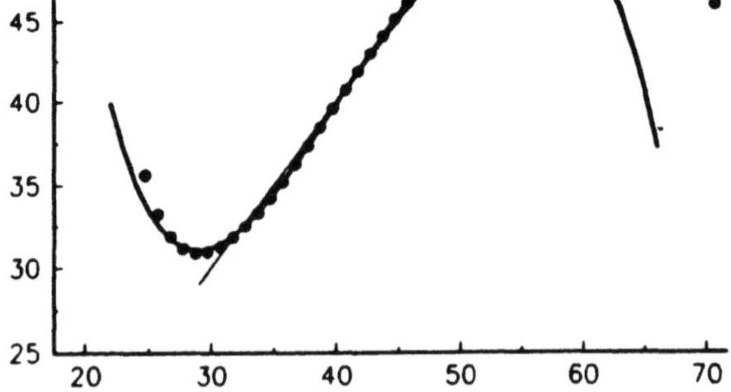

Figure 2. Lorenz oscillator at r=280. Return map of $x_2(P)$ vs. $x_2(0)$, showing the quintic expansion. The insert shows two anti-symmetric limit cycles.

GENERALIZED FOURIER ANALYSIS FOR STOCHASTIC STABILITY PROBLEMS

R. GRIESBAUM AND W. WEDIG
University of Karlsruhe
D - 76128 Karlsruhe, BRD

Abstract:

To obtain top Lyapunov exponents of dynamical systems the densities of associated invariant measures are calculated. They are determined by stationary solutions of Fokker-Planck equations which are solved by means of Fourier expansions. These approaches are investigated for two- and three-dimensional systems subjected to parametric excitations by white noise.

1. Stability problems of interest

Necessary and sufficient conditions for the asymptotic stability of stochastic dynamical systems [1] are obtained by the multiplicative ergodic theorem [2] which allows to determine the exponential growth of non-trivial solutions by means of the top Lyapunov exponents. For this purpose, the invariant measures of the systems are investigated by projecting the linear state processes on the surface of hyperspheres. The Fokker-Planck equations associated to the projection processes are solved by means of Fourier expansions. According to the Furstenberg-Khasminskii formula [3], they give the densities of the invariant measures from which statistical characteristics like Lyapunov numbers or rotation numbers are calculable. These quantities are generalizations [4] of real and imaginary eigenvalue parts of dynamic systems with time-invariant coefficients.

To give a typical application in stochastic structural dynamics let us consider a uniform beam of the mass μ per unit length ℓ. The beam is simply supported at both ends and axially loaded by a time-varying force $P(t)$. As shown in figure 1, $w(x,t)$ describes the lateral beam vibrations depending on space x and time t. The coefficient β gives an external viscous damping per unit length; EA and EI denote axial and bending stiffness,

S. Narayanan and R.N. Iyengar (eds.),
IUTAM Symposium on Nonlinearity and Stochastic Structural Dynamics, 77–94.

respectively. The straight non-bent beam represents the equilibrium position. It is asymptotically stable if the beam vibrations induced by arbitrary initial perturbations decay, exponentially.

Under Kirchoff's assumption [5], that axial end deflections $u(t)$ are proportional to $P(t)$, the beam vibrations are described by the following boundary value problem:

$$EIw_{xxxx} + \beta w_t + \mu w_{tt} - EAu(t)w_{xx}u(t)/\ell = 0, \tag{1}$$
$$w(0,t) = w(\ell,t) = w_{xx}(0,t) = w_{xx}(\ell,t) = 0. \tag{2}$$

Herein, sub x and t denote partial derivatives with respect to space and time, respectively. The equation (1) is solved by the modal expansion

$$w(x,t) = \sum_{n=1}^{\infty} X_n(t) \sin n\pi x/\ell, \qquad 0 \leq x \leq \ell. \tag{3}$$

It satisfies the boundary conditions (2) and reduces the partial differential equation (1) to the uncoupled ordinary differential equations

$$\ddot{X}_n(t) + 2D_n\omega_n\dot{X}_n(t) + \omega_n^2[1 + u(t)A\ell/(n^2\pi^2 I)]X_n(t) = 0, \tag{4}$$
$$2D_n\omega_n = \beta/\mu, \qquad \omega_n^2 = EI(n\pi)^4/(\mu\ell^4), \quad n = 1,2,3,\ldots \tag{5}$$

to be applied for all $n = 1,2,3,\ldots$. Herein, ω_n and D_n denote natural frequencies and damping measures, respectively.

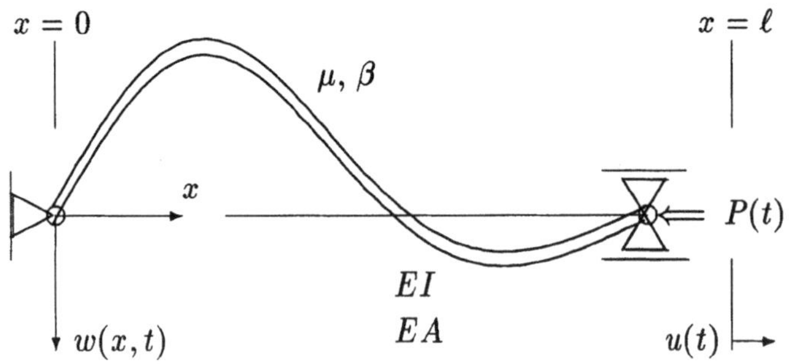

Figure 1. Uniform beam under axial loading $P(t)$

In this paper, we restrict our interest to axial loads $P(t) = EAu(t)/\ell$ which can be described by white noise processes. They are applied to two- and three-dimensional dynamic systems. The densities of the associated invariant measures are explicitly calculated by Fourier expansions. Generalizations are made to perform limiting procedures to delta-shaped densities.

2. Stability of second order systems

According to the multiplicative ergodic theorem [2], the top Lyapunov exponent λ of a linear stochastic system is determined by

$$\lambda = \lim_{t\to\infty} \frac{1}{t} \log ||\vec{X}(t, \vec{X}_0)||, \quad \vec{X}^T = (X_1, \ldots, X_n). \tag{6}$$

Herein, $||\cdot||$ denotes the Euclidean norm of the n-dimensional vector process $\vec{X}$ assumed to be started by the initial value $\vec{X}_0$ at time $t = 0$. In the following, the multiplicative ergodic theorem is applied to the system

$$\ddot{X}_t + 2D\omega_1\dot{X}_t + \omega_1(\omega_1 + \sqrt{\omega_1}\sigma\xi_t)X_t = 0, \tag{7}$$

subjected to parametric excitation by white noise ξ_t of intensity σ.

2.1 PROJECTION OF THE STATE PROCESSES

Introducing the linear state processes $X_{1,t}$ and $X_{2,t}$ of displacement and velocity into the oscillator oscillator (7) leads to the first order system

$$dX_{1,t} = \omega_1 X_{2,t}dt, \qquad E(dW_t^2) = dt, \tag{8}$$

$$dX_{2,t} = -\omega_1(2DX_{2,t} + X_1, t)dt - \sigma\sqrt{\omega_1}X_{1,t}dW_t. \tag{9}$$

Herein, dW_t denotes the increment of the scalar Wiener process W_t by which white noise ξ_t is formally replaced. To perform the projection on the unit circle the polar coordinate processes

$$\Psi_t = \arctan(X_{2,t}/X_{1,t}), \qquad A_t = \sqrt{X_{1,t}^2 + X_{2,t}^2} \tag{10}$$

are applied to (8) and (9) yielding the nonlinear Itô equations

$$d\Psi_t = -\omega_1[1 + (D + \tfrac{1}{2}\sigma^2\cos^2\Psi_t)\sin 2\Psi_t]dt - \sigma\sqrt{\omega_1}\cos^2\Psi_t dW_t, \tag{11}$$

$$dL_t = -\omega_1(2D\sin^2\Psi_t - \tfrac{1}{2}\sigma^2\cos^2\Psi_t\cos 2\Psi_t)dt - \tfrac{1}{2}\sigma\sqrt{\omega_1}\sin 2\Psi_t dW_t \tag{12}$$

for the phase process Ψ_t and the natural logarithm $L_t = \log A_t$ of the amplitude A_t or Euclidean norm $A_t = ||\vec{X}_t||$ of $X_{1,t}$ and $X_{2,t}$.

The amplitude equation (12), integrated in $t \geq 0$ for the initial value A_0 at $t = 0$ and inserted into the multiplicative ergodic theorem (6), leads to the top Lyapunov exponent

$$\lambda = -\omega_1 \lim_{T\to\infty} \frac{1}{T}\int_0^T (2D\sin^2\Psi_t - \tfrac{1}{2}\sigma^2\cos^2\Psi_t\cos 2\Psi_t)dt \tag{13}$$

taking into account that in this time average only drift terms remain, meanwhile the diffusion term of (12) is vanishing because of its non-anticipating independence on dW_t. Provided that the modulo phase process is stationary and ergodic, there exists an invariant measure such that the time average (13) can be replaced by the corresponding space average

$$\lambda = -\omega_1 \int_{-\pi/2}^{+\pi/2} (2D\sin^2\psi - \tfrac{1}{2}\sigma^2\cos^2\psi\cos 2\psi)p(\psi)d\psi. \qquad (14)$$

Herein, $p(\psi)$ is the density of the invariant measure defined in the angle range $|\psi| \leq \pi/2$ and determined by the stationary solution of the Fokker-Planck equation

$$\frac{\partial p}{\partial t} = \omega_1 \frac{\partial}{\partial\psi}\{[1 + (D + \tfrac{1}{2}\sigma^2\cos^2\psi)\sin 2\psi]p\} + \tfrac{1}{2}\omega_1\sigma^2\frac{\partial^2}{\partial\psi^2}(\cos^4\psi p). \quad (15)$$

This equation determines the time-dependent density $p(\psi, t)$ of the angle process Ψ_t according to the definition by the Itô equation (11).

2.2 INVARIANT DENSITIES OF THE PROJECTION

In the stationary case $\partial p/\partial t = 0$, the Fokker-Planck equation (15) can once be integrated with respect to the angle variable ψ yielding an ordinary differential equation of the first order form

$$\tfrac{1}{2}\sigma^2\cos^4\psi\frac{dp}{d\psi} + [1 + (D - \tfrac{1}{2}\sigma^2\cos^2\psi)\sin 2\psi]\,p(\psi) = C. \qquad (16)$$

Herein, C is a constant of integration determined by the normalization condition, noted in (17). Following [6], mild solutions $p(\psi)$ of the Fokker-Planck equation (16) are calculated by the Fourier expansion

$$p(\psi) = \frac{1}{\pi}\sum_{n=0}^{\infty}(c_n\cos 2n\psi + s_n\sin 2n\psi), \qquad \int_{-\pi/2}^{+\pi/2} p(\psi)d\psi = 1. \qquad (17)$$

For $c_0 = 1$, the Fourier series satisfy the normalization as well as the stationarity and periodicity condition $p(\psi + \pi) = p(\psi)$, respectively.

The insertion of the Fourier expansion (17) into the Fokker-Planck equation (16) leads to the infinite equation systems

$$\begin{aligned} &c_n + \tfrac{3}{8}\sigma^2 n s_n + \tfrac{1}{16}\sigma^2[(n+1)s_{n+2} + (n-1)s_{n-2}] \\ &+[(2n+1)\tfrac{1}{8}\sigma^2 + \tfrac{1}{2}D]s_{n+1} + [(2n-1)\tfrac{1}{8}\sigma^2 - \tfrac{1}{2}D]s_{n-1} = 0, \end{aligned} \qquad (18)$$

$$\begin{aligned} &s_n - \tfrac{3}{8}\sigma^2 n c_n - \tfrac{1}{16}\sigma^2[(n+1)c_{n+2} + (n-1)c_{n-2}] \\ &-[(2n+1)\tfrac{1}{8}\sigma^2 + \tfrac{1}{2}D]c_{n+1} - [(2n-1)\tfrac{1}{8}\sigma^2 - \tfrac{1}{2}D]c_{n-1} = v_n \end{aligned} \qquad (19)$$

to be applied for all $n = 1, 2, 3, \ldots$ with $s_0 = 0$ and $c_0 = 1$. Herein, the inhomogeneous coefficients v_n are arising because of $c_0 = 1$. The first two coefficients of v_n are calculated by

$$v_1 = \tfrac{1}{8}\sigma^2 - \tfrac{1}{2}D = 2\tau - d, \qquad\qquad v_2 = \tfrac{1}{16}\sigma^2 = \tau \tag{20}$$

and the higher ones are vanishing for all $n = 3, 4, \ldots$. The application of $n = 1, 2, \ldots$ to (18) and (19) leads to the explicit matrix equation

$$\left[\begin{array}{cc|cc|c}
1 & 6\tau & 0 & 6\tau + d & \cdots \\
-6\tau & 1 & -6\tau - d & 0 & \cdots \\
\hline
0 & 6\tau - d & 1 & 12\tau & \cdots \\
-6\tau + d & 0 & -12\tau & 1 & \cdots \\
\hline
0 & 2\tau & 0 & 10\tau - d & \cdots \\
-2\tau & 0 & -10\tau + d & 0 & \cdots \\
\hline
\vdots & \vdots & \vdots & \vdots & \ddots
\end{array}\right]
\left[\begin{array}{c} c_1 \\ s_1 \\ \hline c_2 \\ s_2 \\ \hline c_3 \\ s_3 \\ \hline \vdots \end{array}\right]
=
\left[\begin{array}{c} 0 \\ v_1 \\ \hline 0 \\ v_2 \\ \hline 0 \\ 0 \\ \hline \vdots \end{array}\right] \tag{21}$$

where τ and d are abbreviations, already explained in (20). Notice that the application of $n = 0$ is not needed since (18) and (19) are fulfilled a priori; i.e. for $n = 0$ (19) yields $s_0 = 0$, meanwhile (18) determines the constant of integration introduced in the Fokker-Planck equation (16).

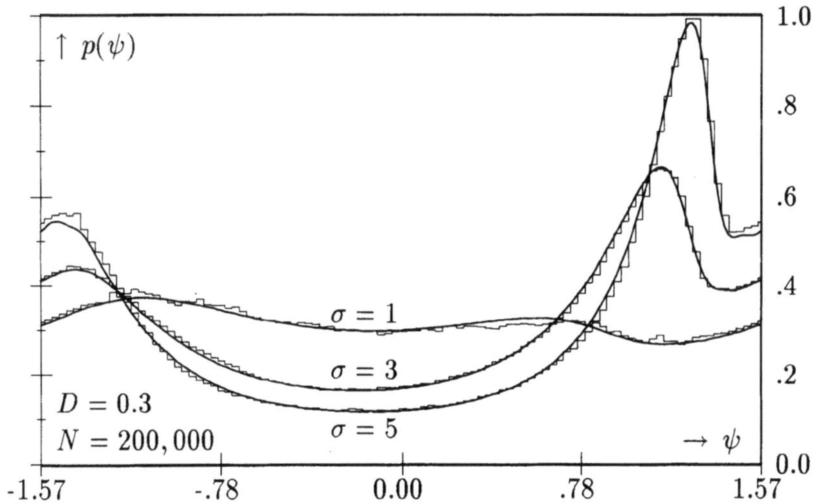

Figure 2. Fourier expansions and Monte-Carlo simulations

If the Fourier series (17) converges in mean square to the exact density $p(\psi)$, the coefficients c_n and s_n are sequences which tend to zero for increasing n. From this it follows that the matrix equation (21) can approximately be solved by truncation at a finite index number $n = N$ applying the closure conditions $c_{N+1} = c_{N+2} = 0$ and $s_{N+1} = s_{N+2} = 0$. Figure 2 shows three typical evaluations for the parameters $D = 0.3$ and $\sigma = 1, 3$ and 5. Hereby, the Fourier coefficients are calculated up to $N = 25$. The density values obtained are drawn by solid lines and overlaid by step lines of corresponding histograms obtained by Monte-Carlo simulations. The comparison of both results show good coincidences in the entire angle range $|\psi| \leq \pi/2$. Finally, the calculated Fourier expansion is inserted into the Furstenberg-Khasminskii formula (14) to obtain the Lyapunov exponent

$$\lambda = -\omega_1[D - \tfrac{1}{8}\sigma^2 - \tfrac{1}{2}(D + \tfrac{1}{4}\sigma^2)c_1 - \tfrac{1}{8}\sigma^2 c_2]. \tag{22}$$

It is completely determined by the first two coefficients c_1 and c_2. To give a first analytical approximation the matrix equation (21) is truncated by $c_2 = c_3 = 0$ and $s_2 = s_3 = 0$ and evaluated to

$$c_1^{(1)} = -\tfrac{3}{8}\sigma^2 \frac{\frac{1}{4}\sigma^2 - D}{1 + (\frac{3}{8}\sigma^2)^2}, \qquad s_1^{(1)} = \frac{\frac{1}{4}\sigma^2 - D}{1 + (\frac{3}{8}\sigma^2)^2}. \tag{23}$$

Higher order analytical approximations are derivable in the same way.

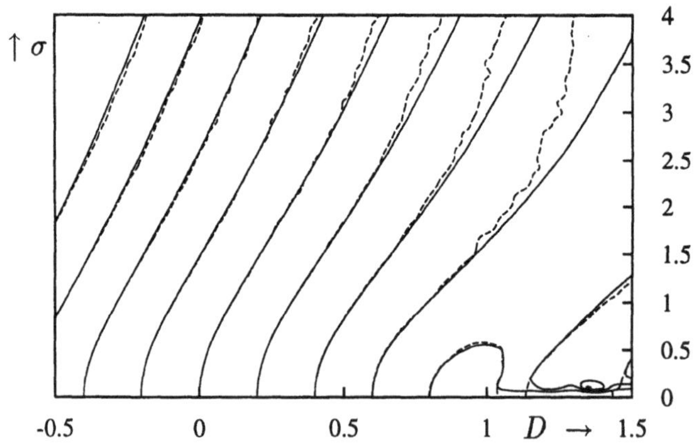

Figure 3. Contour lines of top Lyapunov exponents

In figure 3, solid lines show level contours $\lambda(\sigma, D)/\omega_1$ =const. of the top Lyapunov exponent (22) in the parameter range $0 \leq \sigma \leq 4$ and $-0.5 \leq D \leq 1.5$ evaluated for $N = 20$ Fourier coefficients and the step size $\Delta\lambda/\omega_1 = 0.2$. They can be compared with the interrupted lines of corresponding results obtained by Monte-Carlo simulations. Both results show good coincidences at least for low noise intensities and undercritical dampings $|D| < 1$. Deviations are observable for higher noise intensities where Monte-Carlo simulations become inaccurate at least for the applied

Euler scheme and scan rate of $\Delta t = 0.01$ s. In the overcritical case $D \geq 1$, the density $p(\psi)$ degenerates to delta distributions for vanishing noise intensities $\sigma = 0$. This is the reason why Fourier expansions fail for $D \geq 1$ and $\sigma = 0$ leading to considerable deviations from the exact result that is known by $\lambda = -D\omega_1 + \omega_1\sqrt{D^2 - 1}$ in case of $\sigma = 0$ and $D \geq 1$.

2.3 STATIONARITY OF ANGLE DISTRIBUTIONS

As already mentioned, the density histograms, shown in figure 2, are obtained by Monte-Carlo simulations which are evaluated according to the ergodic theorem for Borel sets A as

$$\mu(A) = \lim_{T\to\infty} \frac{1}{T} \int_0^T 1_A(\psi(t,\omega))dt. \tag{24}$$

Herein, 1_A denotes the indicator function applied to one angle trajectory $\psi(t, \omega)$ which is bounded by the modulo operator $\mathrm{mod}(\pi, \psi)$ to the angle range $|\psi| \leq \pi/2$.

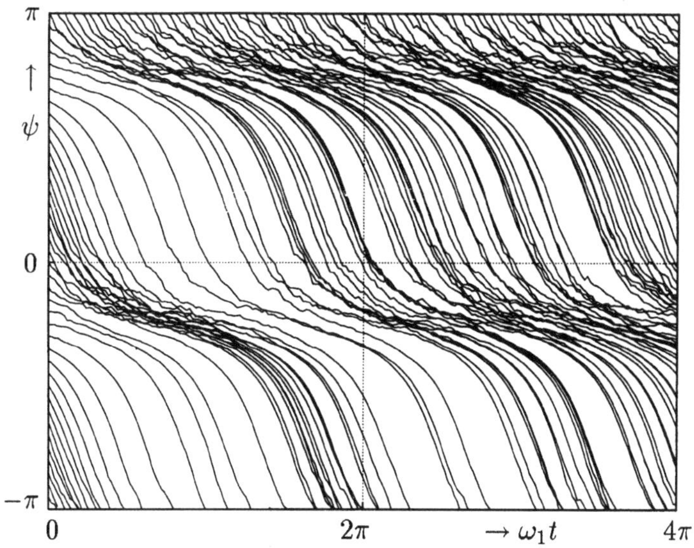

Figure 4. Transient behaviour of angle trajectories

Monte-Carlo simulations of the angle process are performed by means of the angle equation (11) in the discretized form

$$\Delta\Psi_n = -\omega_1[1 + (D + \tfrac{1}{2}\sigma^2 \cos^2\Psi_n)\sin 2\Psi_n]\Delta t - \sigma\sqrt{\omega_1 \Delta t}\cos^2\Psi_n\, R_n \tag{25}$$

to be applied for $\Delta\Psi_n = \Psi_{n+1} - \Psi_n$ and all $n = 0, 1, 2, \ldots$. The recurrence formula (25) is started by the initial angle Ψ_0 and driven by a sequence of normally distributed numbers R_n of zero mean and normalized mean square $E(R_n^2) = 1$. In figure 2, the scan rate selected was $\Delta t = 0.002$ s and applied for $N_S = 2 \cdot 10^6$ simulation steps. The ergodic theorem (24) was evaluated for $K = 100$ angle classes of the step size $\Delta\psi = \pi/K$.

Figure 4 shows some typical angle trajectories generated by the recurrence formula (25) for the parameters $\omega_1 = 1\ \mathrm{s}^{-1}$, $D = 0.8$ and $\sigma = 0.1$. The recursion was started with initial values ψ_0 and t_0 which were uniformly distributed at the left and upper bordering lines of figure 4. Starting from these lines, the angle trajectories possess a certain transient behaviour and become then stationary. For the parameters, applied in figure 4, stationarity was approximately reached already in the lower right part of figure 4 where the density of the angle realizations starts to be time-invariant and π-periodic.

For quantitative investigations of the initially transient behaviour the analysis is extended by the time-dependent Fourier expansion

$$p(\psi, t) = \frac{1}{\pi} \sum_{n=0}^{\infty} [c_n(t) \cos 2n\psi + s_n(t) \sin 2n\psi]. \tag{26}$$

Its insertion into the Fokker-Planck equation (15) yields two sets of ordinary differential equations

$$\begin{aligned} &\tfrac{1}{2n\omega_1}\dot{s}_n + c_n + \tfrac{3}{8}\sigma^2 s_n + \tfrac{1}{16}\sigma^2[(n+1)s_{n+2} + (n-1)s_{n-2}] \\ &\quad + \tfrac{1}{2}[(2n+1)\tfrac{1}{4}\sigma^2 + D]s_{n+1} + \tfrac{1}{2}[(2n-1)\tfrac{1}{4}\sigma^2 - D]s_{n-1} = 0, \end{aligned} \tag{27}$$

$$\begin{aligned} &\tfrac{1}{2n\omega_1}\dot{c}_n - s_n + \tfrac{3}{8}\sigma^2 n c_n + \tfrac{1}{16}\sigma^2[(n+1)c_{n+2} + (n-1)c_{n-2}] \\ &\quad + \tfrac{1}{2}[(2n+1)\tfrac{1}{4}\sigma^2 + D]c_{n+1} + \tfrac{1}{2}[(2n-1)\tfrac{1}{4}\sigma^2 - D]c_{n-1} = -v_n \end{aligned} \tag{28}$$

to be applied for all $n = 1, 2, \ldots$ and $c_0(t) \equiv 1$. The inhomogeneous coefficients v_n are already given in (20). Obviously, these differential equations lead to the stationary density, drawn in figure 2, if the eigensolutions

$$s_n(t) = S_n \exp(\lambda t), \quad c_n(t) = C_n \exp(\lambda t), \qquad n = 1, 2, \ldots \tag{29}$$

are asymptotically stable. Herein, the exponents λ are the eigenvalues of (27) and (28) to be calculated by the theory of infinite determinants.

For first investigations, the couplings to higher and lower order solutions are neglected so that the solutions of (27) and (28) are approximated by the uncoupled differential equations

$$\dot{s}_n(t) + \tfrac{3}{4}\sigma^2 n\omega_1 s_n(t) + 2n\omega_1 c_n(t) = 0, \tag{30}$$

$$\dot{c}_n(t) + \tfrac{3}{4}\sigma^2 n\omega_1 c_n(t) - 2n\omega_1 s_n(t) = 0. \tag{31}$$

Notice that these equations possess the eigenvalues

$$\lambda = -\tfrac{3}{4}\sigma^2 n\omega_1 \pm i2n\omega_1 \qquad n = 1, 2, \dots. \tag{32}$$

For $\sigma > 0$, they have negative real parts. Consequently, the stationary density distribution, calculated by Fourier expansions, is asymptotically stable for parametric excitations with non-vanishing noise intensities.

3. Higher order dimensional systems

The stability analysis of higher dimensional problems is characterized by extended singularities introduced by the projection on hyperspheres. To explain basic features it is sufficient to consider a three-dimensional system consisting on the linear oscillator (7) and a low pass filter coupled by parametric white noise, as follows:

$$\ddot{X}_t + 2D\omega_1\dot{X}_t + \omega_1(\omega_1 X_t + \sqrt{\omega_1}\sigma\xi_t Z_t) = 0, \tag{33}$$

$$\dot{Z}_t + \beta\omega_1 Z_t - \omega_1^{-1/2}\sigma\xi_t\dot{X}_t = 0, \qquad \beta \geq |D|. \tag{34}$$

For vanishing noise intensities $\sigma = 0$, both equations are uncoupled and possess the deterministic eigenvalues

$$\lambda_{1,2} = -D\omega_1 \pm i\omega_1\sqrt{1 - D^2} \qquad\qquad \lambda_3 = -\beta\omega_1. \tag{35}$$

Hence, $\lambda = -D\omega_1$ is the top Lyapunov exponent for $\sigma = 0$ provided that the dampings satisfy $|D| \leq 1$ and $\beta \geq |D|$.

3.1 PROJECTION BY SPHERICAL COORDINATES

Introducing the state processes X_t and Z_t as well as the dimensionless velocity $V_t = \dot{V}_t/\omega_1$, the ordinary differential equations (33) and (34) are written in form of the stochastic Itô equations

$$dX_t = \omega_1 V_t dt, \qquad E(dW_t^2) = dt, \tag{36}$$

$$dV_t = -\omega_1[X_t + (2D + \tfrac{1}{2}\sigma^2)V_t]dt - \sqrt{\omega_1}\sigma Z_t dW_t, \tag{37}$$

$$dZ_t = -\omega_1(\beta + \tfrac{1}{2}\sigma^2)Z_t dt + \sqrt{\omega_1}\sigma V_t dW_t. \tag{38}$$

Herein, dW_t denotes the increment of the scalar Wiener process W_t by which white noise ξ_t is formally replaced. The σ^2-terms in (37) and (38) are correction terms [7] according to the Stratonovich integral. They are needed to interprete white noise perturbations of higher derivatives in the physical sense of broad-band excitations.

The projection of the linear state processes X_t, Z_t and V_t on the unit sphere is performed by means of the spherical coordinates

$$\begin{aligned} z &= a\sin\varphi, & a &= \sqrt{x^2+v^2+z^2}, \\ x &= a\cos\varphi\cos\psi, & \psi &= \arctan(v/x), \\ v &= a\cos\varphi\sin\psi, & \varphi &= \arctan(z/\sqrt{x^2+v^2}). \end{aligned} \tag{39}$$

Applying Itô's calculus the associated transformations are performed leading to the following nonlinear projection equations:

$$d\log A_t = h_1(\Psi_t, \Phi_t)dt + g_1(\Psi_t, \Phi_t)dW_t, \tag{40}$$
$$d\Psi_t = h_2(\Psi_t, \Phi_t)dt + g_2(\Psi_t, \Phi_t)dW_t, \tag{41}$$
$$d\Phi_t = h_3(\Psi_t, \Phi_t)dt + g_3(\Psi_t, \Phi_t)dW_t. \tag{42}$$

Herein, $h_i(\Psi_t, \Phi_t)$ and $g_i(\Psi_t, \Phi_t)$ are abbreviations for corresponding drift and diffusion terms, respectively. For $i = 1, 2, 3$ they are determined by

$$h_1 = -\omega_1(\beta\sin^2\Phi_t + 2D\sin^2\Psi_t\cos^2\Phi_t), \qquad g_1 \equiv 0 \tag{43}$$
$$g_2 = -\sigma\sqrt{\omega_1}\cos\Psi_t\tan\Phi_t, \qquad g_3 = \sigma\sqrt{\omega_1}\sin\Psi_t, \tag{44}$$
$$h_2 = -\omega_1[1 + (D + \tfrac{1}{4}\sigma^2 + \tfrac{1}{2}\sigma^2\tan^2\Phi_t)\sin 2\Psi_t], \tag{45}$$
$$h_3 = -\omega_1[(\beta - 2D\sin^2\Psi_t)\cos^2\Phi_t + \tfrac{1}{2}\sigma^2\cos^2\Psi_t]\tan\Phi_t. \tag{46}$$

Notice, that the two angle equations (41) and (42) are decoupled from the amplitude. Thus, the logarithmic amplitude equation (40) can formally be integrated and adjusted to the initial amplitude value A_0. This integrated form of (40) determines the top Lyapunov exponent

$$\lambda = \lim_{t\to\infty}\frac{1}{t}\log\frac{A_t}{A_0} = \lim_{T\to\infty}\frac{1}{T}\int_0^T h_1(\Psi_t, \Phi_t)dt \tag{47}$$

in accordance to the multiplicative ergodic theorem (6).

Monte-Carlo simulations of the two angle equations (41) and (42) were performed by means of forward Euler schemes applied for the system data $\omega_1 = 1\ \text{s}^{-1}$, $D = 0.2$, $\beta = 1$, $\sigma = 0.5$ and the scan rate $\Delta t = 0.01$ s. The angle trajectories were simulated by $N = 10^8$ time steps and evaluated in the angle ranges $|\psi| \le \pi/2$ and $|\varphi| \le \pi/2$ leading to the joint density distribution $p(\psi, \varphi)$ shown in figure 5. The related top Lyapunov exponent obtained by this Monte-Carlo simulation was $\lambda/\omega_1 = -0.265942$.

Due to the negative mean value of the drift term $h_2(\Psi_t, \Phi_t)$ the angle process Ψ_t of the oscillator rotates clockwise so that the joint density $p(\psi, \varphi)$ becomes periodic and non-vanishing at both bordering lines at $|\psi| = \pi/2$. The angle process Φ_t of the coupled first order system, however, is fluctuating without any rotation. This is a consequence of the singular tan-term in $h_3(\Psi_t, \Phi_t)$. It produces natural boundaries at $|\varphi| = \pi/2$ by repelling Φ_t into the direction of $\varphi = 0$ in a purely deterministic way. This characteristic feature is maintained for all parameter variations of $|D| \leq 1$ and $\beta \geq 1$. For vanishing noise intensities $\sigma = 0$ the joint density $p(\psi, \varphi)$ degenerates to a singular distribution which is delta-shaped and concentrated around the middle line $\varphi = 0$.

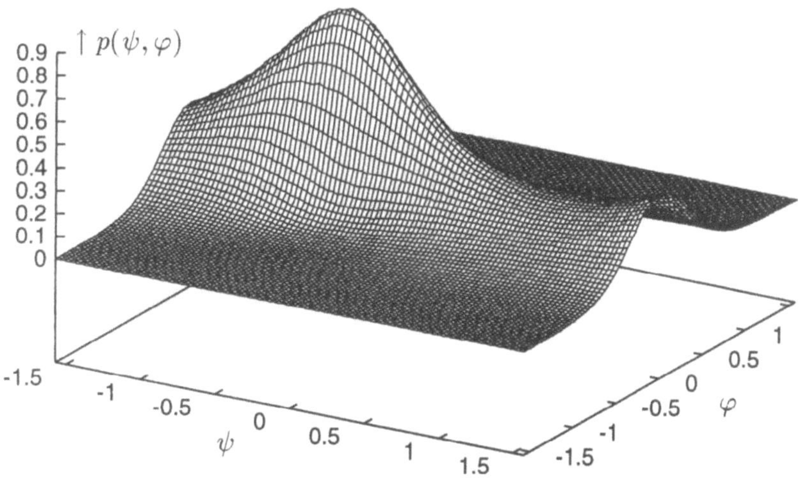

Figure 5. Joint density by Monte-Carlo simulation

3.2 LYAPUNOV EXPONENTS BY AVERAGING

Figure 6 shows level lines of Lyapunov exponents in dependence on the damping D of the oscillator and the noise intensity σ for fixed $\omega_1 = 1\ \mathrm{s}^{-1}$ and $\beta = 1$. The applied level step was $\Delta\lambda = 0.2$. These numerical results were obtained by the angle simulations, mentioned above, and by evaluations of the time average (47) for the scan rate $\Delta t = 0.01$ s and $n = 10^7$ time steps. In the lower left part of figure 6 the Lyapunov exponents are positive indicating an exponential growth of the systems solutions. For increasing noise perturbations, however, the growth behaviour can be inverted so that the solutions of the system become stable. These stabilization effects are observable also in the lower right part of the picture at least for sufficiently small noise intensities of the parametric excitation.

Similar stabilization results are obtained by means of an averaging method [8] which is applied in order to derive analytic approximations. For that purpose the density of the rotation angle Ψ_t is assumed to be uniformly distributed and independent on the fluctuating angle Φ_t. Due to this assumption the rotation angle Ψ_t can be eliminated, and the remaining two processes A_t and Φ_t can approximately be described by two Itô differential equations of the following simplified forms:

$$\begin{aligned} d\log A_t &= -\omega_1(\beta\sin^2\Phi_t + D\cos^2\Phi_t)dt, && (48)\\ d\Phi_t &= \bar{h}_3(\Phi_t)dt + \bar{g}_3 dW_t, \qquad \bar{g}_3 = \sigma\sqrt{\omega_1/2}, && (49)\\ \bar{h}_3(\Phi_t) &= -\omega_1[(\beta - D)\cos^2\Phi_t + \tfrac{1}{4}\sigma^2]\tan\Phi_t. && (50) \end{aligned}$$

The Fokker-Planck equation associated to the averaged angle equation (49) possesses the stationary form (51) and is integrated, as follows:

$$\frac{\partial}{\partial\varphi}[\bar{h}_3(\varphi)\bar{p}(\varphi)] - \frac{1}{2}\frac{\partial^2}{\partial\varphi^2}[\bar{g}_3^2\bar{p}(\varphi)] = 0, \qquad |\varphi| \leq \pi/2, \tag{51}$$

$$\bar{p}(\varphi) = C\cos\varphi\exp(\frac{\beta - D}{\sigma^2}\cos 2\varphi), \qquad \int_{-\pi/2}^{+\pi/2}\bar{p}(\varphi)d\varphi = 1. \tag{52}$$

Herein, C is calculated by the normalization condition, noted in (52).

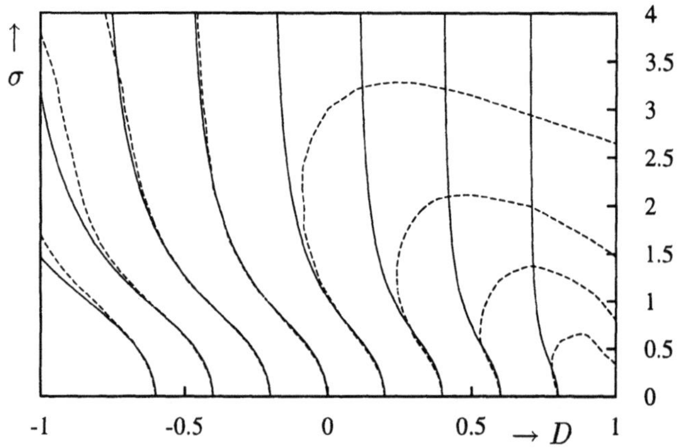

Figure 6. Stabilization effects by increasing noise

The amplitude equation (48) leads to the top Lyapunov exponent

$$\lambda = -\omega_1\int_{-\pi/2}^{+\pi/2}(\beta\sin^2\varphi + D\cos^2\varphi)\bar{p}(\varphi)d\varphi. \tag{53}$$

Figure 6 shows associated level contours by solid lines in comparison with those which are marked by interrupted lines and were obtained by the Monte-Carlo simulations, already described in the beginning of this section. Notice that both results show excellent coincidences near the level line of vanishing Lyapunov exponents. Large deviations of both results are only observable in the upper right part of figure 6 when the damping measures D of the oscillator and the noise intensities σ of its parametric excitation are increasing. In the lower part, however, both results possess a good coincidence again. In case of $|D| \leq 1$, they represent the asymptotics for vanishing noise intensities $\sigma = 0$ where the top Lyapunov exponent is known by $\lambda = -\omega_1 D$.

3.3 TWO-DIMENSIONAL FOURIER EXPANSIONS

To extend the Fourier analysis to the two-dimensional case of the joint angle density $p(\psi, \varphi)$ we consider the Fokker-Planck equation

$$\frac{\partial(h_2 p)}{\partial\psi} + \frac{\partial(h_3 p)}{\partial\varphi} - \frac{1}{2}[\frac{\partial^2(g_2^2 p)}{\partial\psi^2} + 2\frac{\partial^2(g_2 g_3 p)}{\partial\psi\partial\varphi} + \frac{\partial^2(g_3^2 p)}{\partial\varphi^2}] = 0 \qquad (54)$$

which is associated to the stationary angle processes Ψ_t and Φ_t of the equations (41) and (42). The solution $p(\psi, \varphi)$ of this Fokker-Planck equation has to satisfy the two boundary conditions, noted in (55), that the density is vanishing at the border lines $\varphi = \pm\pi/2$, and the periodicity (56), that the density and its derivative is π-periodic with respect to ψ as well as the normalization condition inside the range $|\psi|, |\varphi| \leq \pi/2$, as noted in (55).

$$p(\psi, \varphi = \pm\pi/2) = 0, \qquad \iint_{-\pi/2}^{+\pi/2} p(\psi, \varphi) d\psi d\varphi = 1, \qquad (55)$$

$$\frac{\partial}{\partial\psi} p(\psi + \pi, \varphi) = \frac{\partial}{\partial\psi} p(\psi, \varphi), \qquad p(\psi + \pi, \varphi) = p(\psi, \varphi) \qquad (56)$$

The coefficient functions $h_i(\psi, \varphi)$ and $g_i(\psi, \varphi)$ of the Fokker-Planck equation are already noted in (44), (45) and (46).

As mentioned, these coefficient functions are singular at the line $\varphi = \pm\pi/2$ because of $\tan\varphi$-terms by which the density $p(\psi, \varphi)$ is limited in the angle range $|\varphi| \leq \pi/2$. To eliminate this natural singularity we introduce the formal density substitution [8]

$$p(\psi, \varphi) = \hat{p}(\psi, \varphi) \cos^2\varphi \qquad \text{for } |\psi|, |\varphi| \leq \pi/2 \qquad (57)$$

by which the Fokker-Planck equation (54) takes the form

$$\frac{\partial \hat{h}_2 \hat{p}}{\partial\psi} + \frac{\partial \hat{h}_3 \hat{p}}{\partial\varphi} - \frac{1}{2}[\frac{\partial^2 \hat{g}_2^2 \hat{p}}{\partial\psi^2} + 2\frac{\partial^2 \hat{g}_2 \hat{g}_3 \hat{p}}{\partial\psi\partial\varphi} + \frac{\partial^2 \hat{g}_3^2 \hat{p}}{\partial\varphi^2}] = 0. \qquad (58)$$

Herein, the coefficient functions $\hat{g}_i(\psi, \varphi)$ and $\hat{h}_i(\psi, \varphi)$ are defined by

$$\begin{aligned}
\hat{g}_2(\psi, \varphi) &= -\sigma\sqrt{\omega_1} \cos\psi \sin\varphi, \quad \hat{g}_3(\psi, \varphi) = \sigma\sqrt{\omega_1} \sin\psi \cos\varphi, & (59)\\
\hat{h}_2(\psi, \varphi) &= -\omega_1\{\cos^2\varphi + [(D + \tfrac{1}{4}\sigma^2)\cos^2\varphi + \tfrac{1}{2}\sigma^2 \sin^2\varphi] \sin 2\psi\}, & (60)\\
\hat{h}_3(\psi, \varphi) &= -\omega_3[(\beta - 2D\sin^2\psi)\cos^2\varphi + \tfrac{1}{2}\sigma^2\cos^2\psi] \sin\varphi \cos\varphi. & (61)
\end{aligned}$$

Now, the singularities by $\tan\varphi$ are eliminated; i.e. the Fokker-Planck equation (58) is regular and contains bounded coefficients which are π-periodic in ψ and φ. Hence, the solution $\hat{p}(\psi,\varphi)$ can be approximated by the two-dimensional Fourier expansion

$$\hat{p}(\psi,\varphi) = \sum_{k=-N}^{N} \sum_{\ell=-N}^{N} c_{k,\ell} \exp[i2(k\varphi + \ell\psi)] \tag{62}$$

which is orthogonal with respect to the entire angle range $|\psi|, |\varphi| \leq \pi/2$.

The insertion of the Fourier expansion (62) into the modified Fokker-Planck equation (58) and the comparison of all coefficients of the harmonic functions $\exp[i2(k\varphi + \ell\psi)]$ leads to an infinite system of linear algebraic equations which determine the Fourier coefficients $c_{k,\ell} = a_{k,\ell} + ib_{k,\ell}$ for all indices $k, \ell = \pm 1, \pm 2, \pm 3, \ldots$ dependent on $c_{0,0} = a_{0,0} + ib_{0,0}$. The two coefficients $a_{0,0}$ and $b_{0,0}$ remain indeterminated since the Fokker-Planck equation (58) contains differentiation terms, only. Hence, the zeroth coefficients $a_{0,0}$ and $b_{0,0}$ can be utilized to fulfill the normalization in $|\psi|, |\varphi| \leq \pi/2$ by

$$2a_{0,0} + a_{1,0} + a_{-1,0} = 4/\pi, \qquad 2b_{0,0} + b_{1,0} + b_{-1,0} = 0. \tag{63}$$

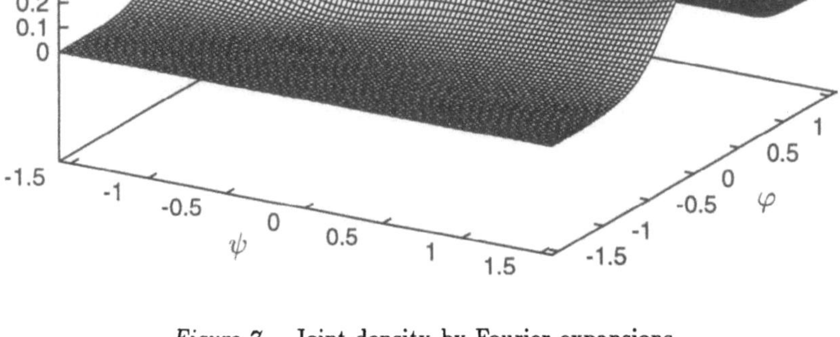

Figure 7. Joint density by Fourier expansions

If the Fourier expansion (62) converges to the exact density $\hat{p}(\psi,\varphi)$ in square mean, the coefficients $c_{k,\ell}$ tend to zero for infinitely increasing indices k and ℓ. Consequently, the coefficient equations can be solved by truncation at finite indices $k, \ell = N$ neglecting the coupling to coefficients with higher indices numbers. These closure conditions lead to the following schema:

$$\begin{array}{ccccc} c_{-2,-2} & c_{-2,-1} & c_{-2,0} & c_{-2,1} & c_{-2,2} \\ c_{-1,-2} & c_{-1,-1} & c_{-1,0} & c_{-1,1} & c_{-1,2} \\ c_{0,-2} & c_{0,-1} & \boxed{c_{0,0}} & c_{0,1} & c_{0,2} \\ c_{1,-2} & c_{1,-1} & c_{1,0} & c_{1,1} & c_{1,2} \\ c_{2,-2} & c_{2,-1} & c_{2,0} & c_{2,1} & c_{2,2} \end{array} \tag{64}$$

It characterizes a possible sequence of approximations by neglecting all Fourier coefficents and their equations outside the rectangular boxes.

Figure 7 shows a typical two-dimensional density distribution $p(psi, \varphi)$ as the result of Fourier expansions obtained for $N = 10$ and the same parameter values $\omega_1 = 1\mathrm{s}^{-1}, D = 0.2, \beta = 1$ and $\sigma = 0.5$, already applied in figure 5. The associated Lyapunov exponent $\lambda = -0.264649$ was calculated by the Furstenberg-Khasminskii formula

$$\lambda = \int_{-\pi/2}^{+\pi/2} \int_{-\pi/2}^{+\pi/2} h_1(\psi, \varphi) \cos^2 \varphi \, \hat{p}(\psi, \varphi) d\psi d\varphi. \tag{65}$$

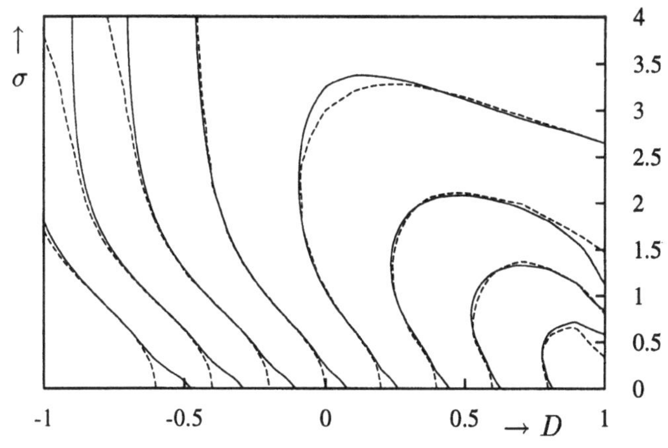

Figure 8. LE by Fourier expansions and MC simulations

Solid lines in figure 8 show further evaluations of the Lyapunov exponent (65) in the parameter range $|D| \leq 1$ and $0 \leq \sigma \leq 4$ for $N = 5$ Fourier terms and for the fixed system parameters $\beta = 1$ and $\omega_1 = 1\mathrm{s}^{-1}$. The comparison with corresponding results of figure 6 obtained by Monte-Carlo simulations show good coincidences. Deviations are only observable for low and high intensities of the parametric excitation. For $\sigma >> 1$, the Monte-Carlo simulations become inaccurate for the applied Euler scheme and the scan rate of $\Delta t = 0.01$ s. If σ tends to zero, the invariant density $p(\psi,\varphi)$ degenerates to singular distributions which are delta-shaped around the angle line $\varphi = 0$. Naturally, such singular densities can not be approximated by classical Fourier series since they become divergent for high gradients in case of $\sigma << 1$.

4. Singular distributions of oscillators

For overcritical dampings $|D| \geq 1$, the unexcited oscillator possesses the top Lyapunov exponent λ_0, noted in (66), and a delta-shaped invariant density which is concentrated around the angle position ψ_0.

$$\psi_0 = -\frac{1}{2}\arcsin(\frac{1}{D}), \qquad \lambda_0 = \omega_1(\sqrt{D^2-1} - D) \tag{66}$$

Such singular distributions can be approximated by nonlinear Fourier expansions as demonstrated in the following.

4.1 OVERCRITICAL OSCILLATOR EQUATIONS

The deterministic top Lyapunov exponent λ_0, noted in (66), is utilized to introduce the substitution

$$X_t = Y_t \exp[\omega_1(\sqrt{D^2-1} - D)t], \quad D \geq 1, \tag{67}$$

and to reduce the parametrically excited oscillator (7) to a simplier second order form, respectively to the following first order system

$$dY_t = \omega_1 V_t dt, \qquad \delta = \sqrt{D^2-1}, \tag{68}$$

$$dV_t = -2\omega_1 \delta V_t dt - \sigma\sqrt{\omega_1}\, Y_t dW_t. \tag{69}$$

It describes the displacement process Y_t and the dimensionless velocity $V_t = \dot{Y}_t/\omega_1$ of the reduced oscillator. The projection on the unit circle is performed by polar coordinate processes Φ_t and A_t leading to

$$\begin{aligned} d\Phi_t &= -\omega_1[\sin^2\Phi_t + (\delta + \tfrac{1}{2}\sigma^2\cos^2\Phi_t)\sin 2\Phi_t]dt - \\ &\quad -\sigma\sqrt{\omega_1}\cos^2\Phi_t dW_t, \qquad \Phi_t = \arctan(V_t/Y_t), \end{aligned} \tag{70}$$

$$\begin{aligned} d\log A_t &= \tfrac{1}{2}\omega_1[\sin 2\Phi_t - 4\delta\sin^2\Phi_t + \sigma^2\cos^2\Phi_t\cos 2\Phi_t]dt - \\ &\quad -\tfrac{1}{2}\sigma\sqrt{\omega_1}\sin 2\Phi_t dW_t, \qquad A_t = \sqrt{Y_t^2 + V_t^2}. \end{aligned} \tag{71}$$

According to (71), the top Lyapunov exponent is given by the formulae

$$\lambda = \tfrac{1}{2}\omega_1 \lim_{T\to\infty}\frac{1}{T}\int_0^T (\sin 2\Phi_t - 4\delta\sin^2\Phi_t + \sigma^2\cos^2\Phi_t\cos 2\Phi_t]dt, \tag{72}$$

$$\lambda = \tfrac{1}{2}\omega_1 \int_{-\pi/2}^{+\pi/2} (\sin 2\varphi - 4\delta \sin^2 \varphi + \sigma^2 \cos^2 \varphi \cos 2\varphi) p(\varphi) d\varphi. \qquad (73)$$

where (72) represents the time average of the multiplicative ergodic theorem. It can be calculated by the corresponding angle average (73) evaluable by the invariant density $p(\varphi)$ of the angle process. In the case $\sigma = 0$, the angle equation (70) possesses the two time-invariant positions

$$\varphi_0 = 0, \qquad \varphi_1 = -\arctan(2\delta) \qquad (74)$$

where the first position is asymptotically stable, and the second one is unstable. Hence for $\sigma = 0$, the time-invariant density of the trivial solution degenerates to a singular delta distribution concentrated at $\varphi_0 = 0$.

4.2 NONLINEAR FOURIER EXPANSIONS

For non-vanishing parametric excitations ($\sigma > 0$) the system angle perform a stationary clockwise rotation. Its time-invariant density is determined by the angle equation (70) and the associated Fokker-Planck equation

$$\tfrac{1}{2}\omega_1 \frac{\partial}{\partial \varphi}\{[1 - \cos 2\varphi + (2\delta + \tfrac{1}{2}\sigma^2) \sin 2\varphi + \tfrac{1}{4}\sigma^4 \sin 4\varphi] p(\varphi)\} +$$
$$+ \tfrac{1}{2}\sigma^2 \omega_1 \frac{\partial^2}{\partial \varphi^2}[\tfrac{1}{8}(3 + 4\cos 2\varphi + \cos 4\varphi) p(\varphi)] = 0, \quad |\varphi| \leq \pi/2. \qquad (75)$$

The solution of (75) is approximated by an exponential function combined with a Fourier expansion, as follows:

$$p(\varphi) = p_0 \exp(\alpha \cos 2\varphi)[1 + \sum_{n=1}^{\infty} (c_n \cos 2\varphi + s_n \sin 2\varphi)]. \qquad (76)$$

The insertion of this set-up into the Fokker-Planck equation (75) and the comparison of all harmonics lead to $2n$ linear equations for the Fourier coefficients c_n and s_n for all $n = 1, 2, \ldots$ and one nonlinear mean value equation for the unknown exponent α. The parameter p_0 represents the normalization constant to be calculated when α, c_n and s_n are known.

For first investigations, it is suitable to neglect higher order Fourier terms by $c_n = s_n = 0$ for $n = 1, 2, \ldots$ applying the zeroth term $c_0 = 1$, only. This closure condition leads to the mean value equation

$$\tfrac{1}{16}\sigma^2(5\alpha^2 + 4\alpha) - \delta\alpha = 0, \qquad \bar{\alpha} = \tfrac{16}{5}(\tfrac{\delta}{\sigma^2} - \tfrac{1}{4}). \qquad (77)$$

The nonlinear equation (77) possesses the two solutions $\alpha_0 = 0$ and $\alpha_1 = \bar{\alpha}$. The non-trivial exponent $\bar{\alpha}$ leads to the normalized angle density

$$p(\varphi) = p_0 \exp(\bar{\alpha} \cos 2\varphi), \qquad p_0^{-1} = \pi I_0(\bar{\alpha}) \qquad (78)$$

where $I_0(\bar{\alpha})$ denotes the modified Bessel function of zeroth order. In the limiting case $\sigma \to 0$ or $\bar{\alpha} \to \infty$, the density $p(\varphi)$, calculated by (78), tends to infinity for $\varphi = 0$ and to zero in $|\varphi| = \pi/2$. Hence, the density (78) degenerates to a singular delta distribution which is concentrated around $\varphi_0 = 0$ and normalized by p_0. Naturally, higher order Fourier terms in (76) are only needed for small but non-vanishing noise intensities $\sigma > 0$.

5. Conclusions

Necessary and sufficient conditions for the asymptotic stability of two- and three-dimensional dynamic systems are investigated by the top Lyapunov exponents and the densities of invariant measures. They are explicitly calculated by Fourier expansions of associated Fokker-Planck equations. These approaches lead to good approximations if the angle distributions are sufficiently smooth without too high gradients.

The Fourier analysis fail if the densities of the invariant measures degenerate to singular delta distributions. In this case, the Fourier expansion is extended by analytical shaping functions which allow to perform the limiting procedure to singular gradients of concentrated densities. The method is applied for overcritically damped oscillators with time-invariant angle concentrations shifted to the zero angle by the top Lyapunov exponent of the unperturbed oscillator.

References

1. Lin, Y.K. and Cai, G.Q. (1995) *Probabilistic Structural Dynamics, Advanced Theory and Applications*, McGraw-Hill Inc., New York.
2. Oseledec, V.I. (1968) A multiplicative ergodic theorem, Lyapunov characteristic numbers for dynamical systems, *Trans. Moscow Math. Soc.* **19**, 197-231.
3. Khasminskii, R.Z. (1974) Necessary and sufficient conditions for asymptotic stability of linear stochastic systems. *Theory Prob. Appl.* **12**, 144-147.
4. Pardoux, E. and Wihstutz, V. (1988) Lyapunov exponent and rotation number of two dimensional linear systems with small diffusion, *SIAM J. Appl. Math.* **48, No. 2**, 441-457.
5. Weidenhammer, F. (1969) Biegeschwingungen des Stabes unter axial pulsierender Zufallslast, *VDI-Berichte* **Nr. 135**, VDI-Verlag, Düsseldorf, p. 101-107.
6. Wedig, W. (1995) Pitchfork and Hopf bifurcations in stochastic systems - effective methods to calculate Lyapunov exponents, in P. Kree and W. Wedig (eds), *Probabilistic Methods in Applied Physics*, Lecture Notes in Physics **451**, Springer, Berlin, 120-148.
7. Arnold, L. (1974) *Stochastic Differential Equations*, Wiley, New York.
8. Griesbaum, R. and Wedig, W. (1999) Ljapunov-Exponent für zwei gekoppelte, lineare Systeme mit weißem Rauschen, GAMM 98 to appear in *ZAMM 99*.

RELIABILITY ANALYSIS OF STRUCTURES SUBJECTED TO DYNAMIC LOADINGS USING NONLINEAR SFEM

ACHINTYA HALDAR
Professor, Dept. Of Civil Engineering and Engineering Mechanics
University of Arizona, Tucson, AZ 85712, USA
JUNGWON HUH
Ph.D. Student, Dept. Of Civil Engineering and Engineering Mechanics
University of Arizona, Tucson, AZ 85712, USA

1. Abstract

An efficient and accurate algorithm is proposed to estimate reliability in the time domain for complicated structural systems subjected to seismic loading considering major sources of nonlinearity. The proposed algorithm intelligently integrates the concepts of the response surface method, the iterative perturbation sensitivity analysis technique, the finite element method, and the first-order reliability method. This leads to the stochastic finite element concept. It has the potential to estimate the risk associated with any linear or nonlinear structure that can be represented by a finite element algorithm subjected to seismic loading or any short duration dynamic loadings. To increase its efficiency, the assumed stress-based finite element algorithm is used. The response surface is developed using the saturated and the central composite design concepts. The algorithm is explained with the help of an example. It is verified using the Monte Carlo simulation technique.

2. Introduction

As the areas of risk and reliability mature, the accurate evaluation of the safety of structures considering their realistic behavior is becoming important to the profession. Since uncertainty can not be avoided in most engineering planning and design, design codes are being developed to reflect this philosophy. From the analysis and design point of view, realistic representation of a structure is essential in terms of structural parameters such as loading conditions, geometric arrangement, material properties, structural behavior (linear or nonlinear), boundary conditions, etc. The presence of uncertainties in the load and resistance-related parameters is well known in the profession. In codified approaches, load and resistance factors are used to address them. However, the mathematical model used to represent the structure may be too idealized to represent realistic structural behavior. For example, structural joints in steel frames are rarely fully restrained (FR); they are essentially partially restrained (PR) with different rigidities. They introduce a source of nonlinearity even when the load is very small. Structural supports are rarely perfectly fixed or pinned. The structural behavior could be quite different if the supports

S. Narayanan and R.N. Iyengar (eds.),
IUTAM Symposium on Nonlinearity and Stochastic Structural Dynamics, 95–106.

and connection conditions are modeled realistically. Most structures will develop nonlinear behavior just before failure. Thus, in the estimation of failure probability, the nonlinear behavior of structures can not be avoided. In spite of significant advances in analytical capabilities and design philosophies, the realistic safety evaluation of structures has yet to be addressed comprehensively.

To evaluate the safety of complicated structures in the presence of different sources of nonlinearity, a finite element method (FEM) -based formulation is desirable, since it is also the first step in a conventional deterministic analysis. In this way, different factors (support and connection conditions, geometric and material conditions, etc.) and the nonlinearity associated with them can be easily incorporated. The use of FEM in the context of uncertainty leads to the concept of stochastic finite element method (SFEM).

There has been a need for efficient and accurate reliability estimation method of nonlinear structures subjected to time-variant loadings, i.e., seismic and other short-duration loadings. Conceptually, the consideration of the uncertainties in dynamic loadings can be introduced in several ways. Two attractive choices are (i) the nonlinear random vibration techniques, and (ii) time history analysis. In the random vibration approach, the cumulant closure, perturbation, Fokker-Planck, stochastic averaging or stochastic equivalent linearization methods can be used; however, they may not be appropriate because most of these methods require too much computational effort to solve problems more complex than a single degree of freedom system. They emphasize the uncertainty in the loadings, but can not estimate reliability in the time domain. Only a few of them inherently deal with the uncertainty of system parameters, and most of them can not consider the uncertainty in the modeling of support and connection conditions. Therefore, the incorporation of uncertainties in the time history analysis is very desirable for the reliability analysis of nonlinear structures subjected to seismic loadings.

There are very few methods available to consider uncertainties in the seismic loading and the nonlinear behavior of structures, as will be elaborated later. The Monte Carlo simulation method (MCS), SFEM, and Response Surface method (RSM) can be used for this purpose. Although a class of Monte Carlo simulation methods may be very appropriate, considering their accuracy and computational efficiency, it may be too costly and cumbersome for the simulation of nonlinear dynamic systems. The SFEM has been successfully developed and applied to structures subjected mainly to static loadings. This method has been proved to be very elegant, economical, and accurate. Using this algorithm, the reliability can be estimated at every time increment of the acceleration of a seismic loading. However, it is very difficult and may be too time-consuming to apply this algorithm to dynamic reliability analysis in the time domain. The RSM has the potential to consider the appropriate mechanical behavior of systems and the uncertainty in the load and resistance-related parameters without compromising the efficiency and accuracy to a great extent. The subject of this study is the possibility of developing such an algorithm, in the context of the SFEM and the first-order reliability method (FORM).

3. Literature Review

As the areas of risk and reliability become an important part of structural analysis or/and design, accurate and efficient ways to estimate them have become a challenging task for the profession. Structural behavior is defined by a limit state function $g(\boldsymbol{X})$, where $\boldsymbol{X}$ is a set of basic random variables. The probability of failure corresponding to a limit state (strength or serviceability performance criterion) can be calculated by evaluating a multidimensional integral (Haldar and Mahadevan 1999), i.e.,

$$p_f = p[g(\mathbf{X}) \leq 0] = \int_D f_x(\mathbf{x})d\mathbf{x} \tag{1}$$

where $f_x(\mathbf{x})$ is the joint PDF for the vector X of basic variables and D is the integration domain of $\mathbf{x}$ where $g(\mathbf{x}) \leq 0$.

This approach is impractical in many cases since the joint probability density function of the basic random variables involved in the limit state may not be available. This led to the development of several reliability methods (Haldar and Mahadevan 1999) with different levels of complexity, including FORM and many different Monte Carlo simulation schemes. Since these areas are well developed, they do not need additional discussion here. However, the implementation potential and accuracy of most of these methods depend on the availability of the limit state functions in terms of the basic random variables involved in the problem. The explicit expressions of limit state functions, in general, are not available for realistic structures, particularly when nonlinear behavior needs to be incorporated in the formulation. Thus, it is not possible to directly calculate the derivatives to obtain the gradient vector, required in search of the design point on the limit state surface using FORM.

In most cases, the structural response considering different sources of nonlinearity has to be computed through a numerical procedure such as finite element analysis. In such cases, the limit state functions are implicit, and their derivatives with respect to the basic random variables are not readily available. Several computational approaches could be pursued for the reliability analysis of structures with implicit limit state functions. They can be broadly divided into three categories, based on their essential philosophy, as (i) Monte Carlo simulation, (ii) response surface approach, (iii) sensitivity-based analysis.

In Monte Carlo simulation, the deterministic analyses need to be performed repeatedly to extract the necessary information on the failure probability. If the deterministic analysis is time-consuming, as is expected for the nonlinear seismic analysis of structures using FEM, the simulation approach may be impractical.

The response surface approach is a set of statistical techniques designed to obtain the best value of the response considering the uncertainty in the input variables. In general, the response surface approach could be inaccurate and inefficient when applied to large complicated structural systems with many random variables, and when nonlinear behavior needs to be considered. It is an approximate technique, and its applicability needs to be determined on a case by case basis. However, its efficiency can be improved, as will be discussed later.

The sensitivity-based reliability approach is more elegant, and in general more efficient than the simulation or response surface approach. Three different types of sensitivity analysis are (i) the finite difference method, (ii) classical perturbation, and (iii) iterative perturbation. The combination of sensitivity analysis and the classical reliability methods (FORM) with implicit performance functions does not suffer the drawbacks of Monte Carlo simulation or the response surface approach. It uses the information about the actual value and the actual gradient of the limit state function at each iteration in the search for the design point, and uses an optimization scheme to converge to the minimum distance point from the origin in the reduced standard normal variable space.

Using iterative perturbation sensitivity analysis, FEM, and FORM incorporating information on distribution of random variables, an elegant approach was proposed by the authors and their associates. It is referred to as the Stochastic Finite Element Method (SFEM). It is only suitable for the static load case. The extension of this method to short-duration dynamic loadings is the subject of this paper.

The efficiency of the deterministic finite element method is important in the implementation of SFEM, since the proposed SFEM algorithm is based on tracking the uncertainty propagating through the steps of deterministic analysis. It has been reported in the literature that the assumed stress-based finite element method has many advantages over the displacement-based method, particularly for nonlinear analysis of large deformation problems. In the assumed stress-based finite element method, the tangent stiffness can be expressed in explicit form, the stresses of an element can be obtained directly, fewer elements are required to describe a large deformation configuration (sometimes as little as half the elements required by the other method), and integration is not required to obtain the tangent stiffness. Different sources of nonlinearity can be incorporated without losing the basic simplicity. It is very accurate and efficient in analyzing nonlinear responses. Since SFEM-based study is expected to be computationally more demanding than the conventional deterministic finite element method, the stress-based finite element method is used in the SFEM proposed here.

4. Proposed Methodology

The proposed algorithm intelligently integrates the concepts of RSM, the iterative perturbation sensitivity analysis technique, FEM, and FORM. To implement the approach, the matrix formulation of FORM is developed. The performance function is expressed in an n-dimensional uncorrelated standard normal space. An iteration algorithm is used to locate the design or checking point (the most probable failure point, MPFP) on the limit state function using the first order approximation. For each iteration, the performance function and its gradient vector are calculated using deterministic finite element models. The distance from the origin to this point in the standard normal space is the reliability index β. Thus, the corresponding probability of failure can be obtained. Some of the salient features of the proposed algorithm are discussed in the following sections.

4.1. RESPONSE SURFACE METHODOLOGY

RSM (Khuri and Cornell, 1996) is an important element of the proposed study. The primary purpose of applying RSM in reliability analysis is to approximate either the original complex and implicit limit state function (Wong, 1984; Rajashekhar and Ellingwood, 1993; Bucher and Bourgund, 1990; Faravelli, 1989; Kim and Na, 1997) or the structural response statistics, i.e. mean and coefficient of variation of response (Yao and Wen, 1996), using a simple and explicit polynomial. In the latter case (Yao and Wen, 1996), since RSM is used to approximate the structural response statistics (mean value and coefficient of variation of the response) rather than the limit state function, some additional approximations are necessary to formulate the limit state function using the response statistics. The implementation potential of these RSMs, in terms of accuracy and efficiency, depends on the form of the polynomial function to be used, the center point, and the sampling points around the center point. They are elaborated further below.

4.1.1. *Polynomials and Experimental Design*

Theoretically, any degree of polynomial with a Taylor series expansion can be used to represent the response function of interest. However, for nonlinear problems under seismic loading, at least a second order polynomial is necessary. This is expected to be more computationally challenging. The selection of the center point and the sampling points or experimental points in the context of the second order polynomials are emphasized in the following discussion.

Second Order Polynomials. In this paper, two types of second order polynomials, i.e., without or with cross terms are used to fit the actual response and are expressed as:

$$\hat{g}(\boldsymbol{X}) = b_0 + \sum_{i=1}^{k} b_i \boldsymbol{x}_i + \sum_{i=1}^{k} b_{ii} \boldsymbol{x}_i^2 \tag{2}$$

$$\hat{g}(\boldsymbol{X}) = b_0 + \sum_{i=1}^{k} b_i \boldsymbol{x}_i + \sum_{i=1}^{k} b_{ii} \boldsymbol{x}_i^2 + \sum_{i=1}^{k} \sum_{j=1}^{i-1} b_{ij} \boldsymbol{x}_i \boldsymbol{x}_j \tag{3}$$

where $\boldsymbol{x}_i$ (i = 1, ····, k) is the ith random variable, and b_0, b_i, b_{ii}, and b_{ij} are unknown coefficients to be determined. The number of coefficients for each polynomial in Eq.(2) and (3), are $p_1 = 2k+1$ and $p_2 = (k+1)(k+2)/2$, respectively.

4.1.2. *Determination of the Center Point*

Conceptually, the center point should be close to the most probable failure point (MPFP) so that the response surface obtained includes most of the failure region with sufficient accuracy. The center point may be selected using a trial and error scheme or an iterative linear interpolation scheme. The success of the trial and error scheme (Yao and Wen, 1996) depends on the experience of the simulator, and may not be simple for every day use.

In the iterative scheme, the center point is initially selected to be the mean value $\boldsymbol{\mu}_i$ of the random variable $\boldsymbol{X}_i$. Then, using values of g($\boldsymbol{X}$) at the sampling points around it, the response surface $\hat{g}(\boldsymbol{X})$ can be generated. Using the information on means, standard deviations, the distributions of $\boldsymbol{X}_i$, $\hat{g}(\boldsymbol{X})$, and FORM, the coordinates of the design point $\boldsymbol{x}_D$ can be estimated. The response can again be evaluated at the design point $\boldsymbol{x}_D$, i.e., $g(\boldsymbol{x}_D)$ and a new center point $\boldsymbol{x}_M$ can be selected using linear interpolation from the mean vector $\boldsymbol{\mu}_x$ to $\boldsymbol{x}_D$ such that g($\boldsymbol{X}$) = 0, i.e., for each random variable:

$$\begin{aligned} &\text{IF } g(\boldsymbol{x}_D) \geq g(\boldsymbol{\mu}_x), \quad x_M = \boldsymbol{\mu}_x + (x_D - \boldsymbol{\mu}_x)\frac{g(\boldsymbol{\mu}_x)}{g(\boldsymbol{\mu}_x) - g(\boldsymbol{x}_D)} \\ &\text{IF } g(\boldsymbol{x}_D) < g(\boldsymbol{\mu}_x), \quad x_M = x_D + (\boldsymbol{\mu}_x - x_D)\frac{g(\boldsymbol{x}_D)}{g(\boldsymbol{x}_D) - g(\boldsymbol{\mu}_x)} \end{aligned} \tag{4}$$

The point $\boldsymbol{x}_M$ is now used as a new center point for the next iteration. This iteration scheme can be repeated until a preselected convergence criterion is satisfied. The center point selection scheme may take only 2 to 3 iterations.

4.1.3. *Experimental Design*

The selection of sampling points around the center point where responses need to be calculated is the next challenge. The available techniques to generate sampling points can be divided into two categories: classical design and saturated design.

Classical Design. In the classical design approach, responses are first calculated at several sampling points and then a regression analysis is carried out to formulate the response surface. In this approach, one of the conceptually simpler designs is a factorial design, in which the response values are estimated for each variable sampled at equal intervals. In order to fit a second order surface for k input variables, the sampling points must have at least three levels for each variable, leading to 3^k factorial design. Box and Wilson (1951) introduced a more efficient approach, known as the central composite design, for fitting second order surfaces. It has been applied to reliability analysis using RSM by several researchers (Faravelli, 1989; Yao and Wen, 1996). On the other hand, saturated design (Bucher et al, 1989; Bucher and Bourgund, 1990; Rajashekhar and Ellingwood, 1993) consists of only as many sample points as the total number of coefficients necessary to define a polynomial. The unknown coefficients are obtained by solving a set of linear equations, without using any regression analysis.

Saturated Design. A design and its corresponding model are called saturated when the number of design points is exactly equal to the number of terms in the model to be fitted. The design, as shown in Figure 1(a), consists of one center point and $2k$ star points (one at +1 and one at -1 for each variable), i.e., the total number of experimental points are $2k+1$, where k is the number of random variables. Therefore, the surface fits exactly at the sampling points and the method needs much fewer sampling points than the central composite design, i.e., this design is the most efficient. However, it may lack the statistical properties (rotatability, orthogonality, etc.) and the sample space between the axes of random variables may not be covered sufficiently, causing inaccuracy in the response surface.

Central Composite Design. The concept behind the cental composite design is shown in Figure 1(b). It consists of a complete 2^k factorial design, n_0 center points ($n_0 \geq 1$), and two axial points on the axis of each design variable at a distance of α from the design center point where $\alpha = \sqrt[4]{2^k}$ in order to make the design rotatable.

The total number of design points is thus $N=2^k+2k+n_0$, which is much more than the number of coefficients to be estimated, $p=(k+1)(k+2)/2$. This design is more accurate than the saturated design and contains several statistical properties, such as the ability to estimate the experimental error and to test the adequacy of the model, orthogonality, and rotatability. Despite all these merits, it requires an unacceptably high level of computational effort with the increase in number of variables for complex systems, and may become more time consuming than simulation. For example, for $k=15$, the experimental points required for the central composite design are 32799 ($N=2^{15}+2\times15+1$) while the saturated design needs only 31 ($N=2\times15+1$). In the proposed algorithm, the saturated design is used for intermediate iterations, and the central composite design is used only at the final iteration.

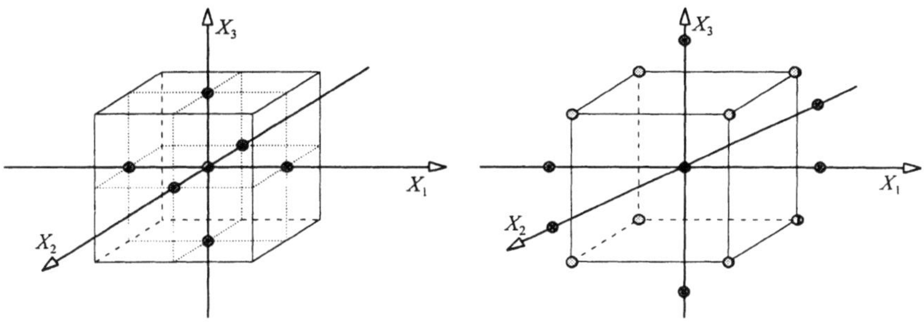

(a) Saturated Design (b) Central Composite Design

Figure 1. Saturated and Central Composite designs for the second order polynomial of k = 3 variables

4.1.4. *Least Square Estimation of RSM*

Using Eq. (3) for a response surface for *k* random variables with cross terms, an actual response g(X) can be expressed by adding an error term, ϵ , as

$$g(X) = b_0 + \sum_{i=1}^{k} b_i x_i + \sum_{i=1}^{k} b_{ii} x_i^2 + \sum_{i=1}^{k} \sum_{j=1}^{i-1} b_{ij} x_i x_j + \varepsilon \tag{5}$$

Equation (5) can be represented in matrix form as

$$\mathbf{Y} = \mathbf{Xb} + \varepsilon \tag{6}$$

where **Y** is the vector of the actual responses, and **X** is a matrix shown in Table 1.

The least squares estimates, $\hat{\mathbf{b}}$, of the elements of **b** in Eq. (6) are

$$\hat{\mathbf{b}} = \left(\mathbf{X}^T\mathbf{X}\right)^{-1}\mathbf{X}^T\mathbf{Y} \tag{7}$$

Analysis of Variance (ANOVA) is used to estimate these coefficients.

Table 1. General representation of the **X** matrix for CCD

x_0	x_1	x_2	....	x_k	x_{11}	....	x_{kk}	x_{12}	x_{13}	....	$x_{k-1,k}$
1	-1	-1	....	-1	1	....	1	1	1	....	1
1	1	-1	....	-1	1	....	1	-1	-1	....	1
:	:	:		:	:		:	:	:		:
1	1	1	....	1	1	....	1	1	1	....	1
1	$-\alpha$	0	....	0	α^2	....	0	0	0	....	0
1	α	0	....	0	α^2	....	0	0	0	....	0
1	:	:		:	:		:	:	:		:
1	0	0	....	$-\alpha$	0	....	α^2	0	0	....	0
1	0	0	....	α	0	....	α^2	0	0	....	0
1	0	0	....	0	0	....	0	0	0	....	0

5. Uncertainties in Time-Variant Seismic and Dynamic Loadings

For discussion clarity, it is important to differentiate between short and long duration loadings. For long duration loadings, the works of Bucher, Chen, and Schüller (1989) and Yao and Wen (1996) are noteworthy. Bucher, Chen, and Schüller (1989) converted the time-variant problem to a time-invariant problem by applying the lifetime maximum effect of combined load processes after evaluating the limit state function. Bucher and Bourgund (1990) and Rajashekhar and Ellingwood (1993) considered short duration loadings. They examined the same dynamic problem, which is a nonlinear SDOF oscillator with random system parameter subjected to a rectangular load pulse with random duration and amplitude. Their procedures can not be used for the reliability analysis of nonlinear structures subjected to seismic loading.

There is no guideline for considering uncertainties in both the amplitude and frequency in the earthquake loading. The uncertainty in the amplitude in the earthquake is successfully considered in the context of RSM in this study. The uncertainty in the frequency content of an earthquake is considered indirectly. Available time histories for the actual earthquake are first normalized with respect to the peak acceleration. Then the reliability of structures is estimated using the algorithm developed in this study. Since all the time histories have a different frequency content, the estimated reliability will indicate the effect of uncertainties in the frequency content.

6. Sensitivity Analysis

The efficiency of the algorithm can be improved significantly by considering some of the random variables to be deterministic, since their influence on the overall reliability analysis could be very marginal. In the context of FORM, the reliability index corresponding limit state, the checking point, and the corresponding direction cosines of each of the random variable are known at this stage. Using this information, the sensitivity index for each of the random variables is evaluated (Haldar and Mahadevan, 2000a and 2000b).

7. Strategy of the Proposed Algorithm and Program RSDRA

The proposed method is an iterative RSM to approximate the limit state function in the failure region. Real earthquake time histories are used for the response calculations. The main steps of the proposed algorithm are as follows:

1. Set the initial center point to start the iteration at the mean values of the random variables.
2. Formulate a response surface model using the second order polynomial without the cross terms, Eq. (2), and fully saturated design in order to keep the algorithm efficient.
3. Find the coefficients b_i of the response surface and develop a closed form expression for the performance function.
4. Determine the design point, $\boldsymbol{x}_D$, by conducting FORM using the performance function obtained in step 3.
5. Find a new center point, $\boldsymbol{x}_M$, for the next iteration using the linear interpolation

scheme, Eq. (4).

6. Repeat this iteration scheme (step 2 ~ step 5) until a preselected convergence criterion is satisfied.
7. Formulate the final response surface model using the full second order polynomial, Eq. (3), and the central composite design.
8. Find the coefficients b_i of the final response surface.
9. Finally, calculate the reliability index and the sensitivity indices using the FORM.

Based on the proposed algorithm, a new structural dynamic reliability analysis program RSDRA is developed by combining the already developed deterministic dynamic analysis program (DYNPR) (Haldar and Gao, 1997) with a new reliability program (RFM).

8. Numerical Example

To elaborate the algorithm further, a single story frame shown in Figure 2(a) is considered and is excited by an actual acceleration(cm/sec^2) time history recorded during the El Centro earthquake in May of 1940 (N-S component), as shown in Figure 2(b).

All the beam and columns of the frame are made of W16×100, and A36 steel is used for this illustrative example. A statistical description of the uncertainties associated with all the random variables is given in Table 2. For the serviceability limit state, the permissible displacement at the top of the frame is assumed to not exceed ***h***/400, where ***h*** is the height of the frame. Thus, for this example, δ_{limit}^{drift} is equal to 1.143 cm, and the limit state is:

$$g(x) = \delta_{limit}^{drift} - y_{max}(x) = 1.143 - y_{max} \tag{8}$$

in which y_{max} is the maximum displacement response of the system.

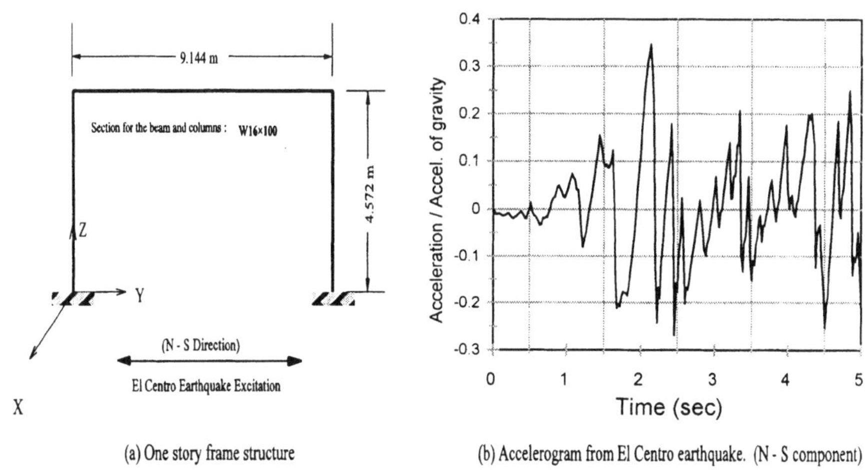

(a) One story frame structure

(b) Accelerogram from El Centro earthquake. (N - S component)

Figure 2. One story frame and earthquake excitation time history

Table 2. Statistical description of the random variables for Case (1) and Case (2)

Random Variable	Mean value	Case (1)		Case (2)	
		C.O.V	Distribution	C.O.V	Distribution
E (kN/m^2)	1.9996×10^8	0.06	Lognormal	0.06	Lognormal
A (m^2)	1.8968×10^{-2}	0.05	Lognormal	-	Constant
I_x (m^4)	6.2018×10^{-4}	0.05	Lognormal	0.05	Lognormal
Z_x (m^3)	3.2446×10^{-3}	0.05	Lognormal	-	Constant
F_y (kN/m^2)	2.4822×10^5	0.10	Lognormal	-	Constant
ξ	0.02	0.15	Lognormal	-	Constant
g_e	1.0	0.20	Type I (EVD)	0.20	Type I (EVD)

g_e is a magnification factor for the amplitude of actual seismic acceleration

Using the proposed nonlinear SFEM, the failure probability, reliability index, and sensitivity indices of the random variables are estimated for the serviceability limit state represented by Eq. (8) for two cases. In Case (1), all seven variables are considered to be

Table 3. Results of the example

Case			Case (1) : 7 random variables			Case (2) : 3 random variables		
MCS*		P_f	0.03222 ($\beta \approx 1.8491$)			0.03233 ($\beta \approx 1.8476$)		
		CPU Time▲	42373 sec			42481 sec		
Type			Type ①	Type ②	Type ③	Type ①	Type ②	Type ③
RSM		β	1.9433	1.8568	1.8539	1.9489	1.8763	1.8745
		P_f	0.02599	0.03167	0.03188	0.02565	0.03031	0.03043
		CPU Time▲	23 sec	78 sec	186 sec	11 sec	15 sec	22 sec
		TNSP•	45	173	429	7	29	45
error			19.34 %	1.71 %	1.07 %	20.65 %	6.26 %	5.87 %
Sensitivity Index	γ	E	-0.4526	-0.4510	-0.4634	-0.4459	-0.4266	-0.4602
		A	0.0403	0.0359	0.0395	-	-	-
		I_x	-0.3977	-0.3722	-0.3934	-0.3913	-0.3582	-0.3809
		Z_x	0.0000	-0.0003	-0.00003	-	-	-
		F_y	0.0000	-0.0006	-0.00005	-	-	-
		ξ	-0.0928	-0.0758	-0.0638	-	-	-
		g_e	0.7917	0.8069	0.7905	0.8050	0.8305	0.8020

*100,000 cycles of simulation

•Total number of sampling points, i.e., deterministic dynamic execution

▲ SGI Origin 2000

random. In Case (2), using the sensitivity analysis in Case (1), only three of them are considered to be random. For each case, 3 types of analyses are conducted; in Type ① only the saturated design is used for all iterations, in Type ② the saturated design is used for the intermediate iterations and the central composite design is used for the final iteration, and in Type ③ the central composite design is used for all iterations. The results are summarized in Table 3. The results are compared with Monte Carlo Simulation results using 100,000 simulations. These results are also shown in Table 3.

Several interesting observations can be made from Table 3. The probability of failure for both cases is very similar to the MCS results, indicating that the proposed algorithm is viable. As expected, Case (2) with fewer random variables is more efficient than Case (1). The use of the central composite design is always desirable. However, the use of Type ②, as proposed in this study, is expected to be very efficient, particularly when a large number of random variables need to be considered, without sacrificing accuracy. For large practical problems, the used of sensitivity analysis with Type ② experimental designs are expected to be more desirable than any other alternative.

9. Conclusions

An efficient and accurate method for nonlinear SFEM is proposed to evaluate reliability in the time domain for frames subjected to short duration dynamic loading. Uncertainties in the seismic excitation and resistance-related parameters are incorporated. An iterative RSM is proposed using two second order polynomials (with and without the cross terms) and a mixture of saturated and central composite designs.

With the help of a simple example, it is shown that the proposed algorithm can be used to estimate the risk of nonlinear structures subjected to short duration dynamic loadings, including seismic loading. The scheme suggested to improve the efficiency of the RSM appears to be effective. Sensitivity analysis can be used to improve the efficiency further. The proposed algorithm is specifically suitable for complicated structural systems with a large number of random variables.

10. Acknowledgments

This paper is based on work partly supported by the National Science Foundation under Grant CMS-9526809. Any opinions, findings, conclusions, or recommendations expressed in this publication are those of the authors and do not necessarily reflect the views of the sponsor.

11. References

Bucher, C.G. and Bourgund, U. (1990) A fast and efficient response surface approach for structural reliability problems, *Structural Safety* **7**, 57-66.

Bucher, C. G., Chen, Y. M., and Schüller, G.I. (1989) Time variant reliability analysis utilizing response surface approach, *Reliability and Optimization of Structural Systems '88: Proc., 2nd IFIP WG7.5 Conf.*, Springer-Verlag, Berlin, Germany, 1-14.

Box, G. E. P. and Wilson, K. B. (1951) On the Experimental Attainment of Optimum Conditions (with discussion), *J. Roy. Statist. Soc.* **B13**, 1-45.

Faravelli, L. (1989) Response-Surface Approach for Reliability Analysis, *Journal of Engineering Mechanics* **115(12)**, *ASCE*, 2763-2781.

Haldar, A. and Gao, L. (1997) Reliability evaluation of structures using nonlinear SFEM, *Uncertainty Modeling in Finite Element, Fatigue, and Stability of Systems,* edited by A. Haldar, A. Guran, and B.M. Ayyub, World Scientific Publishing Co., 23-50.

Haldar, A. and Mahadevan, S. (2000a) *Probability, Reliability and Statistical Methods in Engineering Design,* John Wiley & Sons, New York, N.Y.

Haldar, A. and Mahadevan, S. (2000b) *Reliability Assessment using Stochastic Finite Element Analysis.* John Wiley & Sons, New York, N.Y.

Kim, S. H. and Na, S. W. (1997) Response surface method using vector projected sampling points, *Structural Safety* **19(1)**, 3-19.

Khuri, A. I. and Cornell, J. A. (1996) *Response Surfaces Designs and Analyses*, Marcel Dekker, Inc, New York, N.Y.

Liu, Y. W. and Moses, F. (1994) A sequential response surface method and its application in the reliability analysis of aircraft structural system, *Structural Safety* **16**, 39-46.

Rajashekhar, M.R. and Ellingwood, B.R. (1993) A new look at the response surface approach for reliability analysis, *Structural Safety* **12**, 205-220.

Wong, F. S. (1984) Uncertainties in dynamic soil-structure interaction, *Journal of Engineering Mechanics* **110**, *ASCE*, 308-324.

Yao, T. H.-J. and Wen, Y. K. (1996) Response surface method for time-variant reliability analysis, *J. of Structural Engineering* **122(2)**, *ASCE*, 193-201.

APPLICATION OF CONDITIONAL LINEARIZATION IN THE STUDY OF NONLINEAR SYSTEMS

R.N. Iyengar
Director, Central Building Research Institute
Roorkee 247667, INDIA

D. Roy
Asstt. Professor, Department of Civil Engineering,
Indian Institute of Technology, Kharagpur 721302, INDIA

1. Introduction

Exact solutions in nonlinear random vibrations are hard to come by and are generally limited to single degree of freedom (SDOF) systems driven by white noise in the stationary regime. The most popular and perhaps the simplest available approach is the equivalent linearisation (Lin 1967) technique (ELT). Other approaches include perturbation (Crandall 1973), Fokker-Planck equation (Spanos 1982, Lin and Cai 1995), moment closure (Ibrahim 1978, Iyengar and Dash 1978), and higher order linearisation (Iyengar 1988a) methods. For higher dimensional systems and for systems with non-Gaussian, non-white excitations, however, it is not an easy task to arrive at acceptable, approximate solutions. One such typical higher dimensional problem, which has extensive practical implications also, arises in the study of nonlinear SDOF oscillators under a narrow-banded random excitation, often modeled as a filtered white noise process.

Approximate solutions of a nonlinear oscillator under narrow-banded random inputs have been considered by several researchers. It was, for example, shown by Lyon et al. (1961) through experiments and Iyengar (1988a) through numerical simulations that the response process for the hardening Duffing's oscillator driven by a narrow banded input may, under some conditions, show transitions between two response states. In other words, the associated stationary probability density function (pdf) may be bimodal. Efforts to predict such a response using the ELT has not been successful (Iyengar 1988b). However it has recently been shown (Iyengar and Roy 1996) that using a conditional linearisation procedure, it is possible to derive an integral equation for the pdf of the response amplitude and its solution indeed can show the bimodality in pdf, under an appropriate choice of parameters.

S. Narayanan and R.N. Iyengar (eds.),
IUTAM Symposium on Nonlinearity and Stochastic Structural Dynamics, 107–118.

It thus appears that linearity may be considered as a building block, through which solutions may be attempted even if they were to be approximate. In a related developments, Iyengar and Roy (1998a, 1988b) have recently proposed a method called phase space linearisation (PSL) for investigation of nonlinear oscillators under deterministic inputs. The essence of the PSL technique is to replace a nonlinear dynamical system of equations by a series of linear equations, each valid over a sufficiently short segment of the evolving trajectories in the phase space. The values of the response variables at the beginning of each of these segments serve as initial conditions to these set of linear equations. Thus the reduced set of linear equations, derived from the original system, are conditionally linear given the precise knowledge of these initial values. While in the deterministic case, where one needs to deal only with a single realization of the evolving flow, these initial values are known deterministically, a random nonlinear dynamical system presents a different picture altogether. Since the trajectories, in this case, evolve within an ensemble, the initial values sampled at the beginning of each one of the segments are themselves random and therefore demands a suitable characterization.

With this in view, an attempt is made in the present paper to extend the concepts of PSL to the realm of nonlinear random vibration within a framework of conditional distributions. In the particular case of the response pdf being stationary, an integral equation for the unknown pdf is obtained and is solved using an iterative numerical scheme till convergence. Limited results are presented to verify that the procedure is versatile enough to predict multimodality in the pdf. A few comparisons with the results of Monte Carlo tests are also presented.

2. The Methodology

Let the time axis be ordered such that $t_0 < t_1 < \ldots < t_n < t_{n+1} < \ldots$ and the state variables $X \in R^N$ sampled at each one of these time instants be given by the countably infinite Borel set $S = \{X_0, X_1, \ldots, X_n, X_{n+1}, \ldots\}$. Let the *j*th realization of *S* be denoted as $\overline{S}^{(j)} = \{X_0^{(j)}, X_1^{(j)}, \ldots, X_n^{(j)}, X_{n+1}^{(j)}, \ldots\}$ and all the finite subsets of *S* be denoted as $S_n = \{X_0, X_1, \ldots, X_n\}$. Now, to apply the concept of PSL to this case, one needs to have a linearised representation of the nonlinear equation such that the evolving trajectories are more or less adequately predicted within the Borel set $J_n^{(i)} = \{X(t) \mid X_{n-1}^{(i)}, t \in [t_{n-1}, t_n)\}$ for all i within the ensemble and for $(t_n - t_{n-1})$ sufficiently small. If such a linearisation is possible, then all the elements in the conditional set $J_n^{(i)}$ are jointly Guassian provided that the input forcing function is Gaussian too. To illustrate the procedure further, it would be expedient to consider the example of a hardening Duffing's equation driven by a narrow-banded Gaussian input. The equations are

$$\ddot{x} + 2\eta\omega\dot{x} + \omega^2(1+\alpha x^2)x = f(t)$$
$$\ddot{f} + 2\xi\lambda\dot{f} + \lambda^2 f = w(t) \tag{1}$$
$$< w(t_1)w(t_2) >= K\delta(t_2 - t_1)$$

Noting that the only nonlinear term in the above equation is the cubic term in x, it would be justified to search for an 'equivalent linear' equation of the following form restricted to $J_n^{(i)}$.

$$\ddot{x} + 2\eta\omega\dot{x} + \omega^2(1+\beta_i)x = f(t) \tag{2}$$

Noting that at $t = t_i, X = (x_i, \dot{x}_i)$, the linearised coefficient β_i uniformly valid for the time interval $\tau_i = t_{i+1} - t_i > 0$ sufficiently small, is taken to be of the form

$$\beta_i = \frac{\alpha}{\tau_i}\int_o^{\tau_i} T(x^2, x_i, \dot{x}_i, (t-t_i))dt \tag{3}$$

In the above equation, the integrand T is a stochastic Taylor's expansion (Kloeden and Platen 1992) of $x^2(t)$ centered at $(x_i, \dot{x}_i)$ as a function of the translated time $\tau = t - t_i$. Here it may be mentioned that the stochastic Taylor's expansion differs from the deterministic Taylor's expansion in that use of Ito's formula (Kloeden *et al* 1994) for the associated functional (in this case, $x^2(t)$) leads to certain corrections over the deterministic case. Generally these corrections are of the fractional orders, $(\Delta t)^{n+0.5}$, where n=0,1,2,3..., of the time step Δt. For the present problem, however, one may observe from equation (1) that the white noise comes as an external excitation only in the filter for the narrow banded excitation, *f(t)*. Thus the first stochastic correction to the deterministic Taylor's expansion for $x^2(t)$ is of the order $(\Delta t)^{2.5}$. In other words, if one adopts a three term approximation for the Taylor's series, *T*, then the stochastic expansion is precisely the same as the deterministic one. This leads to the following expression for β_i

$$\beta_i \approx x_i^{\ 2} + x_i\dot{x}_i\tau_i + \mu_i\tau_i^2 \tag{4}$$

where μ_i, the coefficient of τ_i^2, is given by

$$\mu_i = \frac{1}{3}(\dot{x}_i^2 - 2\eta\omega x_i\dot{x}_i \ - \omega^2(x_i^2 + \alpha x_i^4) \tag{5}$$

In expressions (2) through (5), the variable x is to restricted to the set $J_n^{(i)}$, i.e, $x \equiv x \mid x_i$. Since equation (2) is conditionally linear and *f(t)* is Gaussian, one can infer that $x \mid x_i$ is Gaussian for $\mid x - x_i \mid$ sufficiently small. Now let the mean vector of the conditional vector process $(x, \dot{x} \mid x_i, \dot{x}_i)$ be denoted as $(m_i, \dot{m}_i)$. Thus the pdf of this conditional process is completely specified if the second moments can be properly estimated. Towards this, the following decompositions are used.

$$
\begin{aligned}
x \mid x_i &= m_i + z_i \\
\dot{x} \mid \dot{x}_i &= \dot{m}_i + \dot{z}_i
\end{aligned}
\tag{6}
$$

Obviously, $(z_i, \dot{z}_i)$ is a vector Gaussian process with zero mean. Use of equation (6) in the conditionally linear equation (2) leads to the following equations for the mean and variation processes.

$$\ddot{m}_i + 2\eta\omega\dot{m}_i + \gamma_i m_i = 0 \tag{7}$$

$$\ddot{z}_i + 2\eta\omega\dot{z}_i + \gamma_i z_i = f(t) \tag{8}$$

In the above equations, $\gamma_i = \omega^2(1+\beta_i)$. Equation (7), describing the conditional mean vector, needs to be solved using the initial conditions $m_i(t_i) = x_i$, $\dot{m}_i(t_i) = \dot{x}_i$. Noting that the conditionally constant stiffness parameter, γ_i, is always positive and greater than or equal to ω^2, it is clear that (0,0) always remains a stable attractor for this equation. In other words, for nonzero initial conditions, there is a drift of the conditional mean vector towards the attracting fixed point. The solution for the conditional mean vector may be written down in closed form as

$$
\begin{aligned}
m_i(t) &= \exp(-\eta\omega(t-t_i))(C_{1i}\cos\kappa(t-t_i) + C_{2i}\sin\kappa(t-t_i)) \\
\dot{m}_i(t) &= -\eta\omega m_i(t) + \kappa\exp(-\eta\omega(t-t_i))(C_{2i}\cos\kappa(t-t_i) - C_{1i}\sin\kappa(t-t_i)) \\
\kappa &= \omega\sqrt{1+\beta_i-\eta^2}
\end{aligned}
\tag{9}
$$

The arbitrary constants of integration, C_{1i}, C_{2i} needs to be found from the initial conditions $(x_i, \dot{x}_i)$. Now, considering the differential equation (8) in z, one observes that the vector process $(z, \dot{z}, f, \dot{f})$ is jointly Markovian and thus it is possible to obtain all the second moment equations using Ito calculus. In particular one has the following second order moment quantities to solve for

$$s_1 =< z^2 >, s_2 =< z\dot{z} >, s_3 =< zf >,$$
$$s_4 =< z\dot{f} >, s_5 =< \dot{z}^2 >, s_6 =< \dot{z}f >, s_7 =< \dot{z}\dot{f} >, \qquad (10)$$
$$s_8 =< f^2 >, s_9 =< f\dot{f} >, s_{10} =< \dot{f}^2 >$$

$$\begin{aligned}
\dot{s}_1 &= s_2 \\
\dot{s}_2 &= s_5 - 2\eta\omega s_2 - \gamma_i s_1 + s_3 \\
\dot{s}_3 &= s_6 + s_4 \\
\dot{s}_4 &= s_7 - 2\xi\lambda s_4 - \lambda^2 s_3 \\
\dot{s}_5 &= -4\eta\omega s_5 - 2\gamma_i s_2 + 2s_6 \\
\dot{s}_6 &= s_7 - 2\eta\omega s_6 - \gamma_i s_3 + s_8 \\
\dot{s}_7 &= -2\eta\omega s_7 - \gamma_i s_4 + s_9 - 2\xi\lambda s_7 - \lambda^2 s_6 \\
\dot{s}_8 &= 2s_9 \\
\dot{s}_9 &= s_{10} - 2\xi\lambda s_9 - \lambda^2 s_8 \\
\dot{s}_{10} &= -4\xi\lambda s_{10} - 2\lambda^2 s_9 + K
\end{aligned}$$

(11)

The set of ten first order ODEs for determining these quantities, as obtained using Ito's formula for stochastic functionals are as given below.

The last three of equations (11), representing the evolutions of second moment functions of the narrow banded excitation, *f(t)*, are uncoupled from the first seven and hence the last three equations may be solved separately under conditions of stationarity. This leads to

$$s_8 = K / 4\xi\lambda^3, s_9 = 0, \ s_{10} = \lambda^2 s_8 \qquad (12)$$

The first seven equations, however, need not have a stationarity in general. These quantities may even become locally unstable. However, since these equations are all linear, short time evolutions of the associated trajectories may be determined in closed form by determining the conditional eigenvalues of the associated coefficient matrix. One may also solve these equations via some numerical integration scheme over a sufficiently small time interval. In both the cases, the initial conditions for all the conditional moment functions are to be taken as zero.

For the present problem, it may readily be observed the zero vector is a stable fixed point for the vector $S_7 = \{s_1 \quad s_2 \cdots s_7\}^T$ for all possible choices of $(x_i \quad \dot{x}_i)$ and the only inhomogeneous term is s_8, which is the variance of the external force and is a constant. Thus it may be conjectured that the vector S_7 achieves a stationary value of the particular integral quickly enough to be predicted accurately by the conditionally linear equations (11). In such a case, the left hand time derivatives of equations (11) disappears thereby leading to a set of seven algebraic equations for the conditional second moments Thus one can readily show that

$$
\begin{aligned}
s_1 &= \frac{K(1+2C_1\eta\omega)}{8(C+2\eta\omega+C_1\gamma_i)\eta\omega\xi\lambda^3} \\
s_5 &= \lambda^2 s_1 \\
C &= \frac{\lambda^2-\gamma_i}{2(\eta\omega+\xi\lambda)} \\
C_1 &= -(C/\lambda^2)+2\xi/\lambda
\end{aligned}
\tag{13}
$$

$$
q(x,\dot{x}\mid x_i,\dot{x}_i) = \frac{1}{2\pi\sqrt{s_1 s_5 - s_2^2}} \exp\frac{-s_5(x-m_i)^2+2s_2(x-m_i)(\dot{x}-\dot{m}_i)-s_1(\dot{x}-\dot{m}_i)}{2(s_1 s_5 - s_2^2)}
\tag{14}
$$

It may however be stressed here that such a conjecture fails to hold true in general, since the conditional eigenvalues depending on $(x_i, \dot{x}_i)$ may be such that the solution may become locally unstable. In such a case, the locally defined complementary part of the solution plays an important role and has to be obtained (numerically or analytically) over a short time step. In any case, knowledge of these conditional first and second moments enables one to write down the conditionally Gaussian density function as

Given the knowledge of the above transitional density over a short time step, one can write down the integral map for the non-stationary pdf of the oscillator as

$$
p_{i+1}(x_{i+1},\dot{x}_{i+1}) = \int_{-\infty}^{\infty}\int_{-\infty}^{\infty} q(x_{i+1},\dot{x}_{i+1}\mid x_i,\dot{x}_i)p_i(x_i,\dot{x}_i)dx_i d\dot{x}_i
\tag{15}
$$

Now, supposing that a stationary solution exists, one readily sees that in such a case the sequence of functions $\{p_1, p_2, \ldots, p_i, p_{i+1}, \ldots\}$ has a limit, given by p as $i \to \infty$. Thus equation (17) reduces to the following integral equation in p

$$
p(x,\dot{x}) = \int_{-\infty}^{\infty} q(x,\dot{x}\mid\psi,\dot{\psi})p(\psi,\dot{\psi})d\psi d\dot{\psi}
\tag{16}
$$

It would be interesting to note here that the above integral equation, involving a stationary pdf as the unknown, is valid for $(x, \dot{x})$ arbitrarily close to $(\psi, \dot{\psi})$. Moreover since there is no time marching involved in finding the stationary solution, choosing $(x, \dot{x})$ sufficiently close to $(\psi, \dot{\psi})$ establishes the sufficiency of choosing only a three term Taylor's expansion in deriving the conditionally constant coefficient β_i.

3. Solution of the Integral Equation

The present investigation is concerned only with finding a stationary solution if it exists. One can attempt to solve equation (16) in three different ways. The first is via a straightforward iterative approach, till some criteria for convergence are fulfilled. Here one needs an initial guess for the function *p*. In the second method, equation (16) may be considered as Fredholm's integral equation in the associated Banach space with the Gaussian function q as the kernel. Here one can attempt to express p as the sum of orthogonal Hermite polynomials in q with unknown coefficients, thereby reducing the integral equation to a set of linear algebraic equations in terms of these unknown coefficients. The third and possibly the most versatile approach is via a Boundary Element Algorithm, with boundary conditions $p(\pm\infty, \pm\infty) = 0$.

In what follows, an iterative approach is used in the present study to approximately solve the associated integral equation.

4. Numerical Results

In order to solve the integral equation (16) using an iterative approach, one needs to employ an initial guess for the stationary distribution. Here the initial guess for the pdf is taken to be the Gaussian distribution for the associated linear problem. This is given by

$$p^{(0)}(x, \dot{x}) = \frac{1}{2\pi\sqrt{s_{1,l}s_{5,l}}} \exp(-(x^2/2s_{1,l}) - (\dot{x}^2/2s_{5,l})) \tag{17}$$

where

$$\begin{aligned} s_{1,l} &= s_1(\gamma_i = \gamma_{i,l}) \\ s_{5,l} &= s_5(\gamma_i = \gamma_{i,l}) \\ \gamma_{i,l} &= \omega^2 \end{aligned} \tag{18}$$

Thus, the iterative scheme set up is as follows.

$$p^{(i+1)}(x,\dot{x}) = \int_{-\infty}^{\infty}\int_{-\infty}^{\infty} q(x,\dot{x}\,|\,\psi,\dot{\psi})p^{(i)}(\psi,\dot{\psi})d\psi d\dot{\psi}, \quad i=0,1,2,.... \tag{19}$$

The convergence criteria employed is

$$\sup_{x,\dot{x}}(|\,p^{(i+1)}(x,\dot{x}) - p^{(i)}(x,\dot{x})\,|) \le \vartheta \tag{20}$$

where ϑ is the desired upper bound of tolerance. For evaluation of the double integral, Simpson's composite rule (Burden and Faires 1993) has been made use of.

The intensity of the white noise, *K*, is fixed at 2.0 for all the numerical work. The other parameters which are kept fixed are η=0.08, ξ=0.02, α=1.0. The natural frequency parameters ω and λ are then changed to obtain a few unimodal and bimodal pdfs, as reported in figures 1, 2 and 3. Here it may be mentioned that the conditional second moment vector, S_7, has been obtained in the present work by integrating the first seven of equations (11) using a sixth order Runge-Kutta scheme. The numerical integration has been carried out over a small interval of 0.01 sec starting with an initial time of 100 sec. One may observe that even though the pdf as reported in figure 1 is unimodal, it is not Gaussian. Figures 2 and 3 show two typical bimodal pdfs. In all the three cases, convergence was reached within 10 iterations with the upper limit of tolerance ϑ set at 0.001.

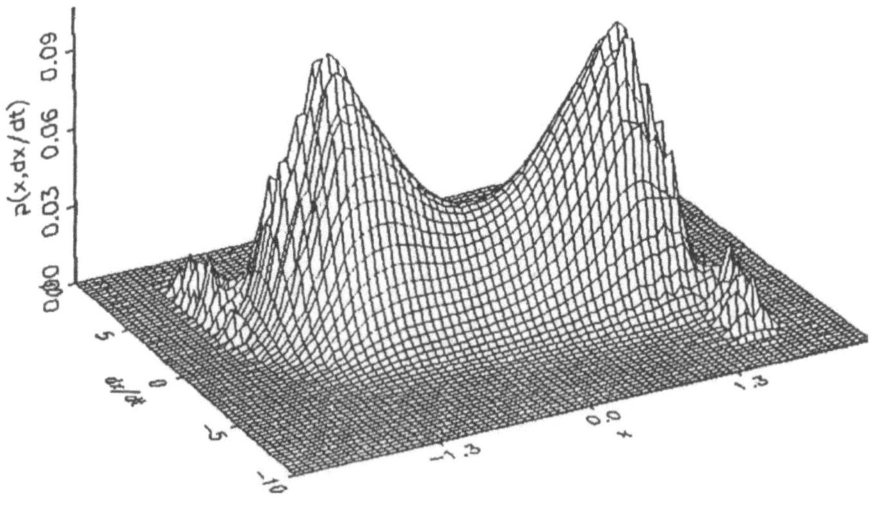

Fig. 1 Stationary pdf of x and $\dot{x}$

$\eta = 0.08$, $\omega = 3.0$, $\alpha = 1.0$, $k = 2.0$, $\xi = 0.02$, $\lambda = 1.5$

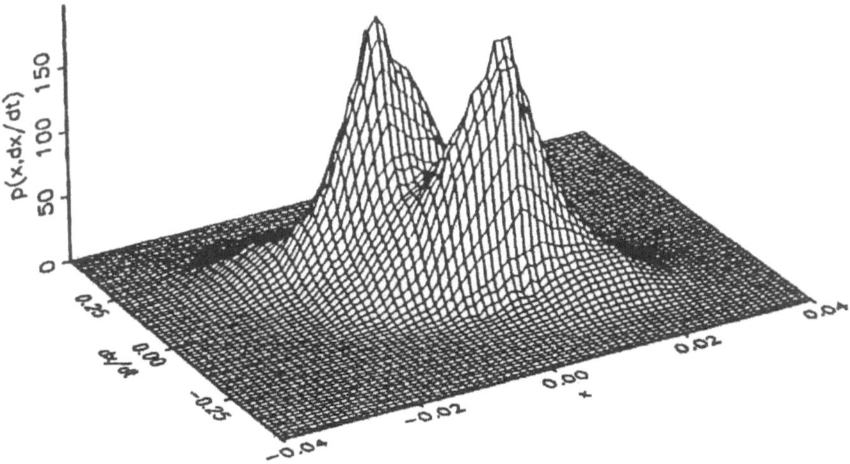

Fig. 2 Stationary pdf of x and $\dot{x}$

$\eta = 0.08$, $\omega = 15.0$, $\alpha = 1.0$, $k = 2.0$, $\xi = 0.02$, $\lambda = 3.0$

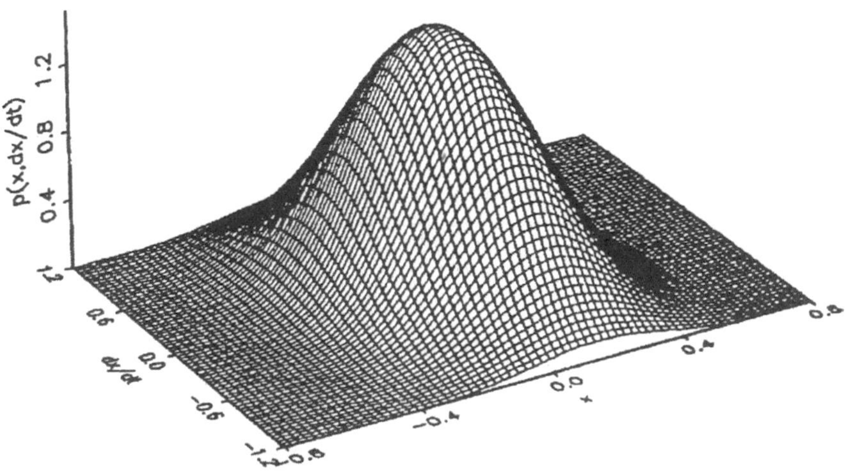

Fig. 3 Stationary pdf of x and $\dot{x}$

$\eta = 0.08$, $\omega = 5.0$, $\alpha = 1.0$, $k = 2.0$, $\xi = 0.02$, $\lambda = 3.0$

By appropriately integrating the 2-dimensional density function, one can readily obtain marginal densities, $p_m(x)$ or $p_m(\dot{x})$. Thus, for example, $p_m(x)$ may be obtained as

$$p_m(x) = \int_{-\infty}^{\infty} p(x,\dot{x})d\dot{x} \tag{21}$$

In figure 4, the marginal bimodal pdf $p_m(x)$, obtained analytically using the present technique, is compared with the numerically simulated marginal pdf. Numerical simulation has been done using a sixth order Runge-Kutta routine with an ensemble size of 10,000 samples. Each of these samples has been numerically integrated for 100 seconds to obtain the approximate stationary histogram. The comparison, as shown in figure 4, appears to be quite good.

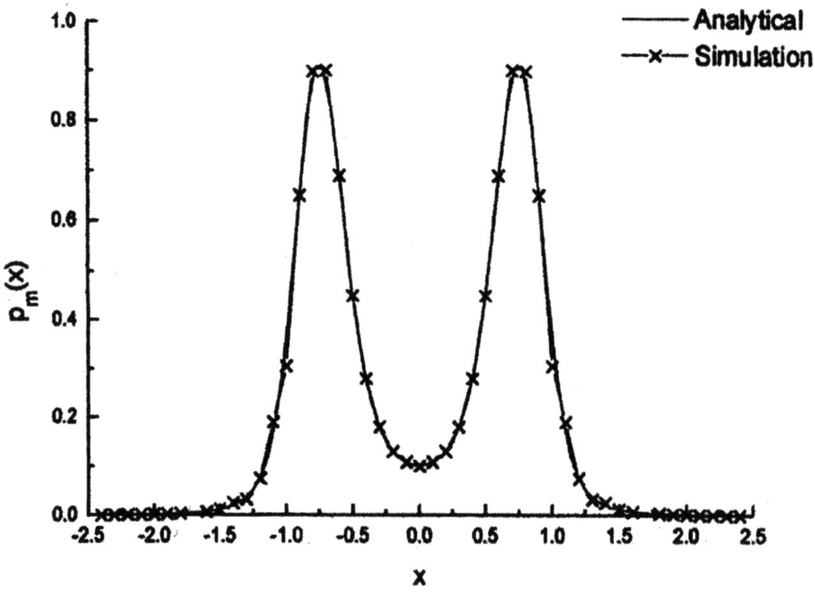

Fig. 4. Comparison of analytical and simulated marginal densities
$\eta = 0.08,\ \omega = 3.0,\ \alpha = 1.0,\ k = 2.0,\ \lambda = 1.5$

5. Summary and Conclusions

The aim of the present investigation has been to explore the possibility of applying the concept of Phase Space Linearization (PSL) to the realms of random vibration. Towards this, the governing nonlinear ODE has been reduced to a set of conditionally linear ODEs, the initial conditions for each of which are known deterministically. To derive these linearized ODEs, the nonlinear terms have been replaced by a set of conditionally linear terms, depending on these initial conditions. These replacements are based on a stochastic Taylor's expansion of the 'nonlinear coefficient' centred at these initial conditions followed by short-time averaging. Next, the conditionally Gaussian pdf for these linearized equations have been obtained by computing the conditional mean and second moment processes in closed form. At this stage, theory of conditional distribution has been made use of to obtain a governing integral map for the time evolution of the nonstationary, non-Gaussian density of the nonlinear oscillator. In the special case of the nonlinear oscillator having a stationary pdf, the integral map is reduced to a Fredholm's integral equation of the second kind with a Guassian kernel. Possible solutions of this integral equation has been attempted for different choices of system parameters using an iterative approach till convergence. The norm for this convergence has been set up as the maximum unsigned 'distance' between two successive approximations of the stationary pdf.

A numerical illustration of the methodology has been provided using the typical example of a hardening Duffing Oscillator under a narrow banded Gaussian excitation. Limited results obtained shows that the present technique is capable of predicting non-Gaussian distributions accurately, including the characteristic bimodal distributions of the hardening Duffing oscillator. In all the cases that have been attempted so far, convergence has been found to reach within a maximum of 10 iterations with a norm of 0.001.

There are however several more issues to be resolved yet. First, the governing integral equation has a double integral, the evaluation of which is rather time consuming. This is especially prohibitive if one attempts to use the technique, as is proposed here, for MDOF oscillators. This limitation may, however, be relieved to some extent if one attempts to work with the amplitude or the energy processes. Moreover, since the conditional distribution over a sufficiently small time step is Gaussian with known mean and second moment processes, it may be feasible to apply the methodology for deriving a principle of faster numerical simulation. Finally, no formal proof for the correspondence of the conditional linearized flow and the originally nonlinear flow is yet available. These issues constitute a few elements of further research on PSL.

6. References

Burden, R.L. and Faires, J.D. (1993) *Numerical Analysis*, PWS Publishing Company, Boston, U.S.A.

Ibrahim, R.A. (1978) Stationary response of a randomly parametric excited non-linear system, *Jl. Applied Mech., Trans. ASME*, **45**, 910-916.

Iyengar, R.N. and Dash, P.K. (1978) Study of random vibration of non-linear systems by Gaussian closure technique, *Jl. Applied Mech., Trans. ASME*, **45**, 393-399.

Iyengar, R.N. and Roy, D. (1996) Conditional linearization in nonlinear random vibration, *Jl. of Engineering Mech.*, 196-200.

Iyengar, R.N. and Roy, D. (1998a) New approaches for the study of nonlinear oscillators, *Jl. Sound and Vib.*, **211**, 843-875.

Iyengar, R.N. and Roy, D. (1998b), Extensions of the Phase Space Linearization method, *Jl. Sound and Vibration*, **211**, 877-903.

Iyengar, R.N. (1988a), Stochastic response and stability of the Duffing oscillator under narrowband excitation, *Jl. Sound and Vib.*, **126**, 255-263.

Iyengar, R.N. (1988b) Higher order linearization in nonlinear random vibration, *Intl. Jl. of Nonlinear Mech.*, **23**, 385-391.

Kloeden, P.E. and Platen, E. (1992) *Numerical solution of stochastic differential equations*, Berlin: Springer-Verlag.

Kloeden, P.E., Platen, E. and Schurz, H. (1994) *Numerical solution of stochastic differential equations through computer experiments*, Berlin: Springer-Verlag.

Lin, Y.K. (1967) *Probabilistic theory of structural dynamics*, New York, McGraw Hill.

Lin, Y.K. and Cai, G.Q. (1995) *Probabilistic structural dynamics : Advanced theory and applications*, McGraw-Hill, N.Y., U.S.A.

Lyon, R.H., Heckl, M., and Hazelgrove, C.B. (1961) Narrow-band excitation of the hard spring oscillator, *Jl. Acoustical Soc. of Am.*, **33**, 1404-1411.

Crandall, S.H. (1973) *Nonlinear vibration problems*, Correlation and spectra of nonlinear system responses, **14**, 39-53.

FINDING RESPONSE CUMULANTS FOR NONLINEAR SYSTEMS WITH MULTIPLICATIVE EXCITATIONS

LOREN D. LUTES
Department of Civil Engineering
Texas A&M University
College Station, TX 77843-3136, USA

COSTAS PAPADIMITRIOU
Division of Engineering and Applied Science
California Institute of Technology
Pasadena, CA 91125, USA

1. Abstract

A relatively straightforward formulation is presented for deriving the differential equations governing the evolution of the response cumulants and moments of a dynamical system. This is a very general framework which applies to linear and nonlinear systems subjected to external and multiplicative non-Gaussian, delta-correlated processes. This formulation provides an alternative to both the partial differential Fokker Planck equation that has sometimes been used in deriving moment or cumulant equations, and the differential (as opposed to derivative) relationships of the Itô calculus. It is believed that many analysts may find the technique used here to be more obvious than the alternatives, since the derivative relationships for the stochastic process are of the same form as in the more familiar ordinary differential equations.

2. Introduction

Methods have been developed over the course of several decades for deriving differential equations which govern the evolution of the moments or cumulants of the response of dynamical systems with stochastic excitations. These so-called state-space techniques provide an alternative to the more traditional methods of using either time-domain or frequency-domain integration to describe the time-history of dynamic response, then using these results to find the values of response moments or cumulants. The original equation of motion is generally used only in determining the coefficients in these state-space equations. The derivation of the state-space equations has sometimes been accomplished by using the Fokker-Planck equation which governs the evolution of the probability distribution of the response process. The authors, though, have primarily focused on the use of procedures whereby the differential equations for moments or cumulants are derived directly from the original equation of motion, without use of the Fokker-Planck equation. In fact, one can consider the derivation of the Fokker-Planck partial differential equation to be an extension of the concept used for deriving these

S. Narayanan and R.N. Iyengar (eds.),
IUTAM Symposium on Nonlinearity and Stochastic Structural Dynamics, 119–132.

ordinary differential equations for moments or cumulants (Lutes and Sarkani 1997). Another approach that is frequently used involves considering the original equation of motion to be an Itô equation describing the differential, rather than the derivative, of the time history.

Liapunov analysis of response covariance for linear systems is one of the earlier uses of the state-space approach. The technique has also been extended to include the calculation of other orders of moments and cumulants, determination of covariance functions, and analysis of various nonlinear systems (Lin 1967, Spanos 1983, Di Paola and Muscolino 1990, Di Paola, Falsone and Pirotta 1992, Lutes and Chen 1992, Hu 1993, Di Paola and Falsone 1994, Falsone 1994, Papadimitriou and Lutes 1994, Hu 1995, Papadimitriou and Lutes 1996, Wojtkiewicz et al. 1996, Di Paola and Falsone 1997a, Di Paola and Falsone 1997b, Hasofer and Grigoriu 1995, Hu 1997, Grigoriu and Waisman 1997). Most of these results are limited to situations in which the excitation is a white-noise, or more generally a delta-correlated, process. In principle, though, it is possible to extend the techniques to include linear or non-linear filtering of the excitation.

For linear or nonlinear systems with only external (or additive) excitations, the various mathematical approaches yield exactly the same equations governing the evolution of moments or cumulants. For multiplicative (or parametric) excitations, though, this is not always the case. In particular, there is the matter of whether integrals are to be interpreted in an "Itô" or "Stratonovich" sense. Essentially this becomes an issue of whether the response of the system at a given instant of time is independent of the delta-correlated excitation at that same instant. Di Paola and Falsone (1993) considered the equation for an increment of the response process and derived moment equations for such a problem which are quite different from those obtained by Itô calculus. The work reported here is somewhat parallel to theirs, but the formulation is different. In particular, the increment of the response in the present work is defined in a straight-forward manner as an integral of the derivative of the process, rather than relying on relationships using differentials and powers of differentials. After reviewing the situation for external excitations and confirming the equations of Di Paola and Falsone for response moments for multiplicative excitations, the method is extended to a direct derivation of equations governing the evolution of response cumulants for nonlinear systems with multiplicative excitations. The primary advantage envisioned for the formulation used here is the apparent simplicity in its interpretation. In particular, the non-Itô integral definition of the response increment allows one to use methods for this stochastic problem that more closely resemble those one would use for a deterministic problem.

3. Moments and Cumulants

The sequence of moment functions for a stochastic process $\{x(t)\}$ is arguably the most natural way to describe the intensity of the process, where the general nth moment function is $E[x(t_1)x(t_2)\cdots x(t_n)]$. This definition is straightforward, and the quantities are easily calculated if the joint probability density functions are known for the $x(t_j)$ random variables. Other alternatives exist, though, for the representation of this information, and they have distinct advantages in some situations. The most commonly considered alternative is the set of cumulant functions, $\kappa_n[x(t_1),x(t_2),\cdots,x(t_n)]$. The general relationship between moments and cumulants can be obtained by noting that the nth moment is a derivative, evaluated at the origin, of the characteristic function, while the corresponding cumulant is a derivative of the log-characteristic function.

One reason that cumulants are sometimes preferred to moments is their relationship to uncertainty about the process. The first cumulant is identical to the first moment, the mean value. Second cumulants give the variance and covariance, providing a measure of the amount that the process varies from its mean value, and the correlation between its values at different times. Similarly, the higher cumulant functions reveal higher-order correlations between the values of the process at different times. In recognition of this, Stratonovich (1963) referred to them as "correlation functions." In addition, of course, cumulant functions for $n \geq 3$ give indications of the extent to which a process is non-Gaussian, since they are identically zero for a Gaussian process.

The focus of this study is the derivation of moment and cumulant functions for the response of dynamical systems governed by stochastic differential equations, with the emphasis being on cumulant functions. In some situations it turns out to be easier to obtain cumulant functions than moment functions, and in other situations the opposite is true (Papadimitriou et al. 1999). In principle, of course, one can derive whichever is simpler, then use the general relationships between the two to obtain the alternate form of the information. This is not a completely satisfactory situation, though, since it is not very convenient to use the general relationships for large values of n. In this regard, though, it may be mentioned that the moment results can be written in a form involving only multiplication and addition of cumulant terms:

$$E[x(t_1)\cdots x(t_n)] = \sum_{v=1}^{n} \sum_{\substack{\{w_1,\cdots,w_v\} \\ w\in\{1,\cdots,n\} \\ \{w_1+\cdots+w_v=n\}}} \sum_{\{j_1,\cdots,j_n\}} \begin{pmatrix} \kappa_{w_1}[x(t_{j_1}),\cdots,x(t_{j_{w_1}})]\cdots \\ \kappa_{w_v}[x(t_{j_n-w_v+1}),\cdots,x(t_{j_n})] \end{pmatrix} \tag{1}$$

Although the several summations in (1) may appear rather confusing, the pattern of the terms is relatively simple. The representative term is the product of v cumulants, with w_l denoting the order of the lth of these cumulants. Each of the $x(t_j)$ arguments appears in one and only one of the cumulants, so that it is necessary that $w_1+\cdots+w_v=n$ in the representative term. The summation, then, is over all possible choices for v (v = 1 to n) and all unique assignments of the t_j arguments to the v cumulants. The term "unique" in this sentence is meant in the sense of unique unordered arrangements, so that (t_2,t_1) is considered to be identical to (t_1,t_2). The explicit relationship for cumulants in terms of moments is not as simple as (1). It contains a corresponding summation of terms, each of which contains a product of moments, but each of the terms in the summation is multiplied by some coefficient which must be determined. A convenient aspect of (1) is the fact that all such coefficients are unity.

Since the current study deals with the evolution of the moments or cumulants of the response, it is crucial to write expressions for the derivatives of these terms. The more obvious situation is for moments, for which one can directly write an expression for the derivative of the kth moment of a process $\{x(t)\}$ in the form of the Riemann definition of a derivative:

$$\frac{d}{dt}E[x^k(t)] \equiv \lim_{\Delta t\to 0}\frac{E[\{x(t^-)+\Delta x\}^k]-E[x^k(t^-)]}{\Delta t}$$
$$= \lim_{\Delta t\to 0}\frac{1}{\Delta t}\sum_{l=1}^{k}\frac{k!}{l!(k-l)!}E[x^{k-l}(t^-)(\Delta x)^l] \quad (2)$$

in which $\Delta t = t^+ - t^-$, and may be arbitrarily small, and $\Delta x = x(t^+) - x(t^-)$. The final form of (2) has been obtained by using the binomial expansion theorem. Similarly, one can write

$$\frac{d}{dt}\kappa_k[x(t),\cdots,x(t)] \equiv \lim_{\Delta t\to 0}\frac{\kappa_k[x(t^-)+\Delta x,\cdots,x(t^-)+\Delta x]-\kappa_k[x(t^-),\cdots,x(t^-)]}{\Delta t}$$
$$= \lim_{\Delta t\to 0}\frac{1}{\Delta t}\sum_{l=1}^{k}\frac{k!}{l!(k-l)!}\kappa_k[\underbrace{x(t^-),\cdots,x(t^-)}_{k-l},\underbrace{\Delta x,\cdots,\Delta x}_{l}] \quad (3)$$

which is based on a linearity property of cumulants (Lutes 1986):

$$\kappa_{j+1}\left(x_1,x_2,\cdots,x_j,\sum_{k=0}^{K}\alpha_k z_k\right) = \sum_{k=0}^{K}\alpha_k\kappa_{j+1}(x_1,x_2,\cdots,x_j,z_k) \quad (4)$$

The difficulty in using (2) or (3) is in evaluating the expectations or cumulants involving both $x(t^-)$ and Δx terms. For this purpose it is convenient to write Δx as the integral

$$\Delta x = \int_{t^-}^{t^+}\dot{x}(\tau)d\tau \quad (5)$$

in which the integrand $\dot{x}(\tau)$ will come directly from the governing equation of motion for the system.

4. Response of a System with Additive Excitation

The new work presented here for systems with multiplicative excitations is a direct extension of earlier investigations of the response cumulants for dynamical systems with external excitations, beginning with linear systems (Lutes and Chen 1992, Papadimitriou and Lutes 1994) and later extended to include nonlinear systems (Papadimitriou and Lutes 1996). Some of the results from that earlier work will be briefly reviewed here, since they form the context for the current work. In particular, the system with external excitation was governed by the equation

$$\dot{x}(t) = f\{x(t)\} + W(t) \quad (6)$$

in which the excitation process, $\{W(t)\}$, is considered to be non-Gaussian and delta-correlated, so that $W(t_1)$ and $W(t_2)$ are independent for $t_1 \neq t_2$ and the jth cumulant is given by

$$\kappa_j[W(t_j),W(t_{j-1}),\cdots,W(t_1)] = e_j(t_j)\delta(t_j-t_1)\delta(t_{j-1}-t_1)\cdots\delta(t_2-t_1) \quad (7)$$

in which $\delta(t)$ denotes the Dirac delta function and $e_j(t)$ is a nonstationary intensity coefficient.

The results for the system of (6) have been shown to be

$$\dot{\kappa}_k[x(t),\cdots,x(t)] = k\kappa_k[\underbrace{x(t^-),\cdots,x(t^-)}_{k-1},f\{x(t)\}] + e_k(t) \tag{8}$$

In order to use (8) for a nonlinear system it is necessary to have a technique for relating a cumulant with a "nonlinear argument" to the ordinary cumulants of the response. A relationship has been developed for this purpose for the special case where $f\{x(t)\}$ is a polynomial in $x(t)$, or can be approximated as such a polynomial. In particular, it has been shown (Papadimitriou and Lutes 1996) that

$$\kappa_{j+1}(x_1,\cdots,x_j,z_1z_2) = \kappa_{j+2}(x_1,\cdots,x_j,z_1,z_2) + \sum_{k=0}^{j}\sum_{\{l_1,\cdots,l_j\}}\kappa_{k+1}(x_{l_1},\cdots,x_{l_k},z_1)\kappa_{j+1-k}(x_{l_{k+1}},\cdots,x_{l_j},z_2) \tag{9}$$

in which there is a summation over all the ways in which $\{1,\cdots,j\}$ can be separated into the two sets $\{l_1,\cdots,l_k\}$ and $\{l_{k+1},\cdots,l_j\}$, as well as over k. Results for (8) with a polynomial $f\{x(t)\}$ can then be found from recursive application of this relationship.

It should be noted that the cumulant equations for the system with a nonlinearity are not uncoupled. In particular, using (8) to obtain an equation for $\dot{\kappa}_j$ generally yields an equation which indicates that the evolution of κ_j also depends on the values of other terms κ_{j+l}. Thus, one must choose some "closure" method to reduce the problem so that the evolution of a finite number of response cumulants can be found by solving a finite set of equations. Best known is cumulant-neglect closure [Crandall 1980, Wu and Lin 1984]. Although closure methods can generally give meaningful results when carefully applied, caution is required. In particular, relatively good results are often obtained with low orders of closure, but there seems to be no assurance that the method will converge as the order is increased [Papadimitriou and Lutes 1996, Wojtkiewicz et al. 1996, Hasofer and Grigoriu 1995].

This form of analysis of systems with external excitations has also been conveniently extended to include vector arguments by using a cumulant notation which is an extension of the Kronecker product. In addition, it has been shown that more efficient approximate evaluations can be obtained by using the symmetry of Kronecker products and truncated summations of eigensolutions (Papadimitriou and Lutes 1994). The corresponding analysis has been carried out for polynomial nonlinear systems with external excitations (Papadimitriou and Lutes 1996), yielding coupled differential equations for the cumulants which agree with those from Itô analysis (Falsone 1994).

5. Response Moments for a System with Multiplicative Excitation

Attention will now be focused on the problem of determining response quantities for a system described by the equation

$$\dot{x}(t) = f\{x(t)\} + g\{x(t)\}W(t) \tag{10}$$

in which the delta-correlated excitation $W(t)$ appears as a multiplier of a response quantity $g\{x(t)\}$, which is generally nonlinear. The restoring force term $f\{x(t)\}$ is also taken to be nonlinear. It should be noted that one could also insert an additive delta-correlated excitation term into (10), but this would add little to the discussion. The preceding section has considered the response to such an additive excitation, and the issues now relate to the additional difficulties in finding the response to the multiplicative excitation. It should also be noted that Di Paola and Falsone (1993) have already determined expressions for the response moments of (10), but the current work uses a formulation which is somewhat different from theirs. In the following section this new formulation will be extended to provide expressions for the response cumulants of this system with a multiplicative excitation.

One can now substitute (10) into (5) to obtain the increment of the process as

$$\Delta x = \mathrm{O}[(\Delta t)^2] + f\{x(t)\}\Delta t + \int_{t^-}^{t^+} g\{x(\tau_1)\}\, W(\tau_1) d\tau_1 \tag{11}$$

One can now use the same procedure to write $g\{x(\tau_1)\}$ in the final integrand as

$$\begin{aligned} g\{x(\tau_1)\} = g\{x(t^-)\} + \int_{t^-}^{\tau_1} g'\{x(\tau_2)\}\dot{x}(\tau_2\} d\tau_2 = \mathrm{O}[(\Delta t)^2] + g\{x(t^-)\} \\ + g'\{x(t^-)\} f\{x(t^-\}\Delta t + \int_{t^-}^{\tau_1} g'\{x(\tau_2)\} g\{x(\tau_2\}\, W(\tau_2) d\tau_2 \end{aligned} \tag{12}$$

Substituting this into (11) then gives

$$\begin{aligned} \Delta x = \mathrm{O}[(\Delta t)^2] + f\{x(t)\}\Delta t + g\{x(t^-)\}\int_{t^-}^{t^+} W(\tau_1) d\tau_1 \\ + \int_{t^-}^{t^+}\int_{t^-}^{\tau_1} g'\{x(\tau_2)\} g\{x(\tau_2\}\, W(\tau_2)\, W(\tau_1) d\tau_2\, d\tau_1 \end{aligned} \tag{13}$$

This procedure can now be repeated for the integrand term $g'\{x(\tau_2)\} g\{x(\tau_2\}$ in (13), etc. The presentation of this recursive process is simplified by using a new notation of

$$g^{(1)}\{x(t)\} = g\{x(t)\}, \qquad g^{(j)}\{x(t)\} = g\,\{x(t)\}[g^{(j-1)}\{x(t)\}]' \text{ for } j \geq 2 \tag{14}$$

Using this notation, the final expression for the increment of the response is

$$\Delta x = \mathrm{O}[(\Delta t)^2] + f\{x(t)\}\Delta t + \sum_{j=1}^{\infty} g^{(j)}\{x(t^-)\}\int_{t^-}^{t^+}\int_{t^-}^{\tau_1}\cdots\int_{t^-}^{\tau_{j-1}} W(\tau_j)\cdots W(\tau_1) d\tau_j \cdots d\tau_2 d\tau_1 \tag{15}$$

In deriving (15), the identification of $\mathrm{O}[\Delta t]$ and $\mathrm{O}[(\Delta t)^2]$ terms has been based on the cumulant relationships in (7). In particular, one sees from (7) that a j-fold integral of a product of j values of $W(\tau_l)$ may be $\mathrm{O}[\Delta t]$ when the range of integration is $\mathrm{O}[\Delta t]$. This then implies that a (j+1)-fold integral of this integrand is $\mathrm{O}[(\Delta t)^2]$, etc. The crucial point, of course, is that there cannot be any $\mathrm{O}[\Delta t]$ cumulant or moment contribution from any term of Δx which is not explicitly listed in (15).

One can now substitute Δx from (15) into the $E[x^m(t^-)(\Delta x)^l]$ term in (2), and evaluate the multiple integrals based on (7). The results are

$$E[x^m(t^-)(\Delta x)^l] = \mathrm{O}[(\Delta t)^2] + \Delta t \sum_{j_1=1}^{\infty}\cdots\sum_{j_l=1}^{\infty} E[x^m(t) g^{(j_1)}\{x(t)\}\cdots g^{(j_l)}\{x(t^-)\}]\frac{e_{j_1+\cdots+j_l}(t)}{j_1!\cdots j_l!} \quad \text{for } l \geq 2 \tag{16}$$

and

$$E[x^m(t^-)(\Delta x)] = \mathrm{O}[(\Delta t)^2] + E[x^m(t)f\{x(t\}]\Delta t + \Delta t \sum_{j_1=1}^{\infty} E[x^m(t^-)g^{(j_1)}\{x(t^-)\}]\frac{e_{j_1}(t)}{j_1!} \tag{17}$$

Note that the $j_1!\cdots j_l!$ denominator in (16) results from the particular limits of integration used. From (7) one can see that a multiple integral from t^- to t^+ of $\kappa_{j_1+\cdots+j_l}[W(\tau_{1,j_1}), \cdots, W(\tau_{1,1}),\cdots, W(\tau_{l,j_l}),\cdots, W(\tau_{l,1})]$ with respect to all the τ_{q,j_m} variables would give a contribution of $e_{j_1+\cdots+j_l}(t)$. From symmetry, though, it follows that the limited range of integration of $t^- \leq \tau_{q,j_q} \leq \cdots \leq \tau_{q,1} \leq t^+$ for $q = 1,\cdots,l$ does reduce the integral by the factor of $j_1!\cdots j_l!$.

Substituting (16) and (17) into (2) now gives

$$\frac{d}{dt}E[x^k(t)] = k\,E[x^{k-1}(t)f\{x(t\}]$$
$$+\sum_{l=1}^{k}\frac{k!}{l!(k-l)!}\sum_{j_1=1}^{\infty}\cdots\sum_{j_l=1}^{\infty}E[x^{k-l}(t)g^{(j_1)}\{x(t)\}\cdots g^{(j_l)}\{x(t)\}]\frac{e_{j_1+\cdots+j_l}(t)}{j_1!\cdots j_l!} \tag{18}$$

This expression is not new, since it is identically the one-dimensional (scalar) version of the equation derived by Di Paola and Falsone (1993). In fact, the moment equation is given here only to confirm that the expression for Δx in (15) is equivalent to that used by Di Paola and colleagues. Their expression looks much different from (15) and involves powers of differential elements of the $\{x(t)\}$ process. To the authors, the multiple integrals of (15) seem to have a more obvious meaning than does the distinction between the Δx increment and a dx differential as is required in Di Paola's formulation

6. Response Cumulants for a System with Multiplicative Excitation

The cumulants of the response to multiplicative excitation can now be investigated in a way exactly parallel to that used in the preceding two sections. In particular, one can substitute (15) into a representative joint cumulant term with $x(t^-)$ and Δx arguments, as in (3), to obtain

$$\kappa_{m+l}[\underbrace{x(t^-),\cdots,x(t^-)}_{(m)},\underbrace{\Delta x,\cdots,\Delta x}_{(l)}] = \mathrm{O}[(\Delta t)^2] + \sum_{j_1=1}^{\infty}\cdots\sum_{j_l=1}^{\infty}\int_{t^-}^{t^+}\cdots\int_{t^-}^{\tau_{1,j_1-1}}\cdots\int_{t^-}^{t^+}\cdots\int_{t^-}^{\tau_{l,j_l-1}}$$
$$\kappa_{m+l}[\underbrace{x(t^-),\cdots,x(t^-)}_{(m)},\underbrace{g^{(j_q)}\{x(t^-)\}W(\tau_{q,j_q})\cdots W(\tau_{q,1})}_{\{q=1,\cdots,l\}}]\underbrace{d\tau_{q,j_q}\cdots d\tau_{q,1}}_{\{q=1,\cdots,l\}} \text{ for } l\geq 2 \tag{19}$$

and

$$\kappa_{m+1}[\underbrace{x(t^-),\cdots,x(t^-)}_{(m)},\Delta x] = \mathrm{O}[(\Delta t)^2] + \kappa_{1+m}[\underbrace{x(t),\cdots,x(t)}_{(m)},f\{x(t)]\Delta t + \sum_{j_1=1}^{\infty}\int_{t^-}^{t^+}\cdots\int_{t^-}^{\tau_{1,j_1-1}}$$
$$\kappa_{m+1}[\underbrace{x(t^-),\cdots,x(t^-)}_{(m)},g^{(j_1)}\{x(t^-)\}W(\tau_{1,j_1})\cdots W(\tau_{1,1})]d\tau_{1,j_1}\cdots d\tau_{1,1} \tag{20}$$

It should be noted that the integrand of (19) has been simplified by introducing a notation of

$$\kappa_{m+l}[\underbrace{x(t^-),\cdots,x(t^-)}_{(m)},\underbrace{g^{(j_q)}\{x(t^-)\}W(\tau_{q,j_q})\cdots W(\tau_{q,1})}_{\{q=1,\cdots,l\}}]\underbrace{d\tau_{q,j_q}\cdots d\tau_{q,1}}_{\{q=1,\cdots,l\}}$$
$$\equiv\kappa_{m+l}[\underbrace{x(t^-),\cdots,x(t^-)}_{(m)},g^{(j_1)}\{x(t^-)\}W(\tau_{1,j_1})\cdots W(\tau_{1,1}),\cdots,$$
$$g^{(j_l)}\{x(t^-)\}W(\tau_{l,j_l})\cdots W(\tau_{l,1})]d\tau_{l,j_l}\cdots d\tau_{l,1},\cdots,d\tau_{1,j_1}\cdots d\tau_{1,1} \tag{21}$$

and that the range of integration is $t^- \le \tau_{q,j_q} \le \tau_{q,j_q-1} \le \cdots \le \tau_{q,2} \le \tau_{q,1} \le t^+$ for $q=1,\cdots,l$.

Because of the multiplicative excitation, each of the cumulants in (19) and (20) contains some arguments which are a product of a response term $g^{(j_q)}\{x(t^-)\}$ and j_q excitation terms $W(\tau_{q,j_m})$. These cumulants with product-arguments, though, can be evaluated by using the same tool as was introduced for the analysis of systems with polynomial nonlinearities. In particular, these product-arguments in (19) and (20) can each be handled by using (9), which applies to any cumulant in which an argument is a product of two random quantities. Elimination of all product-arguments then requires repeated applications of (9). Furthermore, recall that each application of (9) includes all possible ways in which the non-product arguments can be separated into two sets. Clearly, this operation produces a very large number of cumulant terms. It can be shown, though, that $O[\Delta t]$ contributions come only from terms in which all of the $W(\tau_{q,j_m})$ expressions appear as arguments in the same cumulant. Thus, one can neglect all terms with $W(\tau_{q,j_m})$ arguments in two or more different cumulants.

As a first step in the simplification, (9) will be used to separate the $z_1 = g^{(j_1)}\{x(t^-)\}$ response portion from the $z_2 = W(\tau_{1,j_1})\cdots W(\tau_{1,1})$ excitation portion in the z_1z_2 argument ($q=1$) in the cumulant in (19). After some simplification this gives

$$\kappa_{m+l}[\underbrace{x(t^-),\cdots,x(t^-)}_{(m)},\underbrace{g^{(j_q)}\{x(t^-)\}W(\tau_{q,j_q})\cdots W(\tau_{q,1})}_{\{q=1,\cdots,l\}}]=O[(\Delta t)^2]$$
$$+\sum_{r_1=0}^{m}\frac{m!}{r_1!(m-r_1)!}\kappa_{r_1+1}[\underbrace{x(t),\cdots,x(t)}_{(r_1)},g^{(j_1)}\{x(t)\}]\kappa_{m+l-r_1}[\underbrace{x(t^-),\cdots,x(t^-)}_{(m-r_1)},$$
$$W(\tau_{1,j_1})\cdots W(\tau_{1,1}),\underbrace{g^{(j_q)}\{x(t^-)\}W(\tau_{q,j_q})\cdots W(\tau_{q,1})}_{\{q=2,\cdots,l\}}]+\kappa_{m+l+1}[\underbrace{x(t^-),\cdots,x(t^-)}_{(m)},$$
$$g^{(j_1)}\{x(t^-)\},W(\tau_{1,j_1})\cdots W(\tau_{1,1}),\underbrace{g^{(j_q)}\{x(t^-)\}W(\tau_{q,j_q})\cdots W(\tau_{q,1})}_{\{q=2,\cdots,l\}}] \tag{22}$$

Note that the one cumulant in (22) which involves only response quantities has been evaluated at t rather than t^-, but this shift introduces only an $O[\Delta t]$ change in that cumulant. This, in turn, causes an $O[(\Delta t)^2]$ discrepancy in (22), since the $O[\Delta t]$ change multiplies an $O[\Delta t]$ cumulant involving all the excitation terms. It should also be noted that (22) can be considered to give the separation of the same z_1z_2 argument in the cumulant in (20), but the result is simply

$$
\begin{aligned}
&\kappa_{m+1}[\underbrace{x(t^-),\cdots,x(t^-)}_{(m)},g^{(j_1)}\{x(t^-)\}W(\tau_{1,j_1})\cdots W(\tau_{1,1})] = \mathrm{O}[(\Delta t)^2] \\
&+\sum_{r_1=0}^{m}\frac{m!}{r_1!(m-r_1)!}\kappa_{r_1+1}[\underbrace{x(t),\cdots,x(t)}_{(r_1)},g^{(j_1)}\{x(t)\}]\kappa_{m+1-r_1}[\underbrace{x(t^-),\cdots,x(t^-)}_{(m-r_1)}, \\
&W(\tau_{1,j_1})\cdots W(\tau_{1,1})] + \kappa_{m+2}[\underbrace{x(t^-),\cdots,x(t^-)}_{(m)},g^{(j_1)}\{x(t^-)\},W(\tau_{1,j_1})\cdots W(\tau_{1,1})] \qquad (23)
\end{aligned}
$$

since the set $\{q=2,\cdots,l\}$ is now empty. Further simplification of (23) will be deferred until later, after the other terms in (22) have been reduced to a situation where the only product terms are of the form $W(\tau_{q,j_q})\cdots W(\tau_{q,1})$, involving only the excitation.

Next, (9) will be used in exactly the same way to separate the second product-argument ($q=2$) in each of two terms of (22). This gives the terms as

$$
\begin{aligned}
&\kappa_{m+l-r_1}[\underbrace{x(t^-),\cdots,x(t^-)}_{(m-r_1)},W(\tau_{1,j_1})\cdots W(\tau_{1,1}),\underbrace{g^{(j_q)}\{x(t^-)\}W(\tau_{q,j_q})\cdots W(\tau_{q,1})}_{\{q=2,\cdots,l\}}] = \mathrm{O}[(\Delta t)^2] \\
&+\sum_{r_2=0}^{m-r_1}\frac{(m-r_1)!}{r_2!(m-r_1-r_2)!}\kappa_{r_2+1}[\underbrace{x(t),\cdots,x(t)}_{(r_2)},g^{(j_2)}\{x(t)\}]\kappa_{m+l-r_1-r_2}[\underbrace{x(t^-),\cdots,x(t^-)}_{(m-r_1-r_2)}, \\
&\underbrace{W(\tau_{s,j_s})\cdots W(\tau_{s,1})}_{\{s=1,2\}},\underbrace{g^{(j_q)}\{x(t^-)\}W(\tau_{q,j_q})\cdots W(\tau_{q,1})}_{\{q=3,\cdots,l\}}] + \kappa_{m+l-r_1+1}[\underbrace{x(t^-),\cdots,x(t^-)}_{(m-r_1)}, \\
&g^{(j_2)}\{x(t^-)\},\underbrace{W(\tau_{s,j_s})\cdots W(\tau_{s,1})}_{\{s=1,2\}},\underbrace{g^{(j_q)}\{x(t^-)\}W(\tau_{q,j_q})\cdots W(\tau_{q,1})}_{\{q=3,\cdots,l\}}] \qquad (24)
\end{aligned}
$$

and

$$
\begin{aligned}
&\kappa_{m+l+1}[\underbrace{x(t^-),\cdots,x(t^-)}_{(m)},g^{(j_1)}\{x(t^-)\},W(\tau_{1,j_1})\cdots W(\tau_{1,1}),\underbrace{g^{(j_q)}\{x(t^-)\}W(\tau_{q,j_q})\cdots W(\tau_{q,1})}_{\{q=2,\cdots,l\}}] \\
&=\sum_{r_1=0}^{m}\frac{m!}{r_1!(m-r_1)!}\kappa_{r_1+1}[\underbrace{x(t),\cdots,x(t)}_{(r_1)},g^{(j_2)}\{x(t)\}]\kappa_{m+l-r_1+1}[\underbrace{x(t^-),\cdots,x(t^-)}_{(m-r_1)}, \\
&g^{(j_1)}\{x(t^-)\},\underbrace{W(\tau_{s,j_s})\cdots W(\tau_{s,1})}_{\{s=1,2\}},\underbrace{g^{(j_q)}\{x(t^-)\}W(\tau_{q,j_q})\cdots W(\tau_{q,1})}_{\{q=3,\cdots,l\}}] \\
&+\sum_{r_1=0}^{m}\frac{m!}{r_1!(m-r_1)!}\kappa_{r_1+2}[\underbrace{x(t),\cdots,x(t)}_{(r_1)},\underbrace{g^{(j_s)}\{x(t)\}}_{\{s=1,2\}}]\kappa_{m+l-r_1}[\underbrace{x(t^-),\cdots,x(t^-)}_{(m-r_1)}, \\
&\underbrace{W(\tau_{s,j_s})\cdots W(\tau_{s,1})}_{\{s=1,2\}},\underbrace{g^{(j_q)}\{x(t^-)\}W(\tau_{q,j_q})\cdots W(\tau_{q,1})}_{\{q=3,\cdots,l\}}] + \kappa_{m+l+2}[\underbrace{x(t^-),\cdots,x(t^-)}_{(m)}, \\
&\underbrace{g^{(j_s)}\{x(t^-)\}}_{\{s=1,2\}},\underbrace{W(\tau_{s,j_s})\cdots W(\tau_{s,1})}_{\{s=1,2\}},\underbrace{g^{(j_q)}\{x(t^-)\}W(\tau_{q,j_q})\cdots W(\tau_{q,1})}_{\{q=3,\cdots,l\}}] \qquad (25)
\end{aligned}
$$

For the special case of $l=2$, of course, $\{q=3,\cdots,l\}$ is an empty set, so that (24) and (25) have no more cumulants with arguments which are a product of response terms and excitation terms. For $l>2$ one proceeds as above to separate the terms in the $q=3$ argument, etc. In total the procedure using (9) must be performed l times for the $q=l$

argument.

The result, after all the repeated application of (9) has separated the response terms from the excitation terms in the arguments of all cumulants and after some simplification, can be written as

$$\kappa_{m+l}[\underbrace{x(t^-),\cdots,x(t^-)}_{(m)},\underbrace{g^{(j_q)}\{x(t^-)\}W(\tau_{q,j_q})\cdots W(\tau_{q,1})}_{\{q=1,\cdots,l\}}] = O[(\Delta t)^2]$$

$$+\sum_{u=0}^{l}\sum_{v=1}^{u}\sum_{\{w_1,\cdots,w_v\}}\sum_{\{r_1,\cdots,r_v\}}\frac{m!}{r_1!r_2!\cdots(m-r_1-\cdots-r_v)!}\underbrace{\sum_{\{j_{q,s}\}}\kappa_{w_q+r_q}[\underbrace{x(t),\cdots,x(t)}_{(r_q)},\underbrace{g^{(j_{q,s})}\{x(t)\}}_{\{s=1,\cdots,w_q\}}]}_{\{q=1,\cdots,v\}}$$

$$\cdot\kappa_{m+2l-u-r_1-\cdots-r_v}[\underbrace{x(t^-),\cdots,x(t^-)}_{(m-r_1-\cdots-r_v)},\underbrace{g^{(j_q)}\{x(t^-)\}}_{\{q=1,\cdots,l-u\}},\underbrace{W(\tau_{s,j_s})\cdots W(\tau_{s,1})}_{\{s=1,\cdots,l\}}] \quad (26)$$

in which u = the total number of $g^{(j_1)}\{x(t)\}$ values which appear in cumulants involving only response terms; v = the number of subsets into which these $g^{(j_1)}\{x(t)\}$ values are separated; $\{w_1,\cdots,w_v\}$ gives the number of $g^{(j_1)}\{x(t)\}$ values in each subset, with $1\le w_q\le u-v+1$ and $w_1+\cdots+w_v=u$; and $\{r_1,\cdots,r_v\}$ gives the number of $x(t)$ values grouped with each subset of these $g^{(j_1)}\{x(t)\}$ values, with $0\le r_q\le m$ and $r_1+\cdots+r_v=m$. The remaining $q-u$ of the $g^{(j_q)}\{x(t^-)\}$ values and $m-r_1-\cdots-r_v$ of the $x(t^-)$ values appear in the final cumulant, which also includes all the excitation products as arguments. The summation over $\{j_{q,s}\}$ denotes inclusion of all unique assignments of the $\{j_1,\cdots,j_l\}$ elements into the subsets $\{j_{1,1},\cdots,j_{1,w_1}\},\cdots,\{j_{v,1},\cdots,j_{v,w_v}\}$.

Next, consider evaluation of the final cumulant in (26). All the $x(t^-)$ and $g^{(j_q)}\{x(t^-)\}$ terms are independent of any $W(\tau)$ term except for the single point at the lower limit of integration with $\tau=t^-$. Thus, in general, the final cumulant is not $O[\Delta t]$ unless $r_1+\cdots+r_v=m$ and $u=l$, so that the only arguments are excitation terms. This gives

$$\kappa_{m+l}[\underbrace{x(t^-),\cdots,x(t^-)}_{(m)},\underbrace{g^{(j_q)}\{x(t^-)\}W(\tau_{q,j_q})\cdots W(\tau_{q,1})}_{\{q=1,\cdots,l\}}] = O[(\Delta t)^2]+\kappa_l[\underbrace{W(\tau_{s,j_s})\cdots W(\tau_{s,1})}_{\{s=1,\cdots,l\}}]$$

$$\cdot\sum_{v=1}^{l}\sum_{\{w_1,\cdots,w_v\}}\sum_{\{r_1,\cdots,r_v\}}\frac{m!}{r_1!r_2!\cdots r_v!}\underbrace{\sum_{\{j_{q,s}\}}\kappa_{w_q+r_q}[\underbrace{x(t),\cdots,x(t)}_{(r_q)},\underbrace{g^{(j_{q,s})}\{x(t)\}}_{\{s=1,\cdots,w_q\}}]}_{\{q=1,\cdots,v\}} \quad (27)$$

with the restriction that $r_1+\cdots+r_v=m$. Note that there are still product-arguments of the form $W(\tau_{s,j_s})\cdots W(\tau_{s,1})$ in one cumulant in (27). These must also be eliminated by application of (9). Most of the resulting terms drop out, though, since a significant contribution can result only if all the $W(\tau)$ terms appear in the same cumulant. The result after repeated application of (9) is

$$\kappa_{m+l}[\underbrace{x(t^-),\cdots,x(t^-)}_{(m)},\underbrace{g^{(j_q)}\{x(t^-)\}W(\tau_{q,j_q})\cdots W(\tau_{q,1})}_{\{q=1,\cdots,l\}}]$$

$$= \mathrm{O}[(\Delta t)^2] + \kappa_{j_1+\cdots+j_l}[\underbrace{W(\tau_{s,j_s}),\cdots,W(\tau_{s,1})}_{\{s=1,\cdots,l\}}]$$

$$\cdot\sum_{v=1}^{l}\sum_{\{w_1,\cdots,w_v\}}\sum_{\{r_1,\cdots,r_v\}}\frac{m!}{r_1!r_2!\cdots r_v!}\underbrace{\sum_{\{j_{q,s}\}}\kappa_{w_q+r_q}[\underbrace{x(t),\cdots,x(t)}_{(r_q)},\underbrace{g^{(j_{q,s})}\{x(t)\}}_{\{s=1,\cdots,w_q\}}]}_{\{q=1,\cdots,v\}} \quad (28)$$

still with the restriction that $r_1+\cdots+r_v=m$.

One can now integrate the generic cumulant of (28) and use the excitation cumulant relationship of (7) in order to obtain a result corresponding to (19) and (20) in terms of the intensities of the excitation. In particular,

$$\int_{t^-}^{t^+}\cdots\int_{t^-}^{\tau_{1,j_1-1}}\cdots\int_{t^-}^{t^+}\cdots\int_{t^-}^{\tau_{l,j_l-1}}\kappa_{m+l}[\underbrace{x(t^-),\cdots,x(t^-)}_{(m)},\underbrace{g^{(j_q)}\{x(t^-)\}W(\tau_{q,j_q})\cdots W(\tau_{q,1})}_{\{q=1,\cdots,l\}}]$$

$$\underbrace{d\tau_{q,j_q}\cdots d\tau_{q,1}}_{\{q=1,\cdots,l\}} = \mathrm{O}[(\Delta t)^2] + \Delta t\frac{e_{j_1+\cdots+j_l}(t)}{j_1!\cdots j_l!}\sum_{v=1}^{l}\sum_{\{w_1,\cdots,w_v\}}\sum_{\{r_1,\cdots,r_v\}}\frac{m!}{r_1!r_2!\cdots r_v!}$$

$$\cdot\sum_{\{j_{q,s}\}}\kappa_{w_q+r_q}[\underbrace{x(t),\cdots,x(t)}_{(r_q)},\underbrace{g^{(j_{q,s})}\{x(t)\}}_{\{s=1,\cdots,w_q\}}] \quad (29)$$

Choosing $m=k-l$, one can now substitute (29) into (19) and (20), then substitute these into (3) to obtain

$$\frac{d}{dt}\kappa_k[x(t),\cdots,x(t)] = k\kappa_k[\underbrace{x(t),\cdots,x(t)}_{(k-1)},f\{x(t)\}]$$

$$+\sum_{l=1}^{k}\sum_{j_1=1}^{\infty}\cdots\sum_{j_l=1}^{\infty}\sum_{v=1}^{l}\sum_{\substack{\{w_1,\cdots,w_v\}\\ w\in\{1,\cdots,l\}\\ \{w_1+\cdots+w_v=l\}}}\sum_{\substack{\{r_1,\cdots,r_v\}\\ r\in\{0,\cdots,k-l\}\\ \{r_1+\cdots+r_v=k-l\}}}\sum_{\{j_{1,s},\cdots,j_{v,s}\}}\frac{k!}{l!r_1!\cdots r_v!}\frac{e_{j_1+\cdots+j_l}(t)}{j_1!\cdots j_l!}$$

$$\cdot\kappa_{r_1+w_1}[\underbrace{x(t),\cdots,x(t)}_{(r_1)},\underbrace{g^{(j_{1,s})}\{x(t)\}}_{\{s=1,\cdots,w_1\}}]\cdots\kappa_{r_v+w_v}[\underbrace{x(t),\cdots,x(t)}_{(r_v)},\underbrace{g^{(j_{v,s})}\{x(t)\}}_{\{s=1,\cdots,w_v\}}] \quad (30)$$

in which the summation over $\{j_{1,s},\cdots,j_{v,s}\}$ now denotes the inclusion of all unique assignments of the $\{j_1,\cdots,j_l\}$ elements into the subsets $\{j_{1,1},\cdots,j_{1,w_1}\},\cdots,\{j_{v,1},\cdots,j_{v,w_v}\}$. Expression (30) is now the desired relationship. It relates the rate of change of any cumulant of the response process to the intensity functions of the excitation and to other cumulants which depend only on the response of the system.

7. Interpretation of the Response Cumulant Result

It should be noted that most of the response cumulants on the right-hand side of (30) involve the nonlinear functions $f\{x(t)\}$ and $g^{(j)}\{x(t)\}$. Also, the cumulants of different orders are coupled, so that numerical solution of (30) will require the application of some type of closure procedure, just as is necessary for (8) for the response of a nonlinear system with an additive excitation. One approach, in particular, is to approximate all the nonlinear functions as polynomials, use (9) to eliminate product-arguments, then use cumulant-neglect closure to obtain an approximate solution. Expression (30), of course, includes the special case in which $f\{x(t)\}$ and $g\{x(t)\}$ are both linear. In this situation, one can see from (14) that all the $g^{(j)}\{x(t)\}$ are also linear, so that no polynomial approximation is required, and no additional product arguments are encountered. Nonetheless, it is necessary to use a closure scheme since the cumulants of different orders are still coupled.

It is also of interest to note that one of the terms on the right-hand side of (30) is exactly an expected value of a product. In particular, for the special case of $l = k$ one must have all the terms $\{r_1,\cdots,r_v\}$ equal to zero. Then the summations over v and $\{w_1,\cdots,w_v\}$ give the sum of products of cumulants of subsets for all groupings of the members in $\{g^{(j_1)}(t),\cdots,g^{(j_l)}(t)\}$. Thus, these summations in the term for $k = l$ give $E[g^{(j_1)}(t)\cdots g^{(j_l)}(t)]$, so that this term is identical to the corresponding term in (18) for the derivative of moments. This is to be expected, though, since these terms result from the kth moment of the Δx increment in (18) and the kth cumulant of the same increment in (30). In general, the difference between the kth moment and the kth cumulant of such an increment is $O[(\Delta t)^2]$, as can be shown from (1). For $l \neq k$, the summations in (30) do not give an expected value, since they omit all combinations with any subsets having only $x(t)$ values.

In order to illustrate the use of (30), it will now be explicitly applied for the special cases of $k = 1$, $k = 2$ and $k = 3$. These results will also be verified by using (1) to write the derivative of the third moment in terms of the derivatives of the first three cumulants, then showing that this result is identical to that obtained from (18).

For $k = 1$ the only possible combination of the indices in (30) is: $\{l = 1, v = 1, w_1 = 1, r_1 = 0, j_{1,1} = j_1\}$, and this gives

$$\frac{d}{dt}\kappa_1[x(t)] = \kappa_1[f\{x(t)\}] + \sum_{j_1=1}^{\infty}\frac{e_{j_1}(t)}{j_1!}\kappa_1[g^{(j_1)}\{x(t)\}] \equiv E[f\{x(t)\}] + \sum_{j_1=1}^{\infty}\frac{e_{j_1}(t)}{j_1!}E[g^{(j_1)}\{x(t)\}] \tag{31}$$

Similarly, the possible combinations of the indices for $k = 2$ in (30) are $\{l = 1, v = 1, w_1 = 1, r_1 = 1, j_{1,1} = j_1\}$, $\{l = 2, v = 1, w_1 = 2, r_1 = 0, j_{1,1} = j_1, j_{1,2} = j_2\}$, and $\{l = 2, v = 2, w_1 = 1, w_2 = 1, r_1 = 0, r_2 = 0, j_{1,1} = j_1, j_{2,1} = j_2\}$. Using these three terms then combining two of the cumulants into an expectation gives

$$\frac{d}{dt}\kappa_2[x(t),x(t)] = 2\kappa_2[x(t),f\{x(t)\}] + 2\sum_{j_1=1}^{\infty}\frac{e_{j_1}(t)}{j_1!}\kappa_2[x(t),g^{(j_1)}\{x(t)\}]$$
$$+\sum_{j_1=1}^{\infty}\sum_{j_2=1}^{\infty}\frac{e_{j_1+j_2}(t)}{j_1!j_2!}E[\, g^{(j_1)}\{x(t)\}g^{(j_2)}\{x(t)\}\,] \tag{32}$$

For $k = 3$ in (30) there are nine possible combinations of the indices: $\{l = 1, v = 1, w_1 = 1, r_1 = 2, j_{1,1} = j_1\}$, $\{l = 2, v = 1, w_1 = 2, r_1 = 1, w_1 = 2, j_{1,1} = j_1, j_{1,2} = j_2\}$, $\{l = 2, v = 2, w_1 = 1, w_2 = 1, r_1 = 1, r_2 = 0, j_{1,1} = j_1, j_{2,1} = j_2\}$, $\{l = 2, v = 2, w_1 = 1, w_2 = 1, r_1 = 0, r_2 = 1, j_{1,1} = j_1, j_{2,1} = j_2\}$, $\{l = 3, v = 1, w_1 = 3, r_1 = 0, j_{1,1} = j_1, j_{1,2} = j_2, j_{1,3} = j_3\}$, $\{l = 3, v = 2, w_1 = 1, w_2 = 2, r_1 = 0, r_2 = 0,$

$j_{1,1} = j_1,\ j_{2,1} = j_2,\ j_{2,2} = j_3\}$, $\{l = 3,\ v = 2,\ w_1 = 1,\ w_2 = 2,\ r_1 = 0,\ r_2 = 0,\ j_{1,1} = j_2,\ j_{2,1} = j_1,\ j_{2,2} = j_3\}$, $\{l = 3,\ v = 2,\ w_1 = 1,\ w_2 = 2,\ r_1 = 0,\ r_2 = 0,\ j_{1,1} = j_3,\ j_{2,1} = j_1,\ j_{2,2} = j_2\}$, and $\{l = 3,\ v = 3,\ w_1 = 1,\ w_2 = 1,\ w_3 = 1,\ r_1 = 0,\ r_2 = 0,\ r_3 = 0.\ j_{1,1} = j_1,\ j_{2,1} = j_2,\ j_{3,1} = j_3\}$. Note that it would not be appropriate to include in the list the combination $\{w_1 = 2,\ w_2 = 1\}$ for $\{l = 3,\ v = 2\}$, because that would duplicate the terms given by $\{w_1 = 1,\ w_2 = 2\}$. That is, for both $\{w_1 = 2,\ w_2 = 1\}$ and $\{w_1 = 1,\ w_2 = 2\}$ the summation over $\{j_{1,s},\cdots,j_{v,s}\}$ gives the same set of three terms in which the three $g^{(j)}\{x(t)\}$ components are separated into one pair and one single element. Adding these nine terms then using (1) to combine the five cumulant terms with $l = 3$ into one expected value gives

$$\frac{d}{dt}\kappa_3[x(t),x(t),x(t)] = 3\kappa_3[x(t),x(t),f\{x(t)\}] + 3\sum_{j_1=1}^{\infty}\frac{e_{j_1}(t)}{j_1!}\ \kappa_3[x(t),x(t),g^{(j_1)}\{x(t)\}]$$
$$+3\sum_{j_1=1}^{\infty}\sum_{j_2=1}^{\infty}\frac{e_{j_1+j_2}(t)}{j_1!j_2!}\Big(\kappa_3[x(t),g^{(j_1)}\{x(t)\},g^{(j_2)}\{x(t)\}]$$
$$+\kappa_2[x(t),g^{(j_1)}\{x(t)\}]\kappa_1[g^{(j_2)}\{x(t)\}] + \kappa_1[g^{(j_1)}\{x(t)\}]\kappa_2[x(t),g^{(j_2)}\{x(t)\}]\Big)$$
$$+\sum_{j_1=1}^{\infty}\sum_{j_2=1}^{\infty}\sum_{j_3=1}^{\infty}\frac{e_{j_1+j_2+j_3}(t)}{j_1!j_2!j_3!}E[\ g^{(j_1)}\{x(t)\}g^{(j_2)}\{x(t)\}g^{(j_3)}\{x(t)\}\ ] \tag{33}$$

In order to check the consistency of (18) and (30) for $k = 3$, one can now use (1) to write the third moment as $E[x^3(t)] = \kappa_3[x(t),x(t),x(t)] + 3\kappa_2[x(t),x(t)]\kappa_1[x(t)] + \kappa_1^3[x(t)]$. Taking the derivative of this relationship, then simplifying gives

$$\frac{d}{dt}E[x^3(t)] = \frac{d}{dt}\kappa_3[x(t),x(t),x(t)] + 3\kappa_1[x(t)]\frac{d}{dt}\kappa_2[x(t),x(t)] + 3E[x(t)^2]\frac{d}{dt}\kappa_1[x(t)] \tag{34}$$

Substituting (31), (32), and (33) into (34) then combining terms by using (1) gives

$$\frac{d}{dt}E[x^3(t)] = 3E[x^2(t)f\{x(t)] + 3\sum_{j_1=1}^{\infty}\frac{e_{j_1}(t)}{j_1!}\ E[x(t)^2 g^{(j_1)}\{x(t)\}]$$
$$+3\sum_{j_1=1}^{\infty}\sum_{j_2=1}^{\infty}\frac{e_{j_1+j_2}(t)}{j_1!j_2!}\ E[x(t)g^{(j_1)}\{x(t)\}g^{(j_2)}\{x(t)\}]$$
$$+\sum_{j_1=1}^{\infty}\sum_{j_2=1}^{\infty}\sum_{j_3=1}^{\infty}\frac{e_{j_1+j_2+j_3}(t)}{j_1!j_2!j_3!}E\Big[g^{(j_1)}\{x(t)\}g^{(j_2)}\{x(t)\}g^{(j_3)}\{x(t)\}\Big] \tag{35}$$

which is identical to the result from (18) with $k = 3$.

8. Summary

Differential equations have been derived which govern the evolution of the moments and cumulants of the response of nonlinear systems with either external or multiplicative stochastic excitations. The methods used do not involve either the Fokker-Planck equation or Itô calculus. Rather, they rely on defining the increment of the response process in a straight-forward manner as an integral of the derivative of the process. After reviewing the situation for external excitations and confirming existing equations for response moments for multiplicative excitations, the method is extended to a direct derivation of equations governing the evolution of response cumulants for nonlinear systems with multiplicative excitations. These equations have been illustrated by explicit application to cumulants up to the third order. The primary advantage envisioned for the formulation used here is the apparent simplicity in its interpretation. In particular, the non-Itô integral definition of the response increment allows one to use methods for this stochastic problem that closely resemble those one would use for a deterministic problem.

9. References

Crandall, S. H. (1980). "Non-Gaussian closure for random vibration excitation," *International Journal of Non-Linear Mechanics*, **15**, 303-313.
Di Paola, M. and Falsone, G. (1993). "Stochastic dynamics of nonlinear systems driven by non-normal delta-correlated processes," *Journal of Applied Mechanics*, ASME, **60**, 141-148.
Di Paola, M. and Falsone, G. (1997a). "Higher order statistics of the response of MDOF linear systems excited by linearly parametric white noises and external excitations," *Probabilistic Engineering Mechanics*, **12**(3), 179-188.
Di Paola, M. and Falsone, G. (1997b). "Higher order statistics of the response of MDOF systems under polynomials of filtered normal white noises," *Probabilistic Engineering Mechanics*, **12**(3), 189-196.
Di Paola, M., Falsone, G. (1994). "Stochastic dynamics of MDOF structural systems under non-normal filtered inputs," *Probabilistic Engineering Mechanics*, **9**(4), 265-272.
Di Paola, M., Falsone, G., and Pirotta, A. (1992). "Stochastic response analysis of nonlinear systems under Gaussian inputs," *Probabilistic Engineering Mechanics*, **7**, 15-21.
Di Paola, M. and Muscolino, G. (1990). "Differential moment equations of FE modeled structures with geometrical non-linearities," *International Journal of Non-Linear Mechanics*, **25**, 363-373.
Falsone, G. (1994). "Cumulants and correlations for nonlinear systems under non-stationary delta-correlated processes," *Probabilistic Engineering Mechanics*, **9**(3), 157-165.
Grigoriu, M. and Waisman, F. (1997). "Linear systems with polynomials of filtered Poisson processes," *Probabilistic Engineering Mechanics*, **12**(2), 97-103.
Hasofer, A.M., and Grigoriu, M. (1995). "A new perspective on the moment closure method," *Journal of Applied Mechanics*, **62**(2), 527-532.
Hu, S-L.J. (1993). "Responses of dynamic systems excited by non-Gaussian pulse processes," *Journal of Engineering Mechanics*, ASCE, **119**(9), 1818-1827.
Hu, S-L.J. (1995). "Parametric random vibrations under non-Gaussian pulse processes," *Journal of Engineering Mechanics*, ASCE, **121**(12), 1366-1371.
Hu, S-L.J. (1997). "Response cumulant equations for dynamic systems under delta-correlated processes," *Journal of Engineering Mechanics*, ASCE, **123**(2), 174-177.
Lin, Y.K. (1967). *Probabilistic Theory of Structural Dynamics*, McGraw-Hill Book Company, New York. Reprinted in 1976 by Krieger Publishing Co., Melbourne, FL.
Lutes, L.D. (1986). "State space analysis of stochastic response cumulants," *Probabilistic Engineering Mechanics*, **1**(2), 94-98.
Lutes, L.D. and Chen, D.C-K. (1992). "Stochastic response moments for linear systems," *Probabilistic Engineering Mechanics*, **7**(3), 165-173.
Lutes, L.D. and Sarkani, S. (1997). *Stochastic Analysis of Structural and Mechanical Vibrations*, Prentice-Hall, Englewood Cliffs, NJ..
Papadimitriou, C. and Lutes L.D. (1994). "Approximate analysis of higher cumulants for MDF random vibration," *Probabilistic Engineering Mechanics*, **9**(1), 71-82.
Papadimitriou, C. and Lutes L.D. (1996). "Stochastic cumulant analysis of MDOF systems with polynomial-type nonlinearities," *Probabilistic Engineering Mechanics*, **11**(1), 1-13.
Papadimitriou, C. Katafygiotis, L.S., and Lutes, L.D. (1999). Response cumulants of nonlinear systems subject to external and multiplicative excitations, *Probabilistic Engineering Mechanics*, in press.
Spanos, P-T.D. (1983). "Spectral moments calculation of linear system output," *Journal of Applied Mechanics*, ASME, **50**, 901-903.
Stratonovich, R.L. (1963). *Topics in the Theory of Random Noise*. Translated from the Russian by Silverman, R.A., Gordon and Breach, New York.
Wojtkiewicz, S.F., Spencer, B.F., and Bergman, L.A. (1996). "On the cumulant-neglect closure method in stochastic dynamics," *International Journal of Non-Linear Mechanics*, **31**(5), 657-684.
Wu, W.F. and Lin, Y.K. (1984). "Cumulant-neglect closure for non-linear oscillators under random parametric and external excitations," *International Journal of Non-Linear Mechanics*, **19**, 349-362.

NONLINEAR DYNAMICS OF BEAMS WITH STOCHASTIC PARAMETER VARIATIONS

by

C S Manohar & Sayan Gupta
Department of Civil Engineering
Indian Institute of Science
Bangalore 560012 India.

1.0 Introduction

The aim of this paper is to investigate the steady state response of beams under the action of random support motions. The study is of relevance in the context of earthquake response of extended land based structures such as pipelines and long span bridges, and, secondary systems such as piping networks in nuclear power plant installations. The following complicating features are accounted for in the response analysis: (a) differential support motions: this is characterized in terms of cross power spectral density functions associated with distinct support motions, (b) nonlinear support conditions, and (c) stochastically inhomogeneous stiffness and mass variations of the beam structure; questions on non-Gaussian models for these variations are considered. The method of stochastic finite elements is combined with equivalent linearization technique and Monte Carlo simulations to obtain response moments.

2.0 Background to the Literature

Analysis of structures with randomly inhomogeneous properties has been widely studied in recent years: see the state of the art report edited by Schueller (1997) and the recent review by Manohar and Ibrahim (1999) on problems of structural dynamics with stochastic parameter variations. A majority of the existing studies have addressed linear structural behavior. A few studies consider the multi-presence of system nonlinearity and stochasticity. Static problems of this genre were studied by Liu *et al* (1986, 88). Liu *et al.* (1987) considered the dynamic response to a step input of an elasto-plastic beam with isotropic hardening whose plastic modulus was modeled as a Gaussian random field. A mean-centered second-order perturbation method in conjunction with direct numerical integration in time was used to study the time evolution of response moments. The transient response of a transversely loaded stochastically inhomogeneous plate on a random nonlinear elastic foundation was studied by Deodatis and Shinozuka (1988). They used Monte Carlo simulation in conjunction with finite element discretization and time integration techniques. Chang and Yang (1991) considered large amplitude vibration of a beam with randomly varying material and geometric properties. They analyzed the free and forced response to harmonic and random

S. Narayanan and R.N. Iyengar (eds.),
IUTAM Symposium on Nonlinearity and Stochastic Structural Dynamics, 133–145.

excitations. Their analysis involved discretization of the random fields using the method of local averages and a second order perturbation scheme. The forced nonlinear random vibration response was obtained using the equivalent linearization technique. Koyluoglu *et al.*, (1995) employed weighted integral method of random field discretization in the development of a nonlinear stochastic finite element formulation for stochastic plane frame analysis. The stiffness and damping properties were taken to be random in nature, and the excitations were modeled as stationary random processes. They employed mean centered second order perturbation method to treat system uncertainties and the Gaussian closure method to handle the nonlinearities. Klosner *et al.* (1992) considered the response of nonlinear sdof and two-dof systems with stochastic parameters under white noise excitation. For more general classes of problems possessing no exact solutions, Klosner *et al.* employed a statistical linearization technique to determine the conditioned response statistics. It should be noted that statistical linearization gives satisfactory results only if the system does not involve secular terms which give rise to internal resonance conditions. Recently, Manohar and Adhikari (1997) and Adhikari and Manohar (1999) have studied dynamics of linear beams with randomly inhomogeneous system properties. They have employed frequency dependent shape functions to discretize both the system property random fields and the displacement fields. This lead to the response characterization in frequency domain in terms of stochastic dynamic stiffness matrices. In the present study, we extend these earlier studies to study the dynamics of a randomly inhomogeneous Euler-Bernoulli beam which is supported on two cubic nonlinear springs and which is subjected to spatially varying random support motions. We employ the method of equivalent linearization in conjunction with stochastic dynamic stiffness matrix method to obtain approximations to the steady state response. Results on response moments are presented and compared with corresponding results from Monte Carlo simulations. Issues related to non-Gaussian models for the beam properties are also addressed.

3.0 Problem Statement

An Euler-Bernoulli beam supported on two discrete springs and which is subjected to support motions $u(t)$ and $v(t)$ is shown in figure 1. The springs are taken to be of the Duffing type, that is, they possess a cubic force deflection characteristics. The support motions $u(t)$ and $v(t)$ are taken to constitute a vector of zero mean, stationary, Gaussian random processes with the 2×2 PSD matrix denoted by $S(\omega)$. The governing equations of motion for this system is given by

$$[EI(x)\ w''(x,t)]'' + m\ddot{w}(x,t) + c\dot{w}(x,t) = 0 \tag{1}$$

with boundary conditions

$$[EI\ w'']'(0,t) = -k_1[w(0,t) - u] - \alpha[w(0,t) - u]^3$$

$$[EI\ w'']'(L,t) = k_2[w(L,t) - v] + \alpha[w(L,t) - v]^3$$

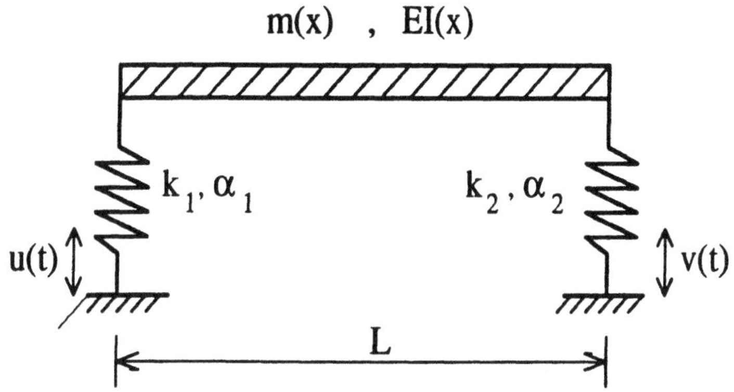

Figure 1. Euler-Bernoulli beam on nonlinear springs

$$w''(0,t) = 0 \quad w''(L,t) = 0 \tag{2}$$

and initial conditions

$$w(x,0) = 0 \quad \dot{w}(x,0) = 0 \tag{3}$$

Here, a prime denotes differentiation with respect to spatial variable x, a dot denotes derivative with respect to time t, $EI(x)$= flexural rigidity, $m(x)$=mass per unit length, c=viscous damping coefficient, k_1, k_2= linear spring rates and α_1, α_2= nonlinear spring rates. The quantities $EI(x)$ and $m(x)$ in this study are modeled as jointly homogeneous random fields and are taken to have the following form

$$EI(x) = EI_0\left[1 + \epsilon_1 f_1(x)\right]; \qquad m(x) = m_0\left[1 + \epsilon_2 f_2(x)\right] \tag{4}$$

Here 0 subscript indicates the mean values, $0 < \epsilon_i << 1$ (i=1,2) are deterministic constants and the random fields $f_i(x)$ are taken to have zero mean, unit standard deviation and covariance $R_{ij}(\xi)$. The following additional restrictions are taken to apply on the random fields $f_i(x)$: (a) $f_i(x)$ (i=1,2) are meansquare bounded; (b) $f_1(x)$ is twice differentiable in a meansquare sense; this requires that $\frac{\partial^4 R_{11}(x_1,x_2)}{\partial x_1^2 \partial x_2^2}$ must exist for all x_1 and x_2 in the interval $(0, L)$ and (c) for a specified deterministic function $g(x)$, which is bounded and continuous in $(0, L)$, integrals of the type $\int_0^L g(x) f_i(x) \mathrm{d}x$ exist in a meansquare sense; this requires that $\int_0^L \int_0^L |g(x_1) g(x_2) R_{ii}(x_1, x_2)| \mathrm{d}x_1 \mathrm{d}x_2 < \infty \quad (i = 1,2)$. The first two conditions ensure that the sample realizations of the beam have sufficiently smooth behavior so that the various stress resultants and boundary conditions (such as those at a free edge) are satisfactorily described. The third condition is needed in the development of the procedure used in this study. It may be noted that the above governing equation of motion constitutes a nonlinear partial differential equation with randomly time varying boundary conditions and randomly inhomogeneous coefficients. The field equation here is linear but the nonlinearity enters through the boundary conditions. Exact solutions to this type of equations are not currently obtainable and one has to resort to digital simulations or to approximate analysis procedures. In this study, the concept of frequency dependent weighted integrals as developed by Manohar and Adhikari (1997) is comboned with the well known method of equivalent linearization to obtain estimates of steady state response.

4.0 Equivalent Linearization

In this study, first, an approximate solution to equations 1-3 using the well known equivalent linearization technique (Socha and Soong 1991) is constructed. For this purpose the cubic nonlinear springs are replaced by equivalent linear springs with equivalent spring stiffnesses β_1 and β_2 for the left and the right springs respectively. The field equation governing the behavior of the equivalent linear system is identical to equation (1). However the boundary conditions now get linearized and are modified to read

$$\begin{aligned} EI\ w'''(0,t) &= -\beta_1\ [w(0,t)-u]; \quad & EI\ w'''(L,t) &= \beta_2\ [w(L,t)-v]; \\ w''(0,t) &= 0; & w''(L,t) &= 0. \end{aligned} \tag{5}$$

To determine the equivalent linear spring rates β_1 and β_2 it is demanded that the average steady state potential energy stored in the nonlinear system and the linearized system are the same. That is

$$\lim_{t\to\infty} < E^l_{NL}(t) >=< E^l_L(t) >; \quad \lim_{t\to\infty} < E^r_{NL}(t) >=< E^r_L(t) > \tag{6}$$

where, $E_{NL}(t)$ and $E_L(t)$ are the potential energies stored in the nonlinear and equivalent linear spring respectively. The superscripts l and r denote respectively the left and right springs. These quantities are given by

$$\begin{aligned} < E^l_{NL} > &= \tfrac{k_1}{2} < (w(0,t)-u)^2 > + \tfrac{\alpha_1}{4} < (w(0,t)-u)^4 >; \\ < E^r_{NL} > &= \tfrac{k_2}{2} < (w(L,t)-v)^2 > + \tfrac{\alpha_2}{4} < (w(0,t)-v)^4 >; \\ < E^l_L > &= \tfrac{\beta_1}{2} < (w(0,t)-u)^2 >; \\ < E^r_L > &= \tfrac{\beta_2}{2} < (w(L,t)-v)^2 > \end{aligned} \tag{7}$$

In these equations $\lim_{t\to\infty}$ has been omitted with an understanding that the expectations in the steady state are being considered. It follows from the above equations that

$$\beta_1 = k_1 + \frac{\alpha_1}{2}\frac{<(w(0,t)-u)^4>}{<(w(0,t)-u)^2>}; \quad \beta_2 = k_2 + \frac{\alpha_2}{2}\frac{<(w(L,t)-v)^4>}{<(w(L,t)-v)^2>} \tag{8}$$

Since the terms appearing on the right hand side of the above equations are themselves functions of β_1 and β_2, it follows that these equations represent a pair of nonlinear transcendental equations for β_1 and β_2. The questions on the determination of these constants will be considered after a scheme of discretizing the random fields $EI(x)$ and $m(x)$ and $w(x,t)$ is outlined. However, at this stage it may be noted that the given nonlinear problem has now been linearized and, hence, it can

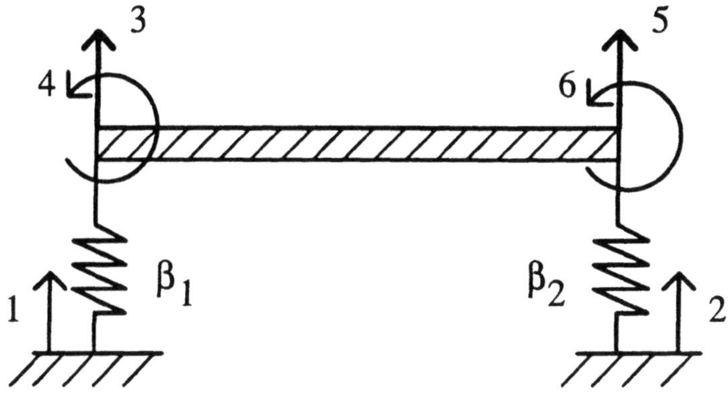

Figure 2. Discretization of equivalent linear beam

now be solved using methods of linear analysis. This reduced problem, is again not amenable for an exact analysis due to the presence of random fields $EI(x)$ and $m(x)$ in the governing field equations. We adopt the direct stochastic dynamic stiffness matrix approach, as developed by Manohar and Adhikari (1997) and Adhikari and Manohar (1999), to proceed further.

5.0 Stochastic dynamic stiffness matrix for the linearized system

After linearization, the governing equation of motion consists of the field equation as given in equation 1 and the linearized boundary conditions as given in equation 5. We find the steady state response of this system by using the direct dynamics stiffness matrix method. For this purpose, the beam structure is discretized as shown in figure 2. The dynamic stiffness matrix for the random beam element has been derived in Appendix A. The dynamic stiffness matrix for a spring element coincides with its static stiffness matrix. Using these facts, the reduced dynamic stiffness matrix for the structure can be shown to be given by

$$[\mathcal{D}^*(\omega)] = \begin{bmatrix} \beta_1 + \mathcal{D}_{11} & \mathcal{D}_{12} & \mathcal{D}_{13} & \mathcal{D}_{14} \\ \mathcal{D}_{21} & \mathcal{D}_{22} & \mathcal{D}_{23} & \mathcal{D}_{24} \\ \mathcal{D}_{31} & \mathcal{D}_{32} & \beta_2 + \mathcal{D}_{33} & \mathcal{D}_{34} \\ \mathcal{D}_{41} & \mathcal{D}_{42} & \mathcal{D}_{43} & \mathcal{D}_{44} \end{bmatrix} \tag{9}$$

Here $\mathcal{D}_{ij}$ refers to the elements of the beam dynamic stiffness matrix derived in Appendix A. The determination of the unknown displacements at nodes 3,4,5 and 6 requires the inversion of the above dynamic stiffness matrix. For this purpose, the method of Neumann's expansion is adopted; see the paper by Shinozuka and Yamazaki (1988) for a discussion on application of Neumann's method for inverting random static stiffness matrices.

6.0 Inversion of the dynamic stiffness matrix

The equilibrium equation in the frequency domain can be shown to be given by

$$\mathcal{D}_0(\omega)\Delta(\omega) = P(\omega) \tag{10}$$

where

$$\Delta(\omega)^T = [\ \delta_3(\omega) \quad \delta_4(\omega) \quad \delta_5(\omega) \quad \delta_6(\omega)\ ]; \quad P^T(\omega) = [\ \beta_1 U(\omega) \quad \beta_2 V(\omega) \quad 0 \quad 0\ ] \tag{11}$$

It may be noted that $\mathcal{D}_0(\omega)$ is a complex valued matrix with $\mathcal{D}_{0ij}(\omega) = \mathcal{D}_{0ji}(\omega)$. For a fixed value of ω, $\mathcal{D}_{0ij}(\omega)$, $(i,j = 1,\cdots,4)$, are functions of the weighted integrals $X_l(\omega)$, $(l = 1,\cdots,10)$. As has been noted in Appendix A, these weighted integrals are themselves complex valued random variables, and, consequently, $\mathcal{D}_{0ij}(\omega)$, $(i,j = 1,\cdots,4)$, are also complex valued random variables. The quantities $U(\omega)$ and $V(\omega)$ are, respectively, the Fourier transforms of the functions $u_0(t)$ and $v_0(t)$ where $u_0(t) = u(t)$, $v_0(t)$ for $0 < t < T_0$ and $u_0(t) = 0$, $v_0(t) = 0$ for $t > T_0$; here $T_0 < \infty$ is a large value of time t. To characterize $\Delta(\omega)$, we need to invert the matrix $\mathcal{D}_0(\omega)$. To do this we write $\mathcal{D}_0(\omega) = \bar{\mathcal{D}}_0(\omega) + \mathcal{D}_{0r}(\omega)$, where $\bar{\mathcal{D}}_0(\omega)$ is the mean component of $\mathcal{D}_0(\omega)$ and $\mathcal{D}_{0r}(\omega)$ is the random component. Following the Neumann expansion rule we get

$$\Delta(\omega) = \Delta_0(\omega) - Q\Delta_0(\omega) + Q^2\Delta_0(\omega) - Q^3\Delta_0(\omega) + \cdots \tag{12}$$

where $Q = [\bar{\mathcal{D}}_0(\omega)][\mathcal{D}_{0r}(\omega)]$ and $\Delta_0 = [\bar{\mathcal{D}}_0(\omega)]^{-1}P$. The matrix of power spectral density functions of the nodal displacements $\delta_i(t)$ $(i = 3,\cdots,6)$, conditioned on the weighted integrals $X_l(\omega)$, denoted by $S_\Delta(\omega)$, can be found by using the definition

$$S_\Delta(\omega) = \lim_{T_0\to\infty} < \Delta(\omega)\Delta^*(\omega) >_c \tag{13}$$

Here * denotes complex conjugation,superscript T is the matrix transpose and the expectation operator $< \cdot >_c$ here refers to the expectation conditioned on the weighted integrals X_l $(l = 1,\cdots,10)$. This would mean that, for a fixed value of ω, the elements of the matrix $S_\Delta(\omega)$ constitute a set of random variables. In the limit of $\epsilon_1 = \epsilon_2 \to 0$, that is, for the limiting case of the beam being deterministic, the above psd matrix represents the response psd due to the random excitations $u(t)$ and $v(t)$ of a homogeneous beam. Combining equation 13 with the Neumann expansion given in equation 12, we can get a series expansion for the elements of the power spectral density matrix $S_\Delta(\omega)$ as function of the weighted integrals X_l $(l = 1,\cdots,10)$. This enables the evaluation of the mean and standard deviation of elements of $S_\Delta(\omega)$.

Remarks

- If a first order approximation given by $\Delta(\omega) = \Delta_0(\omega) - Q\Delta_0(\omega)$ is employed, the mean of $S_\Delta(\omega)$ can be computed given the knowledge of covariance matrix of the weighted integrals. This, in turn, requires description of $f_1(x)$

and $f_2(x)$ up to their respective covariance functions. The computation of variance of $S_\Delta(\omega)$, however, requires the knowledge of moments of the weighted integrals up to fourth order which in turn requires the knowledge of higher order moments of $f_1(x)$ and $f_2(x)$ up to order four.

- If higher order terms are also retained in the expansion $\Delta(\omega) = \Delta_0(\omega) - Q\Delta_0(\omega) + Q^2\Delta_0(\omega) - Q^3\Delta_0(\omega) + \cdots$, the computation of even the mean of $S_\Delta(\omega)$ requires the knowledge of higher order moments of the weighted integrals (and hence of the random fields $f_1(x)$ and $f_2(x)$).

- It may be emphasized that the computation of moments of $S_\Delta(\omega)$ does not require any specific assumption to be made on the probability distribution of the random fields $f_1(x)$ and $f_2(x)$. However the knowledge of higher order moments of $f_1(x)$ and $f_2(x)$ become essential. If one makes an assumption that $f_1(x)$ and $f_2(x)$ are Gaussian random fields, the calculations get significantly simplified. This assumption, however, is not *essential* to apply the procedure described.

7.0 Evaluation of the Equivalent linear parameters

As has been already noted, equation 8 represents a pair of transcendental equations for the parameters β_1 and β_2. To determine these parameters we interpret the expectations appearing in equation 8 as being conditioned on the weighted integrals X_l $(l = 1, \cdots, 10)$. Since the response of the equivalent linearized system, conditioned on X_l $(l = 1, \cdots, 10)$, to the Gaussian excitations $u(t)$ and $v(t)$ is Gaussian, we can simplify equations 8 to read

$$\beta_1 = k_1 + \tfrac{3\alpha_1}{2} < (w(0,t) - u)^2 >_c; \quad \beta_2 = k_2 + \tfrac{3\alpha_2}{2} < (w(L,t) - v)^2 >_c \qquad (14)$$

The determination of β_1 and β_2 is carried out by using an iterative scheme. This consists of the following steps: (a) As a first approximation we take $\beta_1 = k_1$ and $\beta_2 = k_2$. (b) Invert the dynamic stiffness matrix and evaluate $S_{\Delta_{11}}(\omega)$. (c) Evaluate the mean and standard deviation of the conditional response variances σ_{c1}^2 and σ_{c2}^2 given respectively by the area under functions $S_{\Delta_{11}}(\omega)$ and $S_{\Delta_{22}}(\omega)$; these two quantities correspond respectively to the nodal displacements $\delta_3(t)$ and $\delta_5(t)$. (d) Improve upon the estimate of β_1 and β_2 by using in equation 14 the estimates of $< (w(0,t) - u)^2 >$ and $< (w(0,t) - v)^2 >$ using $S_\Delta(\omega)$ in step (b). (e) Repeat the steps above till the estimates of mean and standard deviation of σ_{c1}^2 and σ_{c2}^2 converge.

8.0 Numerical Results

For the purpose of illustration we consider a beam with $l = 8$ m, $EI = 7.538e05$ N m^2, $m = 29.88$ kg/m, $c = 63$ Ns/m, $k_1 = k_2 = 35335$ N/m and $\alpha_1 = \alpha_2 =$1e6 N/m^3. In the absence of spring nonlinearities, the first few natural frequencies are respectively 3, 24, 41, 78 and 202 rad/s. The auto-PSD functions $S_{uu}(\omega)$ and $S_{vv}(\omega)$ for the support accelerations $\ddot{u}(t)$ and $\ddot{v}(t)$ are taken to be given, respectively, by

$$S_{uu}(\omega) = S_o\{\frac{(1+4\eta_{g1}^2(\frac{\omega}{\omega_{g1}})^2)}{(1-(\frac{\omega}{\omega_{g1}})^2)^2+(4\eta_{g1}^2(\frac{\omega}{\omega_{g1}})^2)}\}\{\frac{(\frac{\omega}{\omega_f})^4}{(1-(\frac{\omega}{\omega_f})^2)^2+(4\eta_f^2(\frac{\omega}{\omega_f})^2}\} \quad (15)$$

$$S_{vv}(\omega) = S_o\{\frac{(1+4\eta_{g2}^2(\frac{\omega}{\omega_{g2}})^2)}{(1-(\frac{\omega}{\omega_{g2}})^2)^2)+(4\eta_{g2}^2(\frac{\omega}{\omega_{g2}})^2)}\}\{\frac{(\frac{\omega}{\omega_f})^4}{(1-(\frac{\omega}{\omega_f})^2)^2+(4\eta_f^2(\frac{\omega}{\omega_f})^2}\} \quad (16)$$

In the numerical work the parameters S_o, ω_{g1}, ω_{g2}, η_{g1} and η_{g2} are taken to be 0.01, 20 rad/s, 20 rad/s, 0.6 and 0.4 respectively. The above parameters represent a root mean square acceleration level of input as $0.081g$ and the corresponding zero period acceleration of $0.241g$ for a peak factor of 3.0 where g is acceleration due to gravity. The filter function parameters ω_f and η_f are taken to be 5 rad/s and 0.52 respectively. In the illustrative example considered we take $S_{uv}(\omega)$=0. In the numerical work it was observed that the estimates of β_1 and β_2 converged in about 5 iteration cycles. Figure 4 shows the mean and the standard deviation of psd function of displacement at node 3. In evaluating the standard deviation, the random fields $f_1(x)$ and $f_2(x)$ are taken to be Gaussian. The mean and standard deviation of the conditional variance of displacement at node 3 as a function of excitation parameter ω_g and linear spring stiffness $k = k_1 = k_2$ are shown in figures 5-8. Figures 5-8 also show digital simulation results which have been generated by using a finite element model for system shown in figure 3 with 10 degrees of of freedom. The simulation results are obtained for both Gaussian and non-Gaussian models for $f_1(x)$ and $f_2(x)$. The non-Gaussian model has the same mean and covariance matrix as of the gaussian model but has a first order probability density function which is $exp[-\alpha^2]$. It may be observed that the method outlined in this paper produces results that agree reasonably well with the simulation results. Also, it is seen that the differences in results produced by Gaussian and non-Gaussian models for $f_1(x)$ and $f_2(x)$ are not appreciable.

9.0 Conclusions

A frequency domain approach combined with method of equivalent linearization has been developed to analyze the response of a nonlinearly supported, randomly inhomogeneous beam which is acted upon by differential random support motions. The study employs frequency and damping dependent shape functions to discretize the displacement and the system property random fields. Numerical results illustrating the features of the proposed methodology are presented. Limited comparisons between theoretical results and digital simulation results show reasonably good mutual agreement.

Appendix A : Dynamic stiffness matrix for a random beam element

A beam element subjected to nodal harmonic forces $p_n(t) = P_n \exp[i\omega t]$ which produce nodal harmonic displacements $\delta_n(t) = \Delta_n \exp[i\omega t]$, $(n = 1, \cdots, 4)$ is shown in figure 3. The beam behavior is taken to be linear and, consequently, we take the displacement field to be of the form $w(x,t) = W(x)\exp[i\omega t]$. The dynamic stiffness matrix

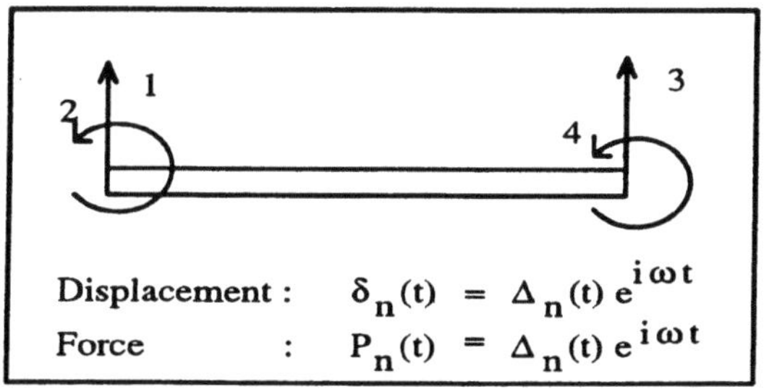

Figure 3. Linear beam element under harmonic loads

$\mathcal{D}(\omega)$ relates the amplitudes of nodal displacements and nodal forces through the relation $P(\omega) = \mathcal{D}(\omega)\Delta(\omega)$ The field equation governing the function $W(x)$ can be shown to be given by

$$\left[EI(x)W''\right]'' + \left[-m(x)\omega^2 + ci\omega\right] W = 0 \tag{17}$$

The derivation of the dynamic stiffness matrix involves solution of the above field equation in conjunction with two separate sets of boundary conditions, one representing the nodal displacement conditions and the other representing the nodal force conditions. In our study we obtain approximate solution to this pair of boundary problems by employing the finite element method. The steps for formulating the dynamic stiffness matrix closely follows the method outlined by Manohar and Adhikari (1997) but with an important difference. In the earlier study we had employed shape functions that were derived from undamped field equations and the influence of damping on dynamic stiffness coefficients was taken into account in an approximate manner. In the present study we refine this procedure by including damping terms into the definition of the shape functions. The following sequence of steps leads to the deduction of $\mathcal{D}(\omega)$.

Step 1 *Derivation of shape functions:*

Obtain the shape functions using the deterministic damped field equation. That is, solve the equation $W'''' - b^4 W = 0$ under a set of 'binary' boundary conditions. The parameter b appearing in this equation is given by $b^4 = \frac{m_0\omega^2 - i\omega c}{EI_0}$. The shape functions $\{\mathbf{N}(x,\omega)\}$ can be shown to be given by $\{\mathbf{N}(x,\omega)\} = [\mathbf{\Gamma}(\omega)]\{\mathbf{s}(x,\omega)\}$, where

$$[\mathbf{\Gamma}(\omega)] = \begin{bmatrix} \frac{1}{2}\frac{cS+Cs}{cC-1} & -\frac{1}{2}\frac{1+sS-cC}{cC-1} & -\frac{1}{2}\frac{cS+Cs}{cC-1} & \frac{1}{2}\frac{cC+sS-1}{cC-1} \\ \frac{1}{2}\frac{cC+sS-1}{b(cC-1)} & \frac{1}{2}\frac{-Cs+cS}{b(cC-1)} & -\frac{1}{2}\frac{1+sS-cC}{b(cC-1)} & -\frac{1}{2}\frac{-Cs+cS}{b(cC-1)} \\ -\frac{1}{2}\frac{S+s}{cC-1} & \frac{1}{2}\frac{C-c}{cC-1} & \frac{1}{2}\frac{S+s}{cC-1} & -\frac{1}{2}\frac{C-c}{cC-1} \\ \frac{1}{2}\frac{C-c}{b(cC-1)} & -\frac{1}{2}\frac{S-s}{b(cC-1)} & -\frac{1}{2}\frac{C-c}{b(cC-1)} & -\frac{1}{2}\frac{S-s}{b(cC-1)} \end{bmatrix} \tag{18}$$

and

$$\{\mathbf{s}(x,\omega)\} = [\sin bx,\ \cos bx,\ \sinh bx,\ \cosh bx]^t \tag{19}$$

is the array of basis functions. Here $C = \cosh bl$, $c = \cos bl$, $S = \sinh bl$ and $s = \sin bl$ and superscript t represents matrix transpose.

Step 2 *Derivation of Element Equation of Motion:*

The displacement field within the element is expressed in the form

$$Y(x,t) = \sum_{j=1}^{4} d_j(t) N_j(x,\omega) \tag{20}$$

Here $d_j(t)$ (j =1,4) are the generalized coordinates representing the nodal displacements. The governing equations for the generalized coordinates $d_j(t)$ can now be obtained by using the condition

$$\frac{\mathrm{d}}{\mathrm{dt}}\left[\frac{\partial \mathcal{L}}{\partial \dot{d}_j}\right] - \frac{\partial \mathcal{L}}{\partial d_j} = 0; \qquad j = 1,4 \tag{21}$$

where $\mathcal{L}(t) = T(t) - V(t)$ is the Lagrangian.

Step 3 *Damped Element Stiffness Matrix:*

Since the motion is harmonic at frequency ω, it follows that $d_j(t) = A_j \exp[\mathrm{i}\omega t]$. Consequently, the damped dynamic stiffness matrix $\mathcal{D}$ can be shown to be given by

$$\mathcal{D}(\omega) = \left[-\omega^2 I_{ij}(\omega) + J_{ij}(\omega)\right]_{(4\times4)} \tag{22}$$

with

$$I_{ij}(\omega) = \int_0^L m(x) N_i(x,\omega) N_j(x,\omega) \mathrm{d}x;$$

$$J_{ij}(\omega) = \int_0^L \left[EI(x)\frac{\mathrm{d}^2}{\mathrm{d}x^2}N_i(x,\omega)\frac{\mathrm{d}^2}{\mathrm{d}x^2}N_j(x,\omega)\right]\mathrm{d}x. \tag{23}$$

Substituting the expression of $m(x)$ and $EI(x)$ from equation (2) in equations (11) and after separating the random and deterministic parts, the damped element dynamic stiffness matrix can be written as

$$\mathcal{D}(\omega) = \bar{\mathcal{D}}(\omega) + \sum_{l=1}^{10}\left[\alpha^l(\omega)\right] X_l(\omega) \tag{24}$$

In the above expression $\bar{\mathcal{D}}(\omega)$ is the deterministic damped element stiffness matrix which can be shown to be given by

$$\bar{\mathcal{D}}(\omega) = \begin{bmatrix} \frac{EI\,(c\,S+C\,s)\,b^3}{-1+c\,C} & -\frac{EI\,s\,b^2\,S}{-1+c\,C} & \frac{EI\,(S+s)\,b^3}{-1+c\,C} & -\frac{EI\,b^2\,(C-c)}{-1+c\,C} \\ -\frac{EI\,s\,b^2\,S}{-1+c\,C} & \frac{EI\,(-C\,s+c\,S)\,b}{-1+c\,C} & \frac{EI\,b^2\,(C-c)}{-1+c\,C} & -\frac{EI\,(S-s)\,b}{-1+c\,C} \\ \frac{EI\,(S+s)\,b^3}{-1+c\,C} & \frac{EI\,b^2\,(C-c)}{-1+c\,C} & -\frac{EI\,(c\,S+C\,s)\,b^3}{-1+c\,C} & \frac{EI\,s\,b^2\,S}{-1+c\,C} \\ -\frac{EI\,b^2\,(C-c)}{-1+c\,C} & -\frac{EI\,(S-s)\,b}{-1+c\,C} & \frac{EI\,s\,b^2\,S}{-1+c\,C} & \frac{EI\,(-C\,s+c\,S)\,b}{-1+c\,C} \end{bmatrix} \tag{25}$$

Furthermore, X_l (l=1,10) are random in nature and are given by

$$X_1 = W_{11};\ X_2 = W_{12};\ X_3 = W_{13};\ X_4 = W_{14};\ X_5 = W_{22};$$

$$X_6 = W_{23};\ X_7 = W_{24};\ X_8 = W_{33};\ X_9 = W_{34};\ X_{10} = W_{44} \tag{26}$$

with

$$W_{kr} = \int_o^L \left[\{-m_0\omega^2\epsilon_1 f_1(x)\}s_k(x)s_r(x) + EI_0\epsilon_2 f_2(x)\frac{\mathrm{d}^2 s_k(x)}{\mathrm{dx}^2}\frac{\mathrm{d}^2 s_r(x)}{\mathrm{dx}^2}\right]\mathrm{d}x. \tag{27}$$

Also, $\left[\alpha^l(\omega)\right]$ (l=1,10) are 4 × 4 symmetric matrices of deterministic functions of ω. The following observations may be mad eon the nature of $X_l(\omega)$: (a) $X_l(\omega)$ ($l = 1,\cdots 10$), are random processes evolving in the frequency parameter ω. Thus, for a fixed value of driving frequency ω, these quantities are random variables and herein they are termed as *Dynamic Weighted Integrals*, since they arise as 'weighted integrals' of the random fields $f_1(x)$ and $f_2(x)$. (b) Since the shape functions are dependent on damping, they are complex valued. This would mean that the weighted integrals $X_l(\omega)$ $(l = 1, \cdots 10)$ are complex valued random quantities. (c) X_l are linear functions of the random fields $f_1(x)$ and $f_2(x)$ and, therefore, if $f_1(x)$ and $f_2(x)$ are modeled as jointly Gaussian random fields, it follows that X_l are also jointly Gaussian. Furthermore, since the dynamic stiffness coefficients are linear function of X_l, it follows that these coefficients in turn are also Gaussian distributed.

References

1. Adhikari, S. and Manohar, C.S. (1999) Dynamical analysis of framed structures with statistical uncertainties, *International Journal of Numerical Methods in Engineering*, **44**, 1157-1178.
2. Chang, C.C. and Yang, H.T.Y. (1991) Random vibration of flexible uncertain beam element, *Journal of Engineering Mechanics, ASCE*, **117**(10) , 2329-2349.
3. Deodatis, G. and Shinozuka, M. (1988) Stochastic FEM analysis of nonlinear dynamic problems, *Stochastic Mechanics* **III**, Shinozuka M (ed), Dept. of Civil Engineering and Operations Res., Princeton University.
4. Klosner, J.M., Haber, S.F. and Voltz,P. (1992) Response of nonlinear systems with parameter uncertainties. *Int. J. Nonlinear Mech.* **27**(4), 547-563.
5. Koyluoglu, H.U., Nielsen, S.R.K. and Cakmak, A.S. (1995) Stochastic dynamics of nonlinear structures with random properties subject to stationary random excitation. *Paper No 133*, Dept. of Building technology and structural engineering, Aalborg University, Denmark.
6. Liu, W.K., Belytschko, T. and Mani, A. (1986) Random field finite elements. *Int. J. Numer. Meth. Eng.* **23**, 1831-1845.
7. Liu, W.K., Belytschko, T. and Mani, A. (1987) Applications of probabilistic finite element methods in elastic/plastic dynamics. *ASME J. Eng. for Indus.* **109**, 2-8.
8. Liu, W.K., Besterfield, G.H. and Belytschko, T. (1988) Variational approach to probabilistic finite elements. *ASCE J. Eng. Mech.* **114**(12), 2115-2133.
9. Manohar, C.S. and Ibrahim, R.A. (1999) Progress in structural dynamics with stochastic parameter variations: 1987 to 1998, *ASME Applied Mechanics Reviews*, **52**(5), 177-197.
10. Manohar, C.S. and Adhikari, S. (1997) Dynamic stiffness of randomly parametered beams, *Probabilistic Engineering Mechanics*, **13**(1), 39-51.
11. Schueller, G.I., (Guest editor), (1997) A state-of-art report on computational stochastic mechanics, *Probabilistic Engineering Mechanics*, **12**(4), 197-321.
12. Socha, L and Soong, T.T. (1991) Linearization in the analysis of nonlinear stochastic systems, *Applied Mechanics Reviews*, **44**, 399-422.
13. Shinozuka, M. and Yamazaki, F. (1988) Stochastic finite element analysis: an introduction, in *Stochastic structural dynamics: Progress in theory and applications*, Edited by Ariaratnam S T, Schueller G I and Elishakoff I, Elsevier Applied Science, London.

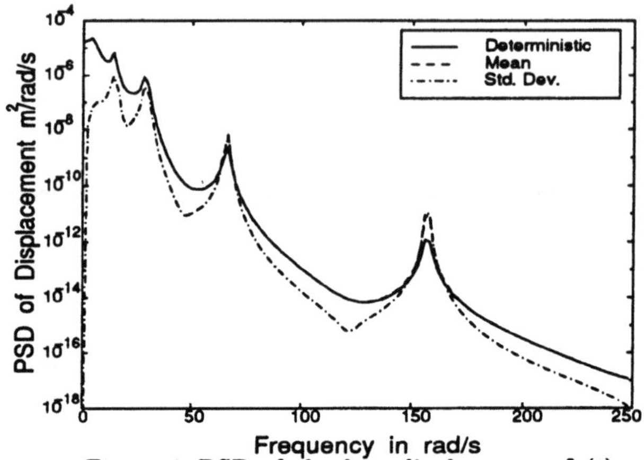

Figure 4. PSD of absolute displacement $\delta_3(t)$.

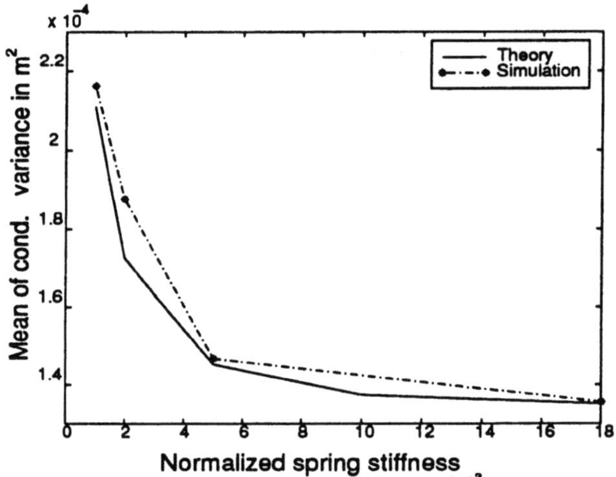

Figure 5. Mean of σ_{cl}^2 Vs $\frac{k_1 l^3}{12EI}$.

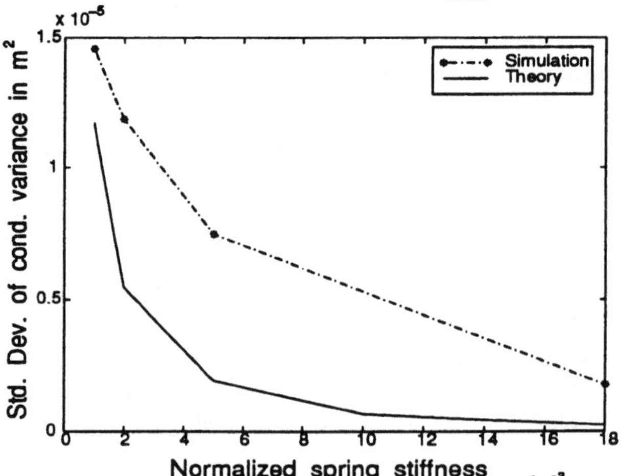

Figure 6. Standard deviation of σ_{cl}^2 Vs $\frac{k_1 l^3}{12EI}$.

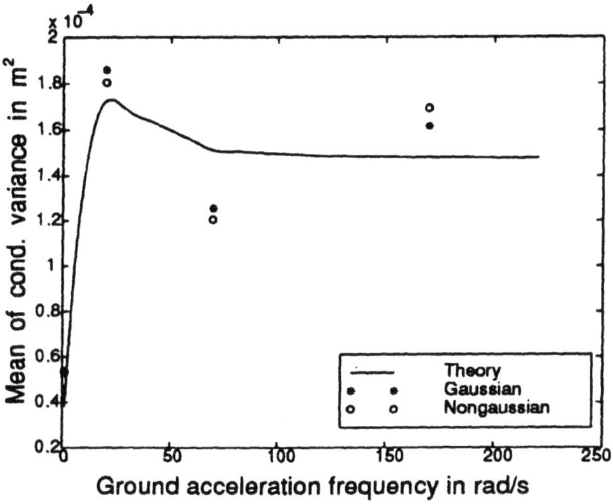

Figure 7. Mean of σ_{cl}^2 Vs ω_g^2 .

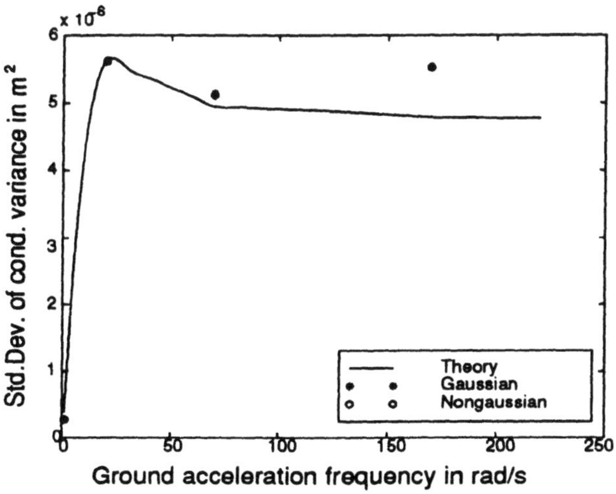

Figure 8. Standard deviation of σ_{cl}^2 Vs ω_g^2 .

NUMERICAL SOLUTION OF THE STATIONARY FPK EQUATION FOR A NONLINEAR OSCILLATOR

S. McWILLIAM
Dynamics Group
School of Mechanical, Materials, Manufacturing Engineering and Management
University of Nottingham
Nottingham, UK

1. Introduction

Many situations occur in engineering in which the loading on structures and the resulting response are random. In the design of these structures it is important that the maximum stresses and fatigue life can be predicted and, for this purpose, it is necessary to employ probabilistic techniques to characterise the statistical properties of the response. For the situation when the system is linear and the excitation is Gaussian, the response statistics are well-known [1]. However, most engineering structures are nonlinear to some extent and these nonlinearities can have a significant influence on the response statistics of the system. This is especially the case at the "tails" of the response distribution which are used to predict the probability of failure.

For linear and non-linear systems subjected to white-noise excitation it is well-known that the state space vector associated with the response is a Markov vector and the probability density function (pdf) is governed by the Fokker-Planck-Kolmogorov (FPK) equation [1]. For other systems, a FPK equation can be written provided that the response is a higher order Markov process. In both these cases the problem of predicting the response statistics reduces to solving the FPK equation for the probability density function. Although there has been some success in finding exact solutions [2], it has been necessary to adopt approximate solution procedures, such as "stochastic averaging" (see [3]), for some practical applications. Alternatively, a numerical solution to the FPK equation can be sought. It is this final class of solution which is considered in the present work.

The most commonly used numerical methods are the Galerkin technique [4], the Finite Element method [5] and the path-integral method [6]. Detailed reviews of each of these numerical methods can be found in the papers by Dunne and Ghanbari [7] and Schueller [8]. Here, a novel approach to calculating the stationary solution to the FPK equation for a non-linear oscillator is presented. It is based on the idea that the probability density function (pdf) can be expressed as a convolution integral involving the Dirac delta function. Provided that an analytic, differentiable representation of the Dirac delta function is available, the FPK equation can be re-written as an integral equation. A

S. Narayanan and R.N. Iyengar (eds.),
IUTAM Symposium on Nonlinearity and Stochastic Structural Dynamics, 147–158.

numerical solution of this equation is obtained by discretising the response domain and numerically solving the resulting linear algebraic equations. The pdf of the response is then obtained from its convolution representation. Although developed independently, the approach is essentially identical to the Distributed Approximating Function (DAF) approach developed by Kouri and Hoffman [9,10], which has been used recently to solve the transient FPK equation for a 1-D system [11]. From this respect, the work reported here can be viewed to be a generalisation of their work to a 2-D FPK equation.

2. Equation of motion and FPK equation

Consider the non-linear system with the following equation of motion:

$$\dot{x}_1 = x_2, \quad \dot{x}_2 = -h(x_1,x_2) + f(t), \tag{1}$$

where x_1 is the displacement and x_2 is the velocity, $h(x_1,x_2)$ is a non-linear function of displacement and velocity, and $f(t)$ is assumed to be a stationary, zero mean, Gaussian white-noise excitation with spectral value S_0.

Many references [1,2] give details of the derivation of the FPK equation for the determination of the transient joint probability density function (jpdf) of non-linear systems subjected to white-noise excitation. Here, the stationary FPK corresponding to the system defined by equation (1) is simply stated as follows:

$$-x_2\frac{\partial p(x_1,x_2)}{\partial x_1} + \frac{\partial h}{\partial x_2}p(x_1,x_2) + h\frac{\partial p(x_1,x_2)}{\partial x_2} + \pi S_0\frac{\partial^2 p(x_1,x_2)}{\partial x_2^2} = 0, \tag{2}$$

where $p(x_1,x_2)$ is the stationary jpdf of the displacement and velocity. In what follows a novel numerical method for solving equation (2) is presented.

3. Convolution based method

The basis of the proposed method is to represent $p(x_1,x_2)$ as a two-dimensional convolution integral as follows:

$$p(x_1,x_2) = \int_{-\infty}^{\infty}\int_{-\infty}^{\infty}\delta(x_1-y_1,x_2-y_2)q(y_1,y_2)dy_1dy_2, \tag{3}$$

where $\delta(z_1,z_2)$ is the two-dimensional Dirac delta function [12], such that:

$$\int_{-\infty}^{\infty}\int_{-\infty}^{\infty}\delta(z_1,z_2)dz_1dz_2=1 \qquad \delta(z_1,z_2)=\begin{cases}0 & \text{when } z_1\neq 0 \text{ and } z_2\neq 0\\ \infty & \text{otherwise}\end{cases} \quad . \qquad (4,5)$$

From the definition of the Dirac delta function and equation (3), q is formally identical to p. However, for clarity of presentation, the notation of equation (3) is retained. Further, it should be noted that the definition of the Dirac delta function provided by equations (4,5) is not suitable for developing a numerical method, and for this reason an approximate representation of the Dirac delta function is used.

To gain some physical insight into the proposed representation it is convenient to discretise the y_1-y_2 domain and re-write the integral terms appearing in equation (3) using summation notation. Using a uniform mesh, equation (3) is re-written as:

$$p(x_1,x_2)=\sum_i\sum_j \delta(x_1-y_{1i},x_2-y_{2j})q(y_{1i},y_{2j})\Delta y_1\Delta y_2 \qquad (6)$$

where Δy_1 and Δy_2 are the uniform spacings of the mesh in the y_1 and y_2 directions respectively, and y_{1i} and y_{2j} are the coordinates of the grid points.

Provided that a suitable approximation to the Dirac delta function is available, it can be seen from equation (6) that an approximate, continuous representation of p can be calculated provided that q is known, where q are the values of p at certain "grid" points. Thus the discretised version of equation (3) can be interpreted as a means of generating an approximate curve using known grid points. It should be emphasised, however, that the resulting approximate representation of p is *not* based upon interpolation between the grid points. Instead, the curve fitting is based upon a weighting of all known grid points, where the weight is greatest for points nearest to the point of interest. This representation forms the basis of the Distributed Approximating Function (DAF) method proposed by Kouri and Hoffman [9,10]. Further details of the approach as applied to the numerical solution of the 1-D transient FPK are given in Zhang et al. [11].

Having represented the jpdf as a convolution integral representation, the intention of the method is to choose an approximate analytic representation of the Dirac delta function, and then devise a solution procedure to calculate q and then p using equation (6).

Given that equation (2) involves spatial derivatives of p with respect x_1 and x_2, it is necessary to differentiate the right hand side of equation (3). From equation (5) it can be seen that the formal definition of the Dirac delta function can not be used in equation (3), since it is not differentiable. For this reason an analytic, continuous, twice differentiable function is used to *approximate* the two-dimensional Dirac delta function; the adopted form will be discussed later. Consequently, the first and second spatial derivatives of $p(x_1,x_2)$ wrt x_i (i=1,2) are obtained from equation (3) as follows:

$$\frac{\partial^m p(x_1,x_2)}{\partial x_i^{\ m}} = \int_{-\infty}^{\infty}\int_{-\infty}^{\infty} \frac{\partial^m \delta(x_1-y_1,x_2-y_2)}{\partial x_i^{\ m}} q(y_1,y_2)dy_1dy_2, \qquad m=1,2. \quad (7)$$

Substituting equation (7) into equation (2), the stationary FPK equation is re-written as:

$$\int_{-\infty}^{\infty}\int_{-\infty}^{\infty}\left[-x_2\frac{\partial\delta(x_1-y_1,x_2-y_2)}{\partial x_1}+\frac{\partial h}{\partial x_2}\delta(x_1-y_1,x_2-y_2)\right.$$
$$\left.+h\frac{\partial\delta(x_1-y_1,x_2-y_2)}{\partial x_2}+\pi S_0\frac{\partial^2\delta(x_1-y_1,x_2-y_2)}{\partial x_2^2}\right]q(y_1,y_2)dy_1dy_2=0. \quad (8)$$

Assuming that $\delta(x_1-y_1,x_2-y_2)$ is a known analytic function, equation (8) represents a two-dimensional integral equation in terms of y_1 and y_2, which is valid for all values of x_1 and x_2. Before considering the numerical solution of this equation it is convenient to consider the representation of δ adopted throughout this work.

A suitable approximation to a Dirac delta function which naturally satisfies the normalisation condition given by equation (4) can be obtained using a weighted Hermite polynomial expansion, such that:

$$\delta(z_1,z_2)=\frac{1}{2\pi\sigma_1\sigma_2}e^{-\frac{1}{2}(\frac{x_1}{\sigma_1})^2-\frac{1}{2}(\frac{x_2}{\sigma_2})^2}\sum_{i=0}^{N}\sum_{j=0}^{N}H_{2i}(0)H_{2j}(0)H_{2i}(\frac{x_1}{\sigma_1})H_{2j}(\frac{x_2}{\sigma_2}),(9)$$

where H_k is the k'th Hermite polynomial, σ_k (k=1,2) are constants (to be decided), and:

$$H_{2k}(0)=1 \qquad k=0$$
$$=\frac{(-1)^k(2k-1)!}{2^{k-1}(k-1)!(2k)!}. \qquad k=1,2,3,... \quad (10)$$

Although in principle $N\to\infty$ in equation (9), a practical numerical scheme requires that N takes a finite value. If $N=0$, $\delta(z_1,z_2)$ is a two-dimensional Gaussian distribution in which z_1 and z_2 are statistically independent and each have zero mean and variance σ_i^2 (i=1,2). The obvious difference between a two-dimensional Gaussian distribution and a Dirac delta function indicates that the accuracy of equation (9) is very much dependent upon the magnitude of the variances, as well as the number of terms in the series. Thus to obtain an accurate representation of a Dirac delta function with only a small to moderate value of N it is necessary to use as small values of variance as possible. (Of course, if

$\sigma_i^2 \to 0$ (i=1,2) then equation (9) will formally tend to a Dirac delta function for all values of N.)

Substituting equation (9) into equation (8) results in an equation of the following form:

$$\int_{-\infty}^{\infty}\int_{-\infty}^{\infty} A(x_1,x_2,y_1,y_2)q(y_1,y_2)dy_1dy_2=0, \qquad (11)$$

which is valid for all values of x_1 and x_2.

In general, σ_i (i=1,2) in equation (9) will have a direct influence on the form of $A(\,)$, where the more desirable, smaller values of σ_i mean that the function becomes more "peaked". The adopted numerical solution of equation (11) is based upon discretising the integration domain into a mesh of uniformly spaced grid points and converting the integrals to summations (see equation (6)). The accuracy and efficiency of this procedure will, of course, be dependent upon the spatial mesh used and how peaked the function A is. There will therefore be some compromise between the values of σ_i used and the spatial mesh of grid points.

A numerical solution of equation (11) is sought by considering only a finite region of the infinite y_1-y_2 domain. This is equivalent to imposing the "boundary condition" that $q(y_1,y_2)=0$ on uniformly spaced grid points outside of this region. The finite region considered is such that $y_{1\min} \leq y_1 \leq y_{1\max}$ and $y_{2\min} \leq y_2 \leq y_{2\max}$, and is divided into a uniform mesh of "elements" so that integral equation (11) can be rewritten using summation notation as follows:

$$\sum_{i=1}^{n_1}\sum_{j=1}^{n_2} A(x_1,x_2,y_{1i},y_{2j})q(y_{1i},y_{2j})\Delta y_1 \Delta y_2=0, \qquad (12)$$

where Δy_1 and Δy_2 are the uniform spacings of the mesh in the y_1 and y_2 directions respectively, and y_{1i} and y_{2j} refer to the co-ordinates of the ij'th grid point along the y_1 and y_2 directions respectively.

Equation (12) is a single equation in terms of the n_1n_2 unknown $q(y_{1i},y_{2j})$ variables. A set of (n_1n_2) equations in these unknowns can be generated by noting that equation (12) is valid for all values of x_1 and x_2. This is achieved by discretising a finite region of the x_1-x_2 domain into a uniform mesh of grid points (in an identical manner to that used for the y_1-y_2 domain) and ensuring that equation (12) is satisfied at each grid point.

Equation (12) can now be re-written as a set of linear algebraic equations such that:

$$\sum_{i=1}^{n_1}\sum_{j=1}^{n_2} A(x_{1k},x_{2l},y_{1i},y_{2j})g(y_{1i},y_{2j})\Delta y_1 \Delta y_2 = 0, \tag{13}$$

where k=1,..,n_1 and l=1,..,n_2 and x_{1k} and x_{2l} are the discretised values of x_1 and x_2. When implementing this discretisation procedure the discretised values of x_1, x_2 ,y_1 and y_2 are selected so that $x_{1i} = y_{1i}$ and x_{2j}=y_{2j} for all i and j, since the chosen representation of the Dirac delta function should ensure that q is identical to p on the same grid points.

Having discretised equation (11) and generated a set of linear algebraic equations it is convenient to express equation (14) using matrix notation such that:

$$\boldsymbol{Aq} = \boldsymbol{0}. \tag{14}$$

All elements in matrix **A** will tend to have non-zero values with some elements being dominant - this is in contrast to the FEM for which the matrix is sparse. The non-trivial solution of equation (14) is obtained by taking into account the normalisation condition, and solving the resulting linear algebraic equations using standard matrix methods.

Recalling equation (3), it follows that $q(x_1,x_2)$ should be equal to $p(x_1,x_2)$. However, as the representation of the Dirac delta function given by equation (9) is approximate, it is necessary to determine p numerically using equation (6).

5. Numerical Examples

The accuracy of the convolution based method to predict the stationary response statistics of a number of non-linear, single degree of freedom systems is considered here, where systems with varying degrees of non-linearity are considered. In all cases the number of terms in the weighted Hermite series defined by equation (9) is set so that N=20; for the cases considered here the choice of N has negligible influence on the accuracy of the results on the graphical scale used [13]. Further, the values of σ_i (i=1,2) used are such that σ_i=2.36Δx_i [11], where Δx_i is determined from knowledge of the finite domain under investigation and the number of grid points n_i.

Two applications of the proposed method are considered in what follows. The first considers two instances of a Duffing oscillator, whilst the second is for a linear system with energy dependent damping. In all cases the calculated jpdf is integrated to give the marginal displacement and velocity pdf's which are more convenient to view. The results are plotted on linear and log-scales so that the accuracy of the proposed method over the main body and "tails" can be examined. The results plotted on log-scales are especially important if the results are to be used in an extreme response calculation.

5.1 APPLICATION 1: DUFFING OSCILLATOR

The first two examples are Duffing oscillators with the following equation of motion:

$$\dot{x}_1 = x_2, \quad \dot{x}_2 = -\alpha_2 x_2 - \alpha_1 x_1 - \gamma x_1^3 + f(t), \tag{15}$$

where $f(t)$ is a zero mean, Gaussian, white-noise excitation with a constant spectral value S_0. For this equation of motion it is known [1] that an exact analytic result exists for the jpdf, where the displacement and velocity are statistically independent. The parameters selected for Example 1 are: $\alpha_1 = 1.0$; $\alpha_1 = 1.0$; $\gamma = 0.5$; $S_0 = 1/\pi$, while those selected for Example 2 are: $\alpha_1 = -0.4$; $\alpha_2 = 1.0$; $\gamma = -0.1$; $S_0 = 1/\pi$.

Example 1 represents a reasonably non-linear system which has mono-modal displacement and velocity pdf's. The stationary marginal distributions for this case are shown in Figures 1 and 2, where the displacement and velocity domains were taken to be: $-3 \le x_1 \le 3$ and $-6 \le x_2 \le 6$, and the grid points are such that $n_1 = n_2 = 31$ in Figure 1 and $n_1 = n_2 = 41$ in Figure 2. In each figure the exact analytic results are compared with those obtained using the proposed technique, the FEM and an equivalent Gaussian distribution. The FEM results are included here, and in Application 2, so that the accuracy of the proposed method can be compared to the Finite Element Method (FEM), with identical grid meshes.

With the exception of the extreme "tails" of the distributions which were calculated as being negative (and therefore do not appear on the log-scale), the proposed method is seen in Figure 1 to give excellent agreement with the exact result with a reasonably course mesh. In contrast, the FEM fails to yield accurate results over the tails of the displacement pdf and main body and tails of the velocity pdf. To overcome the "minor" inaccuracies experienced with the proposed method, an increased number of grid points are used to represent the response domain, and the results are shown in Figure 2. There it can be seen that the proposed method now gives excellent agreement for all values of displacement and velocity, while the FEM gives good agreement over the main body of the distributions, but underestimates the tails.

The second example represents a more non-linear system in which the displacement response shows bi-modal behaviour. Figures 3 and 4 show the displacement and velocity distributions for this case in which $-7 \le x_1 \le 7$ and $-6 \le x_2 \le 6$, and the grid points are such that $n_1 = n_2 = 41$ in Figure 3 and $n_1 = n_2 = 61$ in Figure 4. As for Example 1, the exact analytic results are compared with those obtained using the proposed technique, the FEM and an equivalent Gaussian distribution. For this case both the proposed method and the FEM yield excellent agreement with the exact result over the main body of the distribution in Figure 3. However, although the agreement at the tails is better using the proposed method than the FEM, it should be noted that the predicted values for the FEM and proposed method are negative at the extreme tails. In order to remedy this, the number of grid points are increased, so that $n_1 = n_2 = 61$, and the results are shown in Figure 4. There it can be seen that the proposed method gives excellent agreement with the exact result, even over the tails. Although the performance of the FEM is also seen to improve by increasing the number of elements, the accuracy of the FEM is still not as good as that obtained using the proposed method.

In summary, at least for the Duffing Oscillator systems considered, the proposed method yields numerical results which are in excellent agreement with the exact solution available over the main-body and tails, provided that a sufficient number of elements are used. These results also suggest that the proposed method yields results which are of a higher accuracy than the FEM when identical grid meshes are used, and that a courser mesh may be used with the proposed method to achieve the same level of accuracy as the FEM. Although this may suggest that the proposed method is more efficient than the FEM, it should be recalled that the proposed method yields a matrix equation (equation (14)) with a fully populated matrix, while the FEM yields a sparse matrix. The efficiency of solving these comparable equations will therefore not only be dependent upon the size of the matrix, but will also be dependent upon the storage scheme and solution procedure adopted. In order to provide a consistent comparison of the cpu times required by both methods, NAG routine F04AAF has been used to solve the matrix-vector multiplication problem for the proposed method and the FEM. This routine does *not* incorporate sparse matrix routines which would undoubtedly influence run-time and storage requirements.

Table 1 shows the cpu times for the solution of equation (14) and the corresponding FEM equation for Examples 1 and 2 with different mesh sizes. Also shown is a measure of the accuracy of the displacement pdf, where the measure of error e used is:

$$e = \sum_{i=1}^{n} \left(\frac{p_{exact}(x_i) - p(x_i)}{p_{exact}(x_i)} \right)^2 \Delta x \qquad (16)$$

TABLE 1. Comparison of cpu time and accuracy for Examples 1 and 2 using different meshs.

	CPU time		Accuracy	
Example and mesh size	FEM	Proposed method	FEM	Proposed method
Example 1: 31x31	0.79	16.9	1.18	2.10e-4
Example 1: 41x41	3.10	112.7	0.59	1.44e-5
Example 1: 61x61	24.33	-	0.097	-
Example 2: 41x41	3.17	113.5	1.62	2.20e-1
Example 2: 61x61	21.9	444.0	0.50	4.48e-3

There it can be seen that the cpu time increases dramatically as the mesh size increases, especially for the proposed method, and for matrices of comparable size the FEM results are more efficient. However, the results obtained using the proposed method yield significantly greater accuracy for similar grid meshes. For instance, for Example 1 the proposed method yields results using a 31x31 mesh which are significantly more accurate than those obtained using the FEM with a 61x61 mesh, with a reduced cpu time. However, for Example 2 similar levels of accuracy can be achieved with reduced cpu using the FEM with a 61x61mesh compared to the proposed method with 41x41.

5.2 APPLICATION 2: ENERGY DEPENDENT DAMPED OSCILLATOR

The particular equation of motion to be considered here is given by:

$$\dot{x}_1 = x_2 \qquad \dot{x}_2 = -\beta x_2 - \alpha x_2 (x_1^2 + x_2^2) - x_1 + g^2 f(t), \tag{17}$$

where $f(t)$ is a zero mean, Gaussian, white-noise excitation with a constant spectral value $S_0 = 1/\pi$. For the case considered we let: $\beta = -0.5$, $\alpha = 0.125$ and $g^2 = 0.75$.

The stationary response statistics for this example are calculated using a response domain such that $-4 \le x_1 \le 4$ and $-4 \le x_2 \le 4$, and a grid mesh of points such that $n_1 = n_2 = 31$ and $n_1 = n_2 = 41$. The results for this situation are shown in Figures 5 and 6, where the marginal displacement and velocity pdf's are identical in this case.

From these results it can be seen that the proposed method gives excellent agreement with the exact result, both over the main body of the curve and up to the tails. It should also be noted that the proposed method once again provides superior accuracy compared to the FEM, using fewer elements to achieve higher levels of accuracy.

6. Conclusions

A numerical method for solving the stationary FPK equation has been presented. The accuracy of the method was assessed by comparison with the exact analytic results available for the Duffing Oscillator and a linear oscillator with energy dependent damping. In all cases the proposed method gave excellent results for the marginal distributions provided that an adequate number of elements were used. Further, a comparison between the convolution based method and the Finite Element Method (FEM) was made. For the cases considered, which included bi-modal systems, it was found that the proposed method yielded results which were of a higher accuracy than the FEM, and that fewer "elements" were necessary to achieve the same level of accuracy. A brief investigation into the efficiency of the proposed method compared to the FEM revealed that the FEM was more efficient, even though a greater number of elements than the proposed method were necessary to achieve the same level of accuracy. The main reason for this is that the matrix produced by the proposed method is fully populated, while the corresponding matrix in the FEM is sparse. Efforts to improve the efficiency of the method are currently being pursued by: i) setting those elements of the matrix containing the smallest values to zero, giving the matrix a more appealing "sparse" character; and ii) using other expressions for the Dirac delta function.

7. References

1. Lin, Y.K. *Probabilistic Theory of Structural Dynamics*, McGraw-Hill, New York (1967).
2. Lin, Y.K. and Cai, G.Q. *Probabilistic Structural Dynamics: Advanced theory and applications*, McGraw-Hill, Singapore (1995).
3 Roberts, J.B. A shochastic theory for nonlinear ship rolling in irregular seas, *J. Ship Research*, **26**(4), 229 (1982).
4. Bhandari, R.G. and Sherrer, R.E. Random vibrations in discrete nonlinear dynamical systems, *Journal of Mechanical Engineering Science*, **10**, 168 (1968).
5. Wojtkiewicz, S.F., Bergman, L.A., and Spencer, B.F.Numerical solution of some three-state random vibration problems, *Vibration of nonlinear, random, and time-varying systems*, S.C. Sinha, et al., eds, ASME DE-84-1, 939-947 (1995).
6. Naess, A. and Johnsen, J.M. Response statistics of nonlinear, compliant offshore structures by the path integral solution method, *Probabilistic Engineering Mechanics, 8,* 91 (1993).
7 Dunne, J and Ghanbari, *Journal of Sound and Vibration*, **206**(5), 697-724 (1997).
8. Schueller, G.I., A State-of-the-art report on computational stochastic mechanics, *Probabilistic Engineering Mechanics*, **12**(4), 197-321 (1997).
9. Hoffman, D.K., Nayer, N., Sharafeddin, O.A., and Kouri, D.J. Analytic banded approximation for the discretised free propagator, *J. Phys. Chem.*, 95, 8299 (1991).
10. Hoffman, D.K., and Kouri, D.J. Distributed approximating Function theory: A general, fully quantal approach to wave propagation, *J. Phys. Chem.*, 96, 1179 (1992).
11. Zhang, D.S., Wei, G.W., Kouri, D.J., and Hoffman, D.K. Numerical method for the nonlinear Fokker-Planck equation, *Physical Review E*, **56**, 1 (1997).
12. Davenport, W.B., and Root, W.L. *An introduction to the theory of random signals and noise*, IEEE Press (1987).
13. Knappett, DJ *Response statistics of nonlinear systems*, Internal report, Department of Mechanical Engineering, University of Nottingham, U.K. (1998).

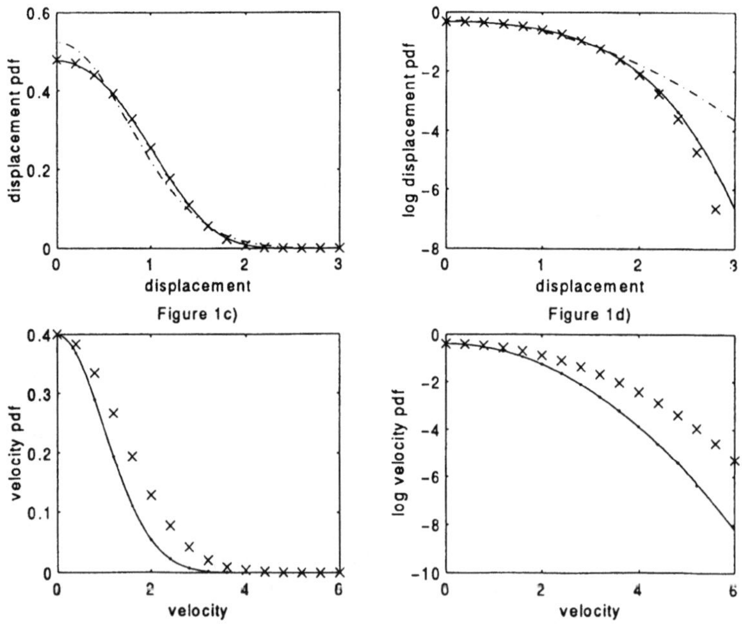

Figure 1: Stationary response statistics for the Duffing Oscillator (equation (15))with α_1=1.0; α_2=1.0; γ=0.5; $S_0=1/\pi$, $n_1=n_2=31$:—, exact solution; • • •, proposed method; ×××, finite element method; - · -, equivalent Gaussian distribution.

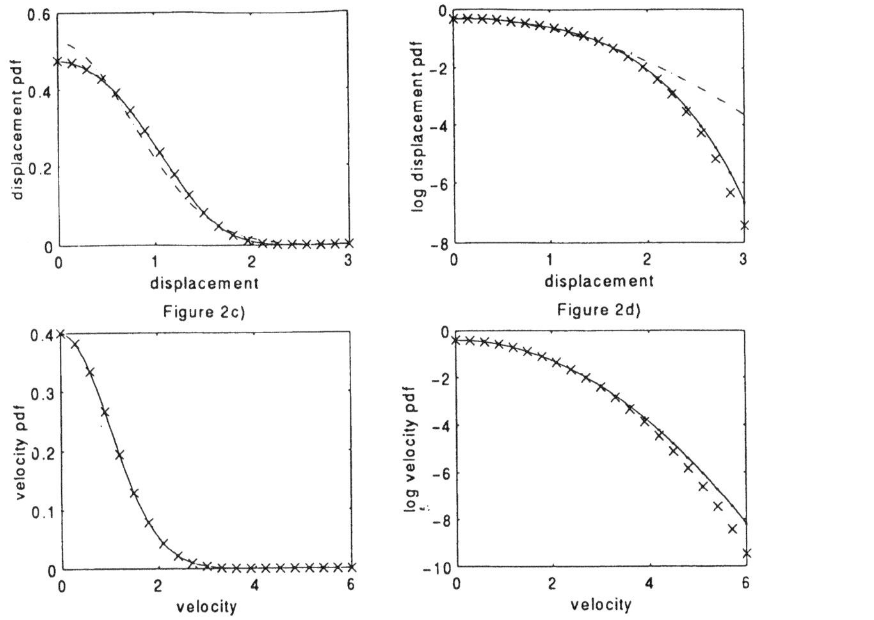

Figure 2: Stationary response statistics for the Duffing Oscillator (equation (16)) with α_1=1.0; α_2=1.0; γ=0.5; $S_0=1/\pi$, $n_1=n_2=21$:—, exact solution; • • •, proposed method; × × ×, finite element method; - · -, equivalent Gaussian distribution.

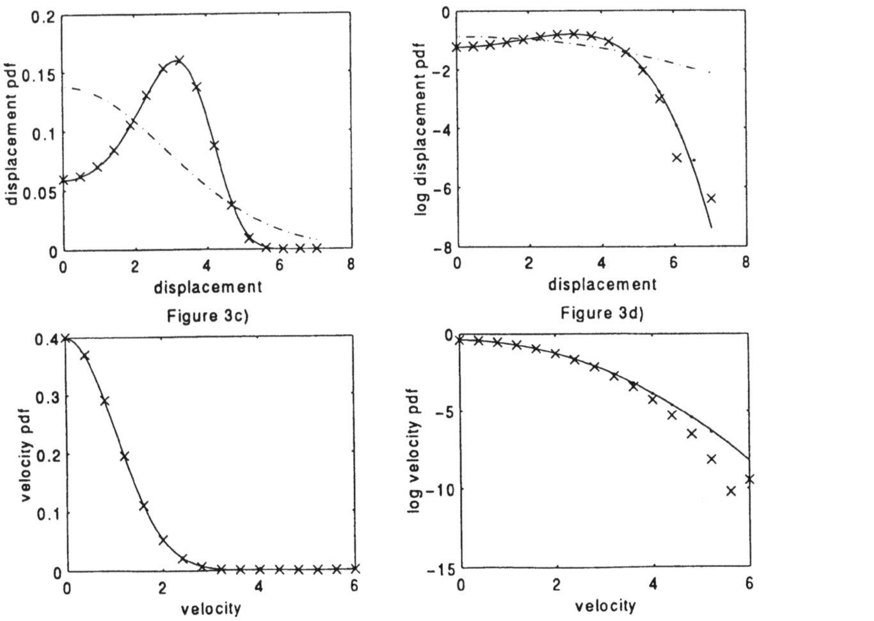

Figure 3: Stationary response statistics for the Duffing Oscillator (equation (16)) with α_1=-0.4; α_2=1.0; γ=-0.1; $S_0=1/\pi$, $n_1=n_2=21$:—, exact solution; • • •, proposed method; × × ×, finite element method; - · -, equivalent Gaussian distribution.

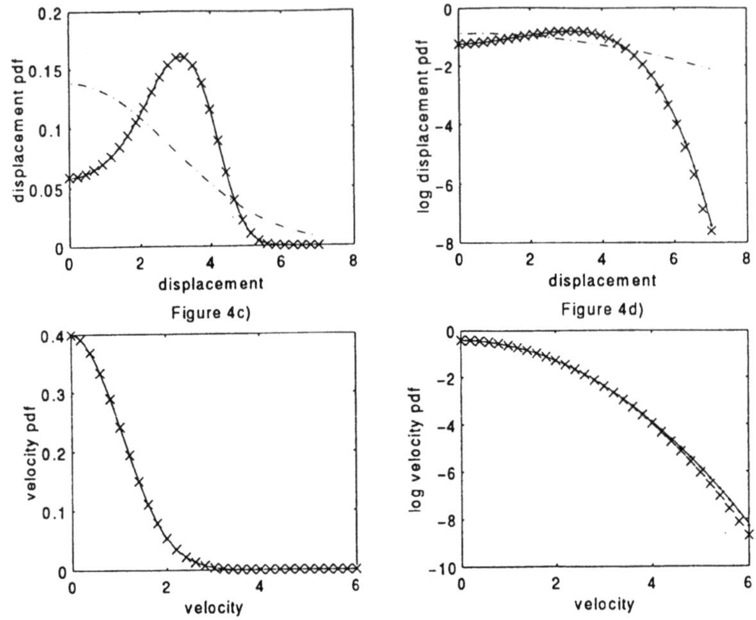

Figure 4: Stationary response statistics for the Duffing Oscillator (equation (16))with α_1=-0.4; α_2=1.0; γ=-0.1; $S_0=1/\pi$, $n_1=n_2=31$: —, exact solution; • • •, proposed method; ×××, finite element method; - · -, equivalent Gaussian distribution.

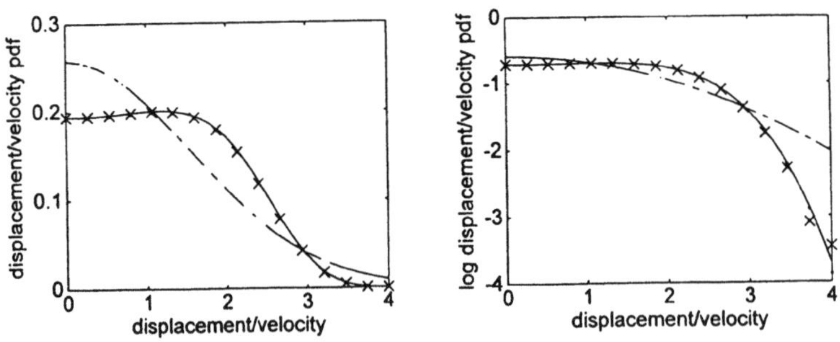

Figure 5: Stationary response statistics for an energy dependent damped oscillator (equation (17)) with $n_1=n_2$ = 31:—, exact solution; • • •, proposed method; - · -, equivalent Gaussian distribution.

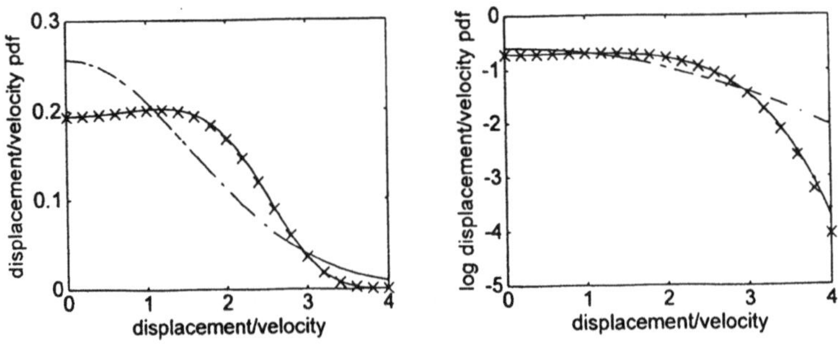

Figure 6: Stationary response statistics for an energy dependent damped oscillator (equation (17)) with $n_1=n_2$ = 41:—; exact solution; • • •, proposed method; - · -, equivalent Gaussian distribution.

PATH INTEGRATION METHODS FOR CALCULATING RESPONSE STATISTICS OF NONLINEAR OSCILLATORS DRIVEN BY α-STABLE LÉVY NOISE

A. NAESS AND C. SKAUG
Department of Structural Engineering
Norwegian University of Science and Technology
R. Birkelands vei 1a, N-7034 Trondheim, Norway

1. Introduction

Systematic exploration and development of the path integration (PI) method as a numerical tool for the solution of stochastic differential equations (SDE) has been going on for more than a decade. Among the first serious efforts in this direction is the work by Wehner & Wolfer (1983; 1983a; 1987). Later, further developments have been published by Naess & Johnsen (1992; 1993), Yu, Cai & Lin (1996) and Naess & Moe (1997).

In this previous work the focus has been on calculating the probability density function (PDF) of the state space vector of nonlinear oscillators driven by stationary Gaussian white noise. Due to an increasing interest in exploring the use of α-stable stochastic processes for modelling physical processes and engineering systems, see Grigoriu (1995), the purpose of the work reported here is to investigate the potential of the PI technique for calculating the PDF of the state space vector of various nonlinear oscillators driven by α-stable Lévy noise. Initial efforts in this direction have been described by Grigoriu (1996) and Naess & Skaug (1998).

The equation of motion of a nonlinear oscillator to be studied is approximated by an Itô stochastic differential equation (SDE) of the following form

$$dY_t = a(Y_t)\,dt + b(Y_t)\,dL_t^{(\alpha)} \tag{1}$$

where Y_t = the (n-dimensional) state space vector process, and $L_t^{(\alpha)}$, $0 < \alpha \leq 2$, is a scalar α-stable Lévy motion process, cf. Samorodnitsky & Taqqu (1994) or Grigoriu (1995). When $\alpha = 2.0$, $L_t^{(\alpha)}$ becomes a Brownian motion.

S. Narayanan and R.N. Iyengar (eds.),
IUTAM Symposium on Nonlinearity and Stochastic Structural Dynamics, 159–169.

$a(Y_t)$ and $b(Y_t)$ are suitable n-dimensional vector functions describing the nonlinear oscillator.

The PI method is based on the property that the state space vector, Y_t, obtained as a solution of an SDE driven by an α-stable Lévy process is a Markov vector process. This fact derives from the basic property that an α-stable Lévy process has independent increments. This makes it possible to use a time stepping procedure to produce the joint PDF $p(y,t)$ of Y_t as a function of time t by exploiting the total probability law

$$p(y,t) = \int_{-\infty}^{\infty} p(y,t|y',t')\, p(y',t')\, dy' \tag{2}$$

As will be shown, for small values of $t-t'$, the TPD $p(y,t|y',t')$ can always be calculated accurately for the nonlinear oscillators under study. Hence, if an initial PDF, $p_0(y) = p(y,t=0)$ say, is given, then equation (2) can be invoked repeatedly to produce the time evolution of $p(y,t)$. If the SDE has an invariant measure, that is, there exists a stationary PDF, $p_s(y)$ say, then eventually $p(y,t)$ will approach this stationary PDF. The number of times equation (2) has to be repeatedly used to reach the stationary situation, depends, of course, on the dynamic system and on the specified initial PDF $p_0(y)$.

In the previous work by Naess and coworkers mentioned above, it was demonstrated how the PI method applied to nonlinear oscillators driven by Gaussian white noise processes could be used to calculate the joint PDF of the state space vector with remarkable accuracy also in the tail regions of the PDFs. In this paper it is demonstrated that the PI method can be implemented numerically so that this feature is preserved for α-stable Lévy noise with an α-parameter equal to 1.0 and 1.5.

For the sake of easy reference, a brief overview over some of the basic features of α-stable Lévy processes that is of relevance here, will be given in the next section.

2. α-STABLE LÉVY PROCESSES

A (real) random variable X is said to have a stable distribution if there are parameters $0 < \alpha \leq 2$, $\sigma > 0$, $-1 \leq \beta \leq 1$, and μ real such that its characteristic function $\psi(\theta) = E[\exp i\theta X]$ assumes the following form (Samorodnitsky & Taqqu, 1994).

If $\alpha \neq 1$:

$$\psi(\theta) = \exp\{-\sigma^\alpha |\theta|^\alpha (1 - i\beta \text{sign}\theta \tan(\pi\alpha/2)) + i\mu\theta\} \tag{3}$$

If $\alpha = 1$:

$$\psi(\theta) = \exp\{-\sigma|\theta|(1 + i\beta(2/\pi)\text{sign}\theta \ln|\theta|) + i\mu\theta\} \tag{4}$$

Here signθ = 1 if $\theta > 0$, = 0 if $\theta = 0$, = -1 if $\theta < 0$. α = the stability parameter, σ = the scale parameter, β = the skewness parameter, and μ = the shift parameter. Since the distribution of X is characterized by these four parameters, the notation $X \sim S_\alpha(\sigma, \beta, \mu)$ is often adopted to denote the situation that X is a stable variable with a specified set of parameters. When X is symmetric α-stable, that is $\beta = \mu = 0$, we shall write $X \sim S\alpha S$.

From equation (2) it follows that the stable distribution is Gaussian when $\alpha = 2$ ($\beta = 0$). Then σ is proportional to the standard deviation, and μ equals the mean value. When $\alpha < 2$, the tails of the probability distributions decay like a power function. Specifically, $\lim_{x\to\infty} x^\alpha Prob\{X > x\} = C_\alpha \sigma^\alpha (1+\beta)/2$ and $\lim_{x\to\infty} x^\alpha Prob\{X < -x\} = C_\alpha \sigma^\alpha (1-\beta)/2$, where C_α is a constant (Samorodnitsky & Taqqu, 1994). Hence it follows that no finite variance exists when $\alpha < 2$. Such distributions are frequently referred to as having heavy tails. Further, a finite mean value is obtained only if $\alpha > 1$. Such distributions therefore leads to much more variability than a Gaussian distribution. This effect will be demonstrated below.

It may also be noted that the probability density functions (PDF) of α-stable random variables exist and are continuous, but unknown in closed form with few exceptions. When $\alpha < 2$ and $\beta = 0$, only one nontrivial case is known in closed form, viz. the Cauchy distribution $S_1(\sigma, 0, \mu)$. If $X \sim S_1(1, 0, 0)$, then its PDF is given as follows

$$p(x) = \frac{1}{\pi(x^2 + 1)} \tag{5}$$

In this paper we shall also investigate the case when $\alpha = 1.5$. A numerical procedure for calculating the PDF is then required. Here we shall use the following result (Samorodnitsky & Taqqu, 1994): $X \sim S_\alpha(1, 0, 0)$ if

$$X = \frac{\sin(\alpha\Pi)}{(\cos\Pi)^{1/\alpha}} \left(\frac{\cos((1-\alpha)\Pi)}{W}\right)^{(1-\alpha)/\alpha} \tag{6}$$

where Π is a random variable uniformly distributed on $(-\pi/2, \pi/2)$ and W is an exponetial random variable with mean 1. The FORTRAN program *rstable* (Samorodnitsky & Taqqu, 1994) is used to generate a very large sample of the random variable $X \sim S_{1.5}(1, 0, 0)$ which is then used to determine the PDF with desired accuracy. Figure 1 shows the PDF used for the calculations to be discussed below.

To be able to change the parameters of the PDFs, the following property will be used later. If $X \sim S_\alpha(1, 0, 0)$, and $\sigma > 0$ and μ are real constants, then $\sigma X + \mu \sim S_\alpha(\sigma, 0, \mu)$.

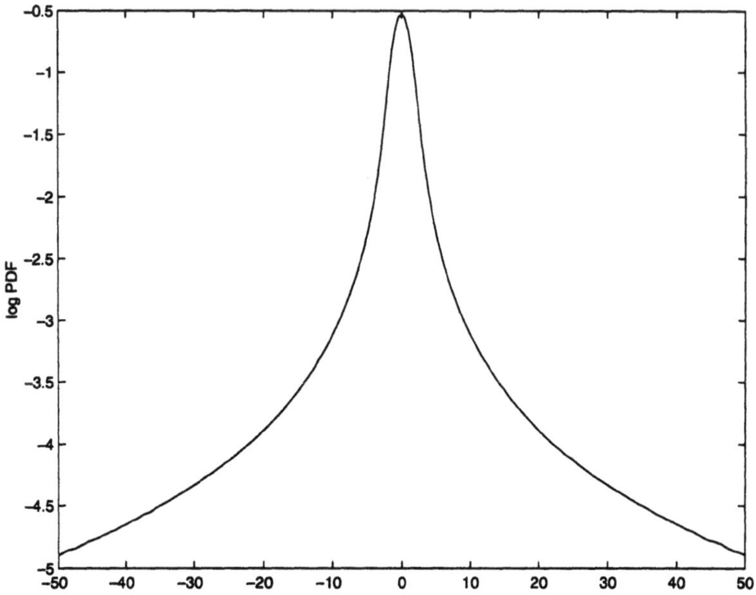

Figure 1. The PDF of $X \sim S_{1.5}(1, 0, 0)$

The stochastic process $L_t^{(\alpha)}$ $(t \geq 0)$ is a (symmetric) α-stable Lévy motion process if 1) $L_0^{(\alpha)} = 0$ (a.s.), 2) $L_t^{(\alpha)}$ has independent increments, 3) $L_t^{(\alpha)} - L_s^{(\alpha)} \sim S_\alpha((t-s)^{1/\alpha}, 0, 0)$ for any $0 \leq s < t < \infty$ and $0 < \alpha \leq 2$.

It follows that such a process has stationary increments, and it becomes a Brownian motion for $\alpha = 2$. Also, for fixed t, $L_t^{(\alpha)} \sim S\alpha S$.

In complete analogy with the formal definition of a Gaussian white noise as the time derivative of a Brownian motion, one can introduce the concept of α-stable Lévy white noise, denoted by $N_t^{(\alpha)}$. Its formal definition is then $N_t^{(\alpha)} = dL_t^{(\alpha)}/dt$. The special case of the 1-stable Lévy white noise will also be referred to as Cauchy white noise.

The distinct difference between a Brownian motion $L_t^{(2)}$ on the one hand and the Cauchy motion $L_t^{(1)}$ and $L_t^{(1.5)}$, on the other, is displayed in Figure 2, which shows part of a realization of each of these processes. In contrast to the Brownian motion, a typical feature of the two other processes is the sudden large excursions, which tend to become more pronounced with decreasing α-parameter. This reflects the much larger variability inherent in e.g. the Cauchy distribution as opposed to the Gaussian law which determine the variability in the Brownian motion.

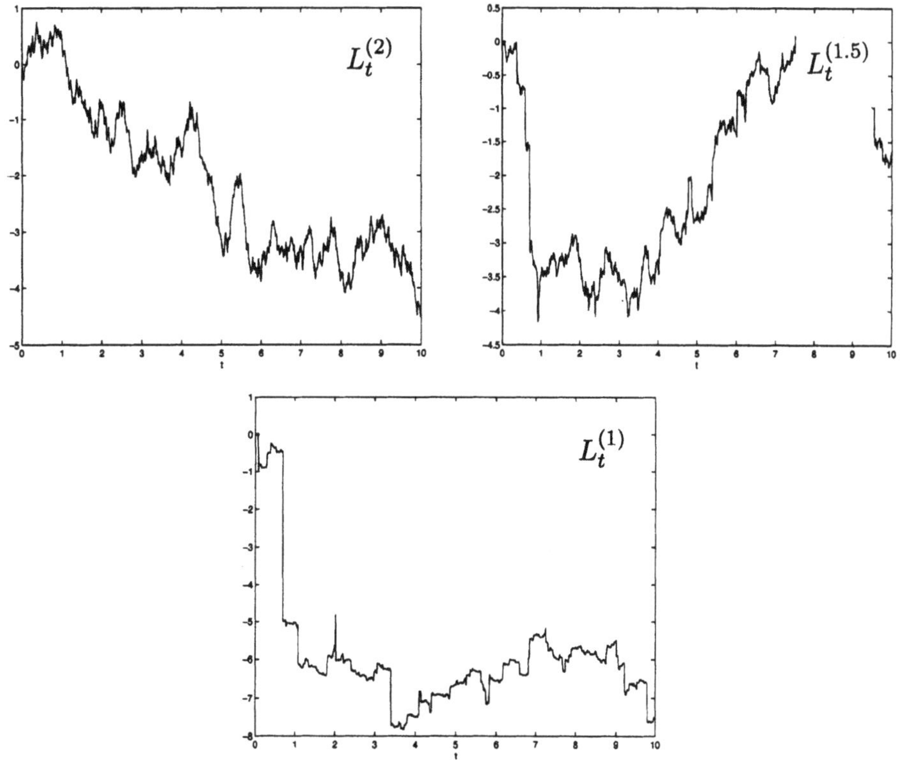

Figure 2. A realization of α-stable Lévy motion $L_t^{(\alpha)}$

3. THE STOCHASTIC DIFFERENTIAL EQUATION

To show how PI can be implemented in numerical solution procedures, we shall discuss a specific class of stochastically excited nonlinear oscillators. Specifically, to avoid certain technicalities concerning general stochastic integrals (Wong & Hajek, 1985; Protter, 1990; Grigoriu, 1996), we shall limit the discussion to nonlinear oscillators whose equation of motion can be expressed as follows

$$\ddot{X}_t + g(X_t, \dot{X}_t) = \gamma N_t^{(\alpha)} \tag{7}$$

where g is a suitable function, γ is a positive constant, and $N_t^{(\alpha)}$ denotes the (standard) α-stable Lévy white noise process defined in the previous section.

Written as an SDE, equation (7) assumes the form

$$dY_t = a(Y_t)\,dt + b\,dL_t^{(\alpha)} \tag{8}$$

where $Y_t = (Y_{1,t}, Y_{2,t})^T = (X_t, \dot{X}_t)^T$ = the 2D state space vector 5process, $a(Y_t) = (Y_{2,t}, a_2(Y_t))^T$ and $b = (0, \gamma)^T$.

For a numerical solution of the SDE (8), a discretization procedure has to be adopted. The basic discretized version of equation (8) is what we shall refer to as the Euler-Maruyama approximation, cf. Kloeden & Platen (1992),

$$Y_t = Y_{t'} + a(Y_{t'})\,\Delta t + b\,\Delta L_{t'}^{(\alpha)} \tag{9}$$

where $\Delta t = t - t'$, and $\Delta L_{t'}^{(\alpha)} = L_t^{(\alpha)} - L_{t'}^{(\alpha)}$.

In the context of Gaussian white noise driven systems, it was pointed out by Naess & Moe (1997), that when the deterministic part of the SDE (8) is considered, equation (9) reduces to the Euler approximation $y_t = y_{t'} + a(y_{t'})\,\Delta t$. As is well known, this aproximation is only accurate to order $O(\Delta t^2)$. To improve the accuracy, a 4th order Runge-Kutta approximation was implemented, which replaced the function $a(y)$ by the corresponding Runge-Kutta approximation, $r(y)$ say. It was found that this allowed for a substantial increase in the the accuracy of the numerical results It is therefore suggested to replace equation (9) by what will be referred to as the Runge-Kutta-Maruyama (RKM) approximation

$$Y_t = Y_{t'} + r(Y_{t'})\,\Delta t + b\,\Delta L_{t'}^{(\alpha)} \tag{10}$$

Since the α-stable Lévy process $L_t^{(\alpha)}$ has independent increments, it follows from equation (10) that the sequence $\{Y_{n\Delta t}\}_{n=0}^{\infty}$ is a Markov chain. For sufficiently small Δt, this Markov chain will approximate the continuous time Markov process solution of the SDE (8). The TPD corresponding to equation (10) can be derived in closed form as follows

$$\begin{aligned} p(y,t|y',t') &= \delta(\,y_1 - y_1' - r_1(y')\Delta t\,) \\ &\quad \cdot \; \tilde{p}(y_2,t|y',t') \end{aligned} \tag{11}$$

where $y = (y_1, y_2)^T$, $\delta(\cdot)$ denotes the Dirac delta function and $\tilde{p}(y_2,t|y',t')$ is the PDF of the conditional random variable $\{Y_{2,t}\,|\,Y_{t'} = y'\} = y_2' + r_2(y')\Delta t + \gamma \Delta L_{t'}^{(\alpha)}$. It then follows from the properties of α-stable variables written down above that this conditional random variable is also an α-stable variable since $\Delta L_{t'}^{(\alpha)}$ is an α-stable variable for every t'. That is, the TPD $p(y,t|y',t')$ can be easily calculated from the normalized version of the PDF of an α-stable variable, that is for $S_\alpha(1,0,0)$, which is already available either as equation (5) or as an accurate empirical PDF.

4. NUMERICAL CALCULATIONS

It has been pointed out above that in each particular case a highly accurate representation of the TPD $p(y,t|y',t')$ is available, either as an analytical formula or a numerical representation. Therefore, to obtain an accurate result for $p(y,t)$ by using equation (1), an accurate representation of $p(y',t')$ is also required. This was achieved by using the procedure described in detail by Naess & Johnsen (1992) or Naess & Johnsen (1993). It combines the discretization of state space into a fairly coarse grid, where the values of $p(y',t')$ have been calculated in the previous time step using equation (1), with an interpolation procedure based on a splines representation of $p(y',t')$.

On the basis of this representation of $p(y',t')$, equation (1) is then used to calculate the values of $p(y,t)$ at the grid points, and the procedure is then repeated. It may be noted that the density of grid points is determined only by the ability to represent the PDF at each time step by a splines approximation in a sufficiently accurate manner. It is also realized that in this approach there is no explicit coupling between the grid size and the time increment, which is solely determined by the accuracy of the discretized dynamic model expressed by equation (10) as compared with the exact formulation of equation (8).

A numerical implementation based on the above principle has been reported to give very accurate results even far into the tails of the PDF (Naess & Johnsen, 1992; Naess & Johnsen, 1993). Later, an alternative procedure has been described (Naess & Moe, 1997), which in some cases can yield more efficient computer codes.

Implicit in the previously proposed approach (Naess & Johnsen, 1992; Naess & Johnsen, 1993; Naess & Moe, 1997) is a combined backward-forward time-stepping procedure. This has been explained in detail by Naess & Moe (1998) for the case of nonlinear oscillators driven by Gaussian white noise. However, the description for the particular case considered in this paper is completely analogous, and need not be repeated. Instead, the interested reader is referred to the given reference for more details.

For illustration, consider the following dynamic model

$$\ddot{X}_t + 2\xi\dot{X}_t + X_t + \lambda X_t^3 = 2\sqrt{\xi}\, N_t^{(\alpha)} \tag{12}$$

which is a Duffing oscillator driven by an α-stable Lévy white noise. It corresponds to the model studied by Naess & Moe (1997) with Gaussian white noise as excitation. The parameter values chosen there were $\xi = 0.25$ and $\lambda = 0.2$. These are also adopted here.

One of the specific problems that one has to cope with in a numerical calculation of the joint PDF of the state space vector of dynamic systems driven by an α-stable Lévy white noise with $\alpha < 2$, is the extension of the domain in state space where the PDF needs to be considered to obtain accurate results by PI. The heavy tails of the PDF of the excitation process is of course reflected in the tails of the PDF of the response process. Compared to the case of Gaussian white noise driven oscillators, in general a much larger domain has to be considered. However, this is compensated by the small gradients associated with the PDF of the response process, which allows a discretization of the state space with a comparable number of grid points as in the case of Gaussian white noise excitation.

4.1. CASE 1: $\alpha = 1.5$

For this case, the choice of domain was $[-30, 30] \times [-30, 30]$ with 31 grid points on each axis. The number of grid points is very similar to what was required for Gaussian white noise excitations while the domain is much larger, cf. Naess & Johnsen (1992). To get a check on the results, extensive Monte Carlo simulations were carried out to obtain accurate estimates of the joint PDF also in the tail region. It is found that reasonable accuracy is obtained over the entire domain with very good accuracy in the tail regions. This is illustrated by the plots in Figures 3 and 4 where the maginal PDF of displacement and velocity response obtained by PI and Monte Carlo simulations together with the relative error $(f^{PI} - f^{MC})/f^{MC}$ between the marginal PDFs are shown. To give an impression of how the heavy tail characteristics of the excitation process is reflected in the tail behaviour of the displacement response, the results for the corresponding Gaussian densities with the same standard deviation are also plotted.

4.2. CASE 2: $\alpha = 1.0$

For the case $\alpha = 1.0$, that is, Cauchy noise, the chosen domain was $[-35, 35] \times [-35, 35]$ also with 31 grid points on each axis. In this case very good agreement was found over the entire domain between the joint PDF obtained by PI and extensive Monte Carlo simulations. Figures 5 and 6 show the maginal PDF of displacement and velocity response obtained by PI and Monte Carlo simulations together with the relative error $(f^{PI} - f^{MC})/f^{MC}$ between the marginal PDFs. The Gaussian densities with the same standard deviation are also included for reference.

It appears that the reason for the higher accuracy obtained by PI for case 2 as compared with case 1 is due to the availability of an analytical formula for the TPD in the former case. This implies that increased accuracy can be expected with an even more accurate estimate of the TPD for case 1.

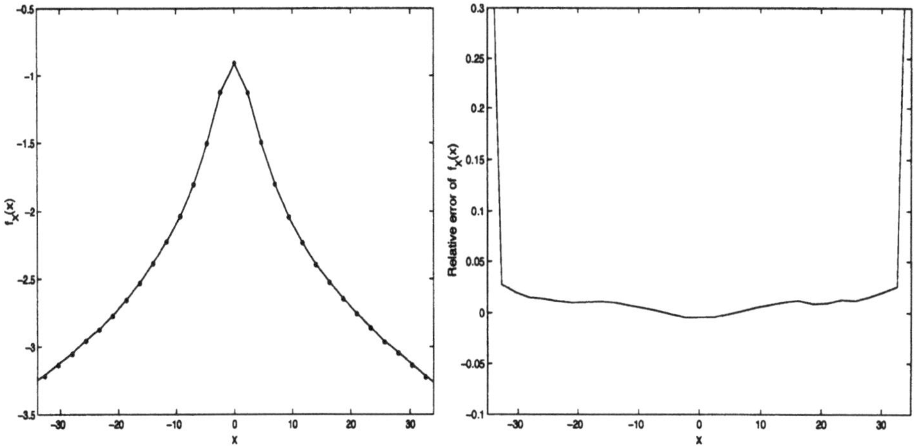

Figure 3. Left: The marginal PDF of displacement response obtained by PI (•)and Monte Carlo simulations (−−) for $\alpha = 1.5$. Right: Relative error.

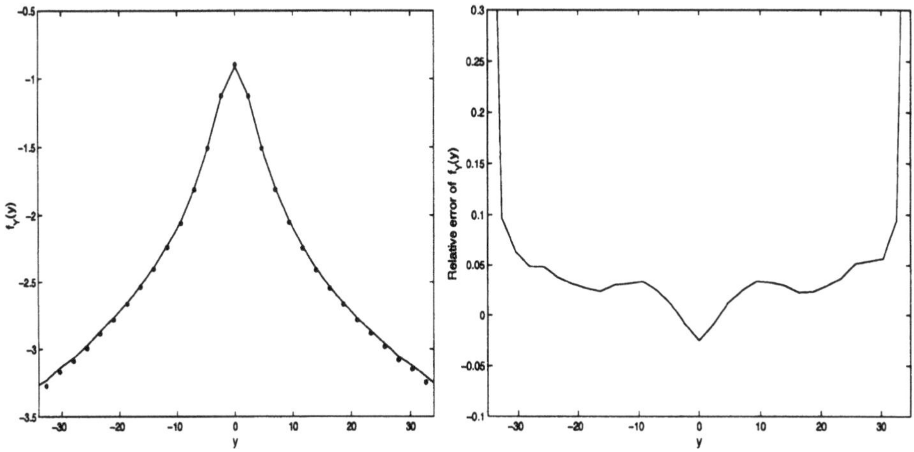

Figure 4. Left: The marginal PDF of velocity response obtained by PI (•)and Monte Carlo simulations (−−) for $\alpha = 1.5$. Right: Relative error.

5. CONCLUDING REMARKS

It has been demonstrated that the path integration (PI) method, which has previously been shown to allow very accurate numerical calculation

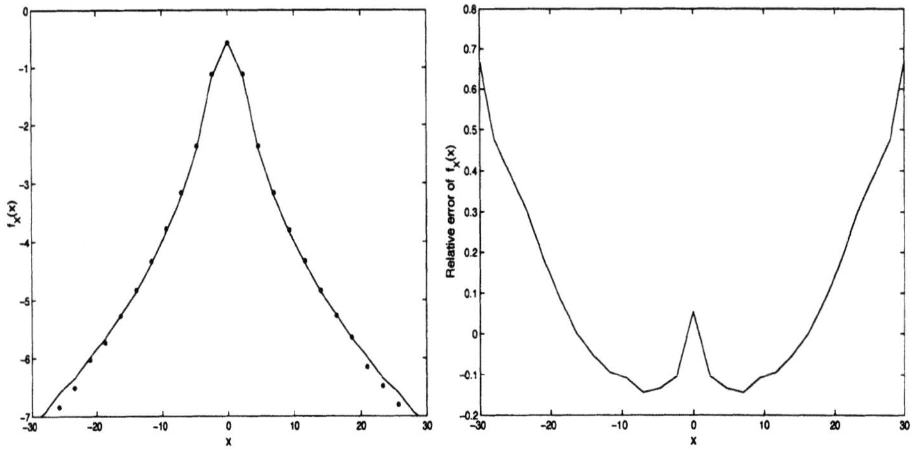

Figure 5. Left: The marginal PDF of displacement response obtained by PI (•)and Monte Carlo simulations (——) for $\alpha = 1.0$. Right: Relative error.

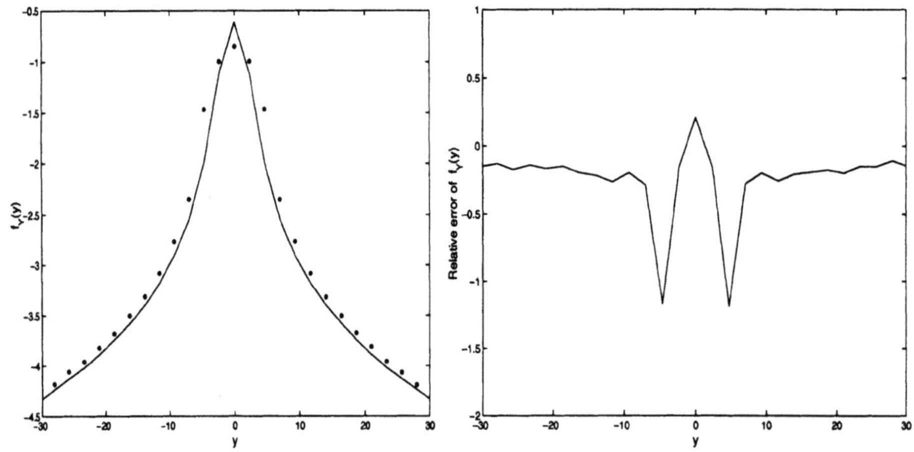

Figure 6. Left: The marginal PDF of velocity response obtained by PI (•)and Monte Carlo simulations (——) for $\alpha = 1.0$. Right: Relative error.

of the joint PDF of the state space vector of nonlinear oscillators driven by Gaussian white noise, can be extended also to the case of nonlinear oscillators driven by an α-stable Lévy white noise. The numerical results, which have been obtained for $\alpha = 1.0$ (Cauchy white noise) and $\alpha = 1.5$ are compared with the results of extensive Monte Carlo simulation, and the conclusion is that good accuracy is still obtained. It has also been observed that the computation time is largely independent of the value of α, even if the domains of calculation have to be extended substantially. The reason being that the number of grid points can be kept practically constant, and the number of grid points is the main factor in determining the computation time.

References

Grigoriu, M. (1995) *Applied Non-Gaussian Processes*, Prentice Hall, Englewood Cliffs

Grigoriu, M. (1996) Parametric Models and Stochastic Integrals, *Proceedings IUTAM Symposium on Advances in Nonlinear Stochastic Mechanics*, Eds. A. Naess and S. Krenk. Kluwer Academic Publishers,Dordrecht

Kloeden, P.E. and Platen, E. (1992) *Numerical Solutions of Stochastic Differential Equations*, Springer-Verlag, New York

Naess, A. and Johnsen, J.M. (1992) Response Statistics of Nonlinear Dynamics Systems by Path Integration, *Proceedings IUTAM Symposium on Nonlinear Stochastic Mechanics*, Eds. N. Bellomo and F. Casciati. Springer-Verlag, Berlin, pp. 401–409

Naess, A. and Johnsen, J.M. (1993) Response Statistics of Nonlinear, Compliant Offshore Structures by the Path Integral Solution Method, *Probabilistic Engineering Mechanics***8**, pp. 91–106

Naess, A. and Moe, V. (1997) New Techniques for Path Integral Solution of the Random Vibration of Non-Linear Oscillators, *Proceedings ICOSSAR'97*, Eds. N. Shiraishi, M. Shinozuka and Y.K. Wen. Balkema, Rotterdam, pp. 795–801

Naess, A. and Moe, V. (1998) Generalized Cell Mapping Versus Path Integration, *Proceedings 3rd International Conference on Computational Stochastic Mechanics*, Balkema, Rotterdam

Naess, A. and Skaug, C. (1998) Extension of the Numerical Path Integration Method to Filtered α-stable Lévy Noise, *Proceedings 3rd International Conference on Computational Stochastic Mechanics*, Balkema, Rotterdam

Protter, P. (1990) *Stochastic Integration and Differential Equations*, Springer-Verlag, Berlin

Samorodnitsky, G. and Taqqu, M.S. (1994) *Stable Non-Gausian Random Processes*, Chapman and Hall, New York

Wehner, M.F. and Wolfer, W.G. (1983) Numerical evaluation of path-integral solutions to Fokker-Planck equations, *Physical Review A*27(5), pp. 2663–2670

Wehner, M.F. and Wolfer, W.G. (1983a) Numerical evaluation of path-integral solutions to Fokker-Planck equations. II. Restricted stochastic processes, *Physical Review A***28**(5), pp. 3003–3011

Wehner, M.F. and Wolfer, W.G. (1987) Numerical evaluation of path-integral solutions to Fokker-Planck equations. III. Time and functionally dependent coefficients, *Physical Review A***35**(4), pp. 1795–1801

Wong, E. and Hajek, B. (1985) *Stochastic Processes in Engineering Systems*, Springer-Verlag, New York

Yu, J.S., Cai, G.Q. and Lin, Y.K. (1996) A New Numerical Scheme for Path Integration Method, *Proceedings 3rd International Conference on Stochastic Structural Dynamics*, San Juan, Puerto Rico

CONTROL OF CHAOS IN NONLINEAR SYSTEMS SUBJECTED TO PARAMETRIC AND STOCHASTIC EXCITATIONS

S. NARAYANAN and M. RAMESH
Machine Dynamics Laboratory
Department of Applied Mechanics
Indian Institute of Technology, Madras 600036, India.

Abstract
The nonfeedback methods of chaos control are well suited for practical applications. This paper discusses the derivation of an analytical method using the generalized Melnikov criterion for determining the control parameters to convert the chaotic motion to a periodic motion for the case of nonfeedback method of chaos control using the addition of a weak periodic force. This is discussed for a nonlinear system subjected to parametric and stochastic excitations.

1. Introduction

Modern design practices lay more emphasis on economical and light weight designs. These conditions require the structures to be more compliant and to operate in highly nonlinear modes. Under certain parameter and initial conditions of the system, the nonlinear response may become chaotic. Control of chaotic motions both by feedback and nonfeedback methods have received the attention of researchers in the last few years [1,2].

In real field environments, purely deterministic (periodic) excitations may seldom exist. There will be random perturbations in addition to the dominant periodic component in modelling of the excitations and the response analysis of real structures. The effectiveness of nonfeedback chaos control schemes of nonlinear oscillators under weak random perturbations has been studied in [3,4]. In [4], the robustness of the nonfeedback methods, viz., constant force addition, weak-periodic force addition, addition of periodic delta and rectangular pulses have been compared and demonstrated for a nonlinear electrical circuit and Duffing-Ueda oscillator. The equation of the Duffing-Ueda oscillator is given by

$$\dot{x}_1 = x_2 \tag{1a}$$

$$\dot{x}_2 = -\alpha\, x_2 - x_1^3 + A\cos(\omega t) + A_c \cos(\omega_c t) + \varepsilon\eta \tag{1b}$$

S. Narayanan and R.N. Iyengar (eds.),
IUTAM Symposium on Nonlinearity and Stochastic Structural Dynamics, 171–184.

Chaotic solution is found for values of $\alpha = 0.1$, $\omega = 1.0$, $A = 11.5$. For $A_c = 1.2$ and $\omega_c = 3.0$ the chaotic motion is converted to a period-two motion where the period is $2\pi/\omega$. Here $A_c \cos(\omega_c t)$ is the weak-periodic control force and $\varepsilon\eta$ is the random noise term where ε is a parameter specifying the intensity of the noise and η is a random variable chosen to be uniformly distributed in the interval [-1, 1] The results of the control performance for the case of weak periodic force addition in the presence of noise for Duffing-Ueda oscillator is shown in Figure 1. The control is lost for a noise intensity $\varepsilon \geq 0.2$ [4].

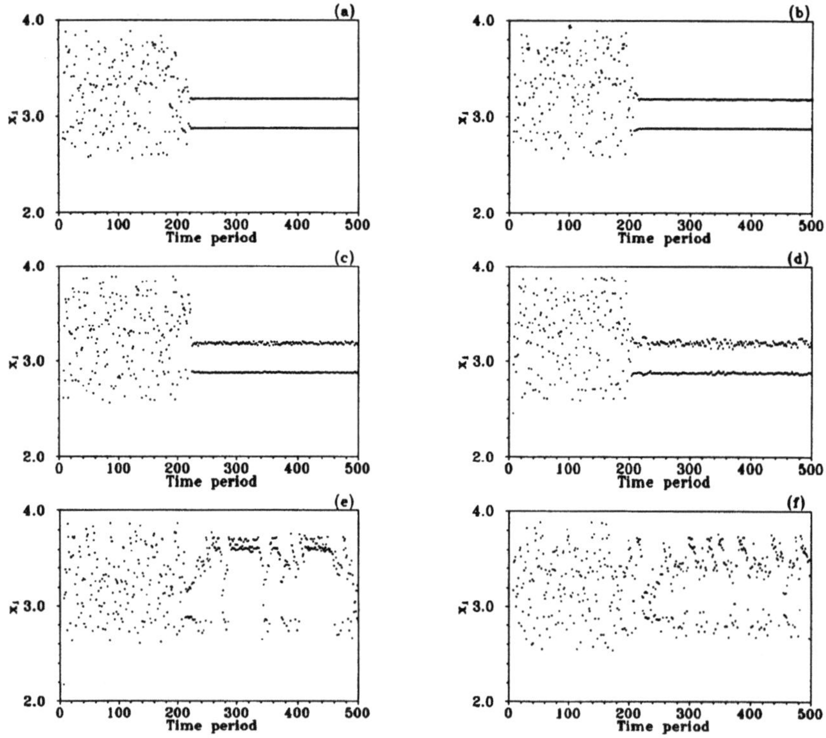

Figure 1. Poincare points of x_1 of eqns.(1a-1b) for $A_c = 1.2$, $\omega_c = 3.0$ (a) without noise, and with noise of intensity (b) 0.01, (c) 0.05, (d) 0.1, (e) 0.2, (f) 0.3. The control is switched on at time period = 200.

With random perturbations included, the response will be chaotic and random [5]. Nonfeedback methods of chaos control are suited for practical applications because of their speed, flexibility, no online monitoring and processing requirements. For applications where none, no real-time, or only highly restricted measurements of the system are available, or where the system behavior is to be altered more drastically, these schemes are quite useful. These methods convert the chaotic motion to *any* arbitrary fixed point or periodic orbit or quasi-periodic orbit. These attributes make them promising for controlling chaotic circuits, fast electro-optical systems, systems in which no parameter is accessible for control, and so on. In nonfeedback methods of chaos control, the existing method of finding the control parameters for controlling or suppressing chaos involves performing numerical simulations over the entire desired parameter space and choosing the control parameters which give a periodic solution. This technique involves an enormous amount of computational time. The aim of this paper is to derive an analytical method based on Melnikov criterion to delineate the chaotic and the periodic regions and thus reduce the computational time for finding the periodic solutions in the control parameter space using numerical simulation. In this paper we consider the suppression of random chaotic motion by the addition of a weak periodic force. In particular we consider a nonlinear system given by the equation of motion

$$\ddot{u} + c\dot{u} + \left(\omega_0^2 + \gamma\cos(\omega t)\right)u + \alpha u^2 + \beta u^3 = 0 \tag{2}$$

The above equation may be considered to represent the first mode vibration of a shallow elastic arch (Figure 2) in nondimensional form or the vibration of a nonlinear mechanism model [6,7].

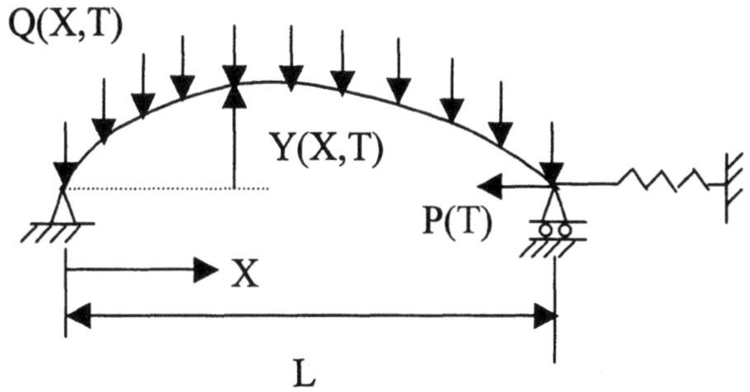

Figure 2. Geometry of the shallow arch.

2. System Equation

The equation of motion of the shallow arch is given by

$$\begin{aligned}\frac{\partial^2 y}{\partial t^2} + c\frac{\partial y}{\partial t} + \frac{\partial^4 y}{\partial x^4} - \frac{d^2 y_0}{dx^4} + (p + h_0)\frac{\partial^2 y}{\partial x^2} - h_0\frac{d^2 y_0}{dx^2} \\ + 2k\frac{\partial^2 y}{\partial x^2}\int_0^1\left[\left(\frac{dy_0}{dx}\right)^2 - \left(\frac{\partial y}{\partial x}\right)^2\right]dx = -q\end{aligned} \tag{3}$$

Introducing the non-dimensional quantities,

$$t=\frac{T}{L^2}\sqrt{\frac{EI}{\mu}},h_0=\frac{H_0L^2}{EI},p=\frac{PL^2}{EI},\quad q=\frac{QL^4}{2EIr},$$

$$r=\sqrt{\frac{I}{A}},\quad \cdot c=\frac{CL^2}{\sqrt{EI\mu}},\quad k=\frac{K}{K+(EA)/L}$$

where T is time, L is the span, E is the Young's modulus, I is the area moment of inertia of the beam, μ is Poisson's ratio, H_0 is the height at the centre, P is the horizontal loading, Q is the distributed loading, r is the radius of gyration, C is the damping, K is the stiffness and assuming $y_0(x) = \lambda \sin(\pi x)$, $q(x,t)=\overline{q}(t)\sin(\pi x)$, $y(x,t)=[\lambda+u(t)]\sin(\pi x)$ and substituting in equation (3) leads to

$$\ddot{u}+c\dot{u}+\pi^2(\pi^2+2\pi^2\lambda^2k-h_0-p)u+3\pi^4k\lambda y^2+\pi^4ku^3=\pi^2\lambda p-\overline{q} \tag{4}$$

For special choice of the parameters, we get the equation (2) from equation (4).

We choose $\omega_0 = 1.0$, $\acute{c} = 0.1$, $\alpha = 3/2$, $\beta = 1/2$. We model the random perturbation to this system in the form of a Gaussian white noise $\eta(t)$ with zero mean and two sided power spectral density $S_0 = 2\pi\kappa$.

We add a weak periodic force $A\cos(\Omega t)$ for purposes of control. With the addition of the stochastic perturbation and weak periodic force equation (2) becomes

$$\ddot{u}+c\dot{u}+\left(1+\gamma\cos(\omega t)\right)u+\alpha u^2+\beta u^3=\eta(t)+A\cos(\Omega t) \tag{5}$$

For $\gamma = 1.42$ and $\omega = 1.0$ and without the presence of noise and without the application of control the system exhibits chaotic behavior. The corresponding phase plane plot and the Poincaré section are shown in Figures 3 and 4 respectively for the initial conditions (0.01, 1.01).

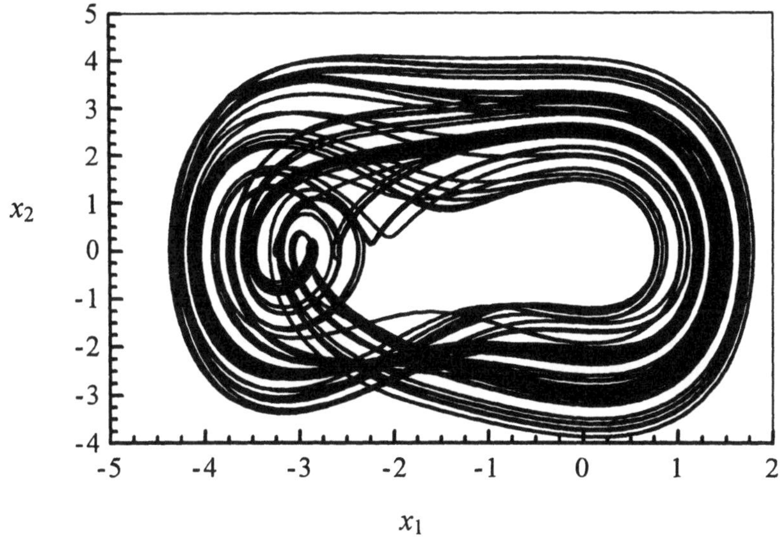

Figure 3. Phase plot for $\gamma = 1.42$, $\omega = 1.0$ (without control and $\eta = 0$)

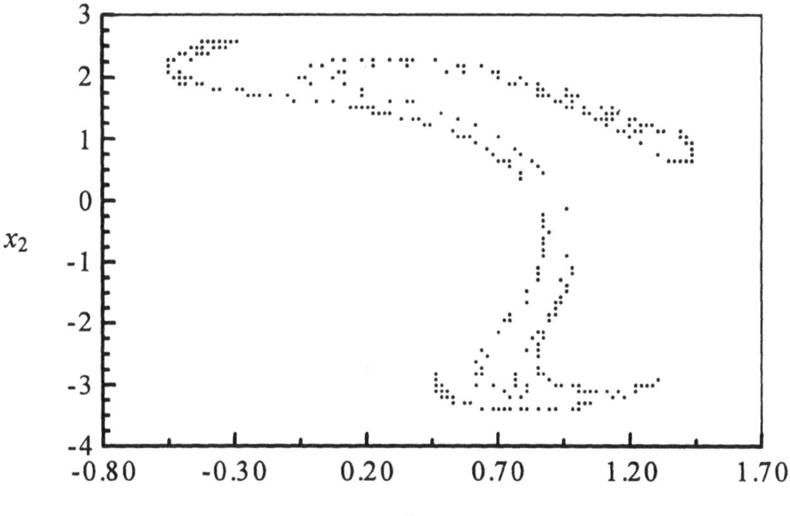

Figure 4. Poincare section for $\gamma = 1.42$, $\omega = 1.0$ (without control and $\eta = 0$)

Equation (5) can be expressed in state space form as

$$\begin{Bmatrix} \dot{x}_1 \\ \dot{x}_2 \end{Bmatrix} = \mathbf{f}(x_1, x_2) + \mathbf{g}(x_1, x_2, t) \tag{6}$$

where $\mathbf{f}(x_1, x_2) = \{x_2, -x_1 - \alpha x_1^2 - \beta x_1^3\}^{\mathrm{T}}$

and $\mathbf{g}(x_1, x_2) = \{0, -c\, x_2 - \gamma\, x_1 \cos(\omega t) + A\cos(\Omega t) + \eta(t)\}^{\mathrm{T}}$

with $u = x_1$ and $\dot{u} = x_2$. Superscript T denotes the transpose.

The system is a two-well potential with saddle at (-1,0). For the unperturbed and undamped system, the Hamiltonian is given by

$$\begin{aligned} \dot{x}_1 &= x_2 = \frac{\partial H}{\partial x_2} \\ \dot{x}_2 &= -x_1 - \frac{3}{2}x_1^2 - \frac{1}{2}x_1^3 = -\frac{\partial H}{\partial x_1} \end{aligned} \tag{7}$$

where

$$H = \frac{1}{2}x_2^2 + \frac{1}{2}x_1^2 + \frac{1}{2}x_1^3 + \frac{1}{8}x_1^4 \tag{8}$$

The level set of H = 1/8 consists of a saddle (-1,0). Solving equation (8) for H = 1/8 for x_2 in terms of x_1 we obtain,

$$x_2 = \pm\sqrt{\frac{1}{4} - x_1^2 - x_1^3 - \frac{x_1^4}{4}} \tag{9}$$

In terms of u and $\dot{u}$ we rewrite equation (9) as

$$u_2 = \pm\sqrt{\frac{1}{4} - u^2 - u^3 - \frac{u^4}{4}} \tag{10}$$

Using separation of variables and substitutions we obtain

$$\pm t + c_1 = -\sqrt{2}\,\mathrm{sech}^{-1}\left(\frac{u+1}{\sqrt{2}}\right) \tag{11}$$

Considering the negative sign, we have

$$u+1=\sqrt{2}\,\mathrm{sech}(\frac{t}{\sqrt{2}}-c_1) \tag{12}$$

Choosing the time origin so that $u=\sqrt{2}-1$ at $t=0$, we have $c_1=0$ and hence

$$u=\sqrt{2}\,\mathrm{sech}(\frac{t}{\sqrt{2}})-1 \tag{13}$$

Consequently, the homoclinic orbit is given by

$$x_1=\sqrt{2}\,\mathrm{sech}(\frac{t}{\sqrt{2}})-1 \text{ and } x_2=-\mathrm{sech}(\frac{t}{\sqrt{2}})\tanh(\frac{t}{\sqrt{2}}) \tag{14}$$

Given the values of the parameters α, β, c, γ and ω and the power spectral density S_0 for which the system exhibits chaotic behavior, the values of A and Ω for which the chaotic motions do not occur (i.e., suppressed) can be identified from the regions in A-Ω parameter space using the generalized stochastic Melnikov criterion derived for this system. This criterion was developed in [5] for the Duffing's oscillator as a mean-square representation to explicitly express the possibility of occurrence of chaotic response in the presence of noise. In this study, we make use of this criterion for estimating the control parameters A and Ω for which the chaotic motions do not occur.

3. Generalized Melnikov criterion

Assuming smallness of the perturbations, the generalized stochastic Melnikov process is given by

$$M_g(\theta_1,\theta_2)=\int_{-\infty}^{\infty} f[x_1(t),x_2(t)]\wedge g[x_1(t),x_2(t),\theta_1,\theta_2]$$

$$=\int_{-\infty}^{\infty} x_2\left(-cx_2-\gamma\, x_1\cos[\omega(t+\theta_1)]+A\cos[\Omega(t+\theta_1)]+\eta(t+\theta_2)\right)dt$$

$$=-c\int_{-\infty}^{\infty}\mathrm{sech}^2(\frac{t}{\sqrt{2}})\tanh^2(\frac{t}{\sqrt{2}})\,dt$$

$$+\gamma\int_{-\infty}^{\infty}\left[\sqrt{2}\,\mathrm{sech}^2(\frac{t}{\sqrt{2}})\tanh(\frac{t}{\sqrt{2}})-\mathrm{sech}(\frac{t}{\sqrt{2}})\tanh(\frac{t}{\sqrt{2}})\right]\cos[\omega(t+\theta_1)]\,dt$$

$$-A\int_{-\infty}^{\infty}\mathrm{sech}(\frac{t}{\sqrt{2}})\tanh(\frac{t}{\sqrt{2}})\cos[\Omega(t+\theta_1)]\,dt$$

$$-\int_{-\infty}^{\infty} \operatorname{sech}(\frac{t}{\sqrt{2}}) \tanh(\frac{t}{\sqrt{2}}) \eta(t+\theta_2)\, dt$$

$$= M_d(\theta_1) + M_r(\theta_2) \tag{15}$$

The first three integrals in equation (15), corresponding to $M_d(\theta_1)$, represent the mean of the stochastic Melnikov process due to the periodic excitations and damping force. The last integral corresponding to $M_r(\theta_2)$, represents the random portion due to white noise. The random portion due to white noise is calculated by treating the white noise process $\eta(t)$ passing through a linear filter given by the homoclinic orbit [8] $x_2 = -\operatorname{sech}(\frac{t}{\sqrt{2}}) \tanh(\frac{t}{\sqrt{2}})$. Because of the linear filtering, the random process $M_r(\theta_2)$ is stationary and of zero mean

$$\langle M_r(\theta_2) \rangle = \int_{-\infty}^{\infty} \operatorname{sech}(\frac{t}{\sqrt{2}}) \tanh(\frac{t}{\sqrt{2}}) \langle \eta(\theta_2) \rangle = 0. \tag{16}$$

Direct multiplication of white noise spectrum $S_\eta(\omega)$ and the transfer function $|F(\omega)|^2$ of the linear filter x_2 gives the corresponding spectrum of the random process.

The frequency response $F(\omega)$ is obtained through Fourier transform

$$F(\omega) = \int_{-\infty}^{\infty} x_2(t)\, e^{-i\omega t}\, dt$$

$$= -\int_{-\infty}^{\infty} \operatorname{sech}(\frac{t}{\sqrt{2}}) \tanh(\frac{t}{\sqrt{2}}) e^{-i\omega t}\, dt = \frac{2i\pi\omega}{\cosh\left(\frac{\pi\omega}{\sqrt{2}}\right)} \tag{17}$$

The variance of $M_r(\theta_2)$ is calculated by integrating its spectrum over the entire frequency range

$$\sigma_{M_r}^2 = \int_{-\infty}^{\infty} |F(\omega)|^2 S_\eta(\omega) d\omega \cong 37.22\kappa \tag{18}$$

By setting equation (15) to zero, we obtain the generalized stochastic Melnikov criterion as

$$M_g(\theta_1, \theta_2) = M_d(\theta_1) + M_r(\theta_2) = 0 \tag{19}$$

The occurrence of possible chaotic response in the presence of noise can be explicitly expressed by an energy interpretation. The criterion for possible chaotic response based on generalized Melnikov process, in the presence of noise, is performed in a mean-square representation

$$\left\langle \left(\frac{2A\pi\Omega \sin(\Omega\theta_1)}{\cosh(\frac{\pi\Omega}{\sqrt{2}})} \right)^2 \right\rangle \geq \left\langle \left(\frac{2\sqrt{2}c}{3} \right)^2 \right\rangle + \left\langle \left(\frac{2\pi\gamma\omega^2 \sin(\omega\theta_1)}{\sinh(\frac{\pi\omega}{\sqrt{2}})} \right)^2 \right\rangle - \left\langle \left(\frac{2\pi\gamma\omega \sin(\omega\theta_1)}{\cosh(\frac{\pi\omega}{\sqrt{2}})} \right)^2 \right\rangle - \left\langle M_r^2(\theta_2) \right\rangle \tag{20}$$

Then the mean-square criterion for possible chaotic response in terms of parameters (A, ω, Ω, c, γ, κ) is given by

$$\left(\frac{A\pi\Omega}{\cosh(\frac{\pi\Omega}{\sqrt{2}})} \right)^2 \geq \frac{2c^2}{9} + \left(\frac{\pi\gamma\omega^2}{\sinh(\frac{\pi\omega}{\sqrt{2}})} \right)^2 - \left(\frac{\pi\gamma\omega}{\cosh(\frac{\pi\omega}{\sqrt{2}})} \right)^2 - 18.61\kappa \tag{21}$$

Equating the expressions in equation (21) gives the mean square representation of the lower bound of possible chaotic domain. In other words, if the values of ω, c, γ, κ are known then the equation (21) delineates the domain of possible occurrence of chaotic response in A-Ω parameter space. Thus the generalized Melnikov criterion provides an analytical way of determining the control parameters A and Ω for which the chaotic motion can be avoided or suppressed.

The bifurcation diagram with the application of control and without noise is shown with the control amplitude A as the parameter and $\gamma = 1.42$ and $\omega = 1.0$ in Figure 5. The Melnikov criterion given by equation (21) is plotted in A-Ω parameter space in Figure 6 without considering the noise ($\kappa = 0$). This plot gives the upper bound for the periodic motion. Hence the values of the control parameters for controlling chaos can be chosen from the region below this curve. The upper bounds for the periodic motion for the cases with and without noise in A-Ω parameter space are also given by the dashed and solid lines respectively in Figure 7. Numerical simulation is used to validate the analytical criterion. Sample realizations of the system response are obtained by integrating equation (5), by a fourth-order Runge-Kutta method. The numerical results are shown by appropriate symbols in Figure 6. White space region in Figure 6 is interspersed with both chaotic and periodic solutions.

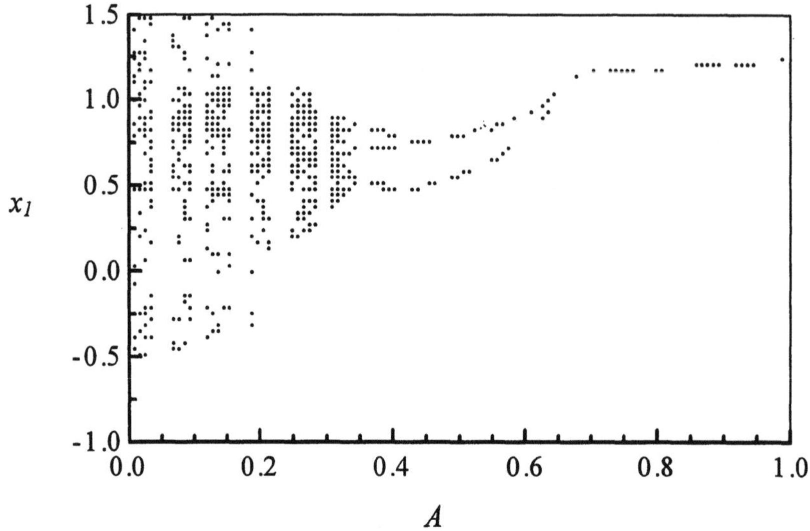

Figure 5. Bifurcation diagram with control applied ($\gamma = 1.42$, $\omega = 1.0$, $\eta = 0$).

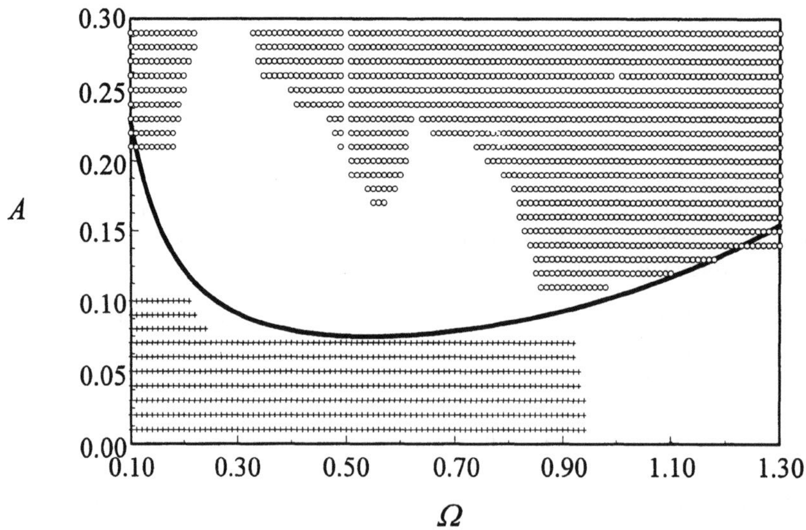

Figure 6. Melnikov criterion (with control for $\gamma = 1.42$, $\omega = 1.0$, $\kappa = 0$). Open circles correspond to chaotic motion and cross symbols correspond to periodic motion.

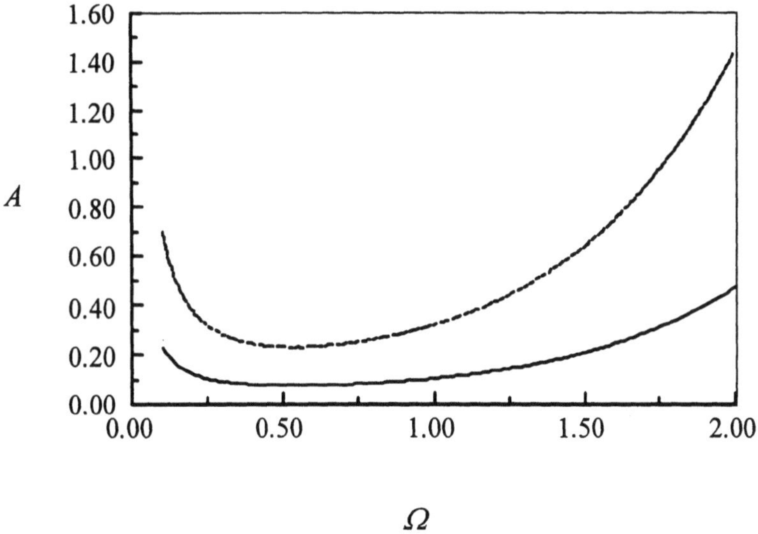

Figure 7. Melnikov criterion with noise (dashed line) and without noise (solid line)

4. Fokker-Planck Equation

The response of the nonlinear system subjected to white noise excitation along with the periodic excitation can be approximated as a Markov process [9]. We further investigate the evolution of the joint probability density function by the Fokker-Planck equation.

The Fokker-Planck equation associated with equation (5) can be expressed as

$$\begin{aligned}\frac{\partial \mathrm{p}(x_1,x_2,t)}{\partial t} &= -\frac{\partial}{\partial x_1}\{x_2 \mathrm{p}(x_1,x_2,t)\} \\ &-\frac{\partial}{\partial x_2}\left\{\left(-cx_2 - x_1 - \alpha x_1^2 - \beta x_1^3 - \gamma x_1 \cos(\omega t) + A\cos(\Omega t)\right)\mathrm{p}(x_1,x_2,t)\right\} \\ &+\pi S_0 \frac{\partial^2}{\partial {x_2}^2}\mathrm{p}(x_1,x_2,t)\end{aligned} \tag{22}$$

The solution to this equation is obtained by a path integral solution procedure. The evolution of the probability density function, which gives a global picture of the system behavior, is studied.

4.1 PATH INTEGRAL SOLUTION

In this method, the travelling path of the probability density function is discretised in terms of infinitesimal segments in probability space [10]. Each segment representing a short time propagation is approximated by a time-dependent Gaussian distribution called the short time probability density function. For the system considered the short time probability density function $\mathbf{G}(x_1', x_2', x_1, x_2, t; \tau)$, is given by

$$\mathbf{G}(x_1', x_2', x_1, x_2, t; \tau) =$$

$$\frac{1}{\sqrt{2\pi\kappa\tau}} \exp\left[-\frac{\tau}{2\kappa}\left(-cx_2 - x_1 - \alpha x_1^2 - \beta x_1^3 - \gamma x_1 \cos(\omega t) + A\cos(\Omega t) - \frac{x_2' - x_2}{\tau} \right)^2 \right] \delta\left(\mathrm{x}_2 - \frac{x_1' - x_1}{\tau} \right) \quad (23)$$

where vectors $[x_1', x_2']^T$ and $[x_1, x_2]^T$ represent the post-state and pre-state, respectively, and the probability density function at the time t is given by

$$p(x_1, x_2, t) = \lim_{\substack{\tau\to 0 \\ N\to\infty \\ N\tau\to t-t_0}} \prod_{i=0}^{N-1} \int .. \int \sum_{j=0}^{N-1} \mathbf{G}((x_1)_{i+1}, (x_2)_{i+1}, (x_1)_i, (x_2)_i, t_j; \tau) \; p((x_1)_0, (x_2)_0, t_0) d(x_1)_i d(x_2)_i \quad (24)$$

where $p((x_1)_0, (x_2)_0, t_0)$ is the probability density function of the initial conditions. The path integral solution converges to the exact solution in the limit as $N \to \infty$ and $\tau \to 0$.

Following the numerical procedure of the path integral based on a discrete lattice representation as given in [11], we have the time evolution of the probability distribution

$$P_{ik}(t+\tau) = \sum_{j=1}^{N} \sum_{l=1}^{N} \mathbf{T}_{ijkl}(\tau) P_{jl}(t) \quad (25)$$

where the propagator matrix $\mathbf{T}_{ijkl}$ is defined as

$$T_{ijkl}(\tau) = \frac{4}{((\Delta x_1)_{i-1} + (\Delta x_1)_i)((\Delta x_2)_{k-1} + (\Delta x_2)_k)} \cdot \int_{(x_1)_i - (\Delta x_1)_{i-1}/2}^{(x_1)_i + (\Delta x_1)_i/2} dx_1 \int_{(x_2)_k - (\Delta x_2)_{k-1}/2}^{(x_2)_k + (\Delta x_2)_k/2} dx_2 \int_{(x_1)_j - (\Delta x_1)_{j-1}/2}^{(x_1)_j + (\Delta x_1)_j/2} \int_{(x_2)_l - (\Delta x_2)_{l-1}/2}^{(x_2)_l + (\Delta x_2)_l/2} dx_1' dx_2' \mathbf{G}(x_1, x_1', x_2 x_2' t; \tau) \quad (26)$$

The time evolution of the joint probability density function for the system considered is shown in Figure 8 with time step $\tau = 0.01$ for 0.05 time units.

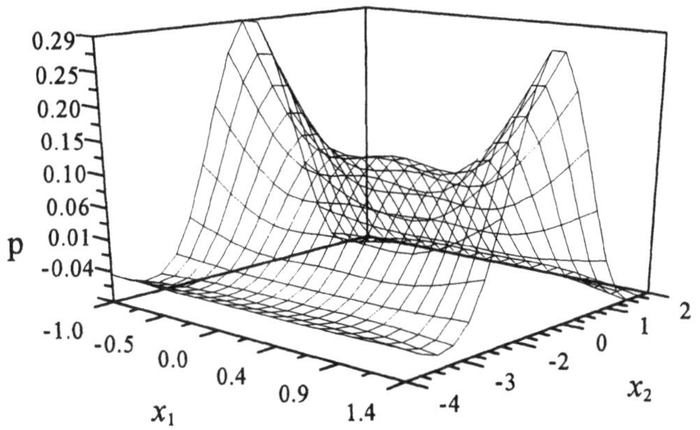

Figure 8. Probability distribution ($\tau = 0.01$; time = 0.05; matrix size = 5×5×5×5).

5. Conclusion

This study discusses the analytical criterion derived using the generalized Melnikov method for determining the control parameters for controlling chaos in a system by the external addition of a weak periodic force. Based on generalized stochastic Melnikov process, a lower bound for possible chaotic region is identified. This delineates the periodic domain from which the control parameters can be chosen. The presence of noise increases the threshold and enlarges the possible periodic domain with the addition of a weak periodic force. The Fokker-Planck equation along with path integral solution gives a time evolution of the probability density function, which characterizes the global behavior of the system.

6. References

1. Ott, E., Grebogi, C. and Yorke, J. (1990) Controlling chaos, *Phys. Rev. Lett.* **64**, 1196-1199.

2. Braiman, Y. and Goldhirsch, I. (1991) Taming chaotic dynamics with weak periodic perturbations, *Phys. Rev. Lett.* **66**, 2545-2548.

3. Rajasekar, S. (1995) Controlling of chaotic motion by chaos and noise signals in a logistic map and a Bonhoeffer-van der Pol oscillator, *Phys. Rev. E* **51**, 775-778.

4. Ramesh, M. and Narayanan, S. (1999) Chaos control by nonfeedback methods in the presence of noise, *Chaos, Solitons and Fractals* **10**(9), 1473-1489.

5. Lin, H. and Yim, S. C. S. (1996) Analysis of a nonlinear system exhibiting chaotic, noisy chaotic, and random behaviors, *J. Applied Mechanics* **63**, 509-516.

6. Lamarque, C. H. and Malasoma, J. M. (1995) Nonlinear behaviour of a nonlinear mechanical system, in Jan Awrejcewicz (ed.), *Nonlinear Dynamics: New Theoretical and Applied Results*, Akademie Verlag.

7. Plaut, R. H. and Hsieh, J. C. (1985) Chaos in a mechanism with time delays under parametric and external excitation, *J. Sound and Vibration* **114**, 73-90.

8. Frey, M. and Simiu, E. (May, 1992) Equivalence between motions with noise-induced jumps and chaos with Smale horseshoes, *Proc. 9th Eng. Mech. Conf. ASCE* Texas A&M University, College Station, TX, 660-663.

9. Kapitaniak, T. (1988) Chaos in systems with noise, *World Scientific*, Singapore.

10. Wissel, C. (1979) Manifolds of equivalent path integral solutions of the Fokker-Planck equation, *Zeit Physik B* **35**, 185-191.

11. Wehner, M.F. and Wolfer, W.G. (1983) Numerical evaluation of path-integral solutions to Fokker-Planck equations, *Phys. Rev. A* 27(5), 2663-2670.

STOCHASTIC STABILITY OF LATERAL-TORSIONAL MOTION OF SLENDER BRIDGES UNDER TURBULENT WIND

M.D. PANDEY and S.T. ARIARATNAM
Department of Civil Engineering
University of Waterloo
Waterloo, ON, Canada, N2L 3G1

Abstract

The paper presents a simulation-based approach to investigate the stability of lateral-torsional motion of slender suspension bridges under stochastic wind turbulence. The bridge motion is represented by a linear, two degrees of freedom oscillator. Stochastic turbulence in wind velocity is considered in the form of a periodic excitation with random phase modulation. The stability condition refers to the asymptotic sample stability, for which the necessary and sufficient condition is that the largest Lyapunov exponent be negative. Monte-Carlo simulation is performed to evaluate the largest Lyapunov exponent, and stochastic differential equations of motion are integrated using Euler's scheme. In contrast with some approximations reported in the literature, the turbulence intensity is shown to have a destabilizing effect on the bridge stability in a sense that an increase in the intensity decreases the critical mean wind velocity below the deterministic flutter velocity.

1 Introduction

The stability of motion of slender suspension bridges subjected to turbulent wind loading has been a topic of considerable research interest in wind engineering. Self-excited wind loads, which depend on the structural motion, are known to govern the condition at which the instability of bridge motion can occur. Furthermore, the stability condition is sensitive to turbulent components in wind velocity which tend to perturb the self-excited loads acting on the bridge. The mechanism of bridge instability in the presence of wind turbulence is not fully understood. Experimental observations suggest that a change in the stability condition may occur due to turbulence (Huston et. al. 1988). In fact, both stabilization and destabilization effects have been noted in section model tests. Therefore, investigation of the effect of random wind turbulence on bridge stability has practical significance.

The instability of bridge motion in many cases appears to be of the flutter type and dominated by the torsional mode. Lin and Ariaratnam (1980) investigated the moment stability of torsional bridge motion excited by a white-noise process, and reported a

S. Narayanan and R.N. Iyengar (eds.),
IUTAM Symposium on Nonlinearity and Stochastic Structural Dynamics, 185–196.

destabilizing effect of turbulence. Bucher and Lin (1988) extended this analysis to include the bending mode in addition to the torsional mode of vibration. The energy transfer from the critical mode to more stable modes was identified as a possible mechanism for the stabilization of bridge motion by turbulence. However, the effect of turbulence may not always be favourable, since the beneficial effect was shown to be sensitive to the aerodynamic parameters of a specific bridge structure (Bucher and Lin 1988). Lin and Li (1993) applied a refined averaging method to investigate the asymptotic sample stability of torsional motion under a non-white turbulence process. Minor stabilization effect of turbulence was reported in cases of wideband and detuned narrowband excitation (Li and Lin 1995). It should be noted that the sample stability is physically more meaningful than stability in moments, which represent ensemble averaging that is difficult to relate to experimental observations.

The purpose of this paper is to present a numerical, simulation-based solutions for the stability of lateral-torsional motion of slender bridges under stochastic wind turbulence. Major advantages of the numerical approach are its conceptual simplicity and versatility (Wedig 1996). For example, the method is able to account for the quadratic noise term in the stability analysis, which was ignored in various studies to avoid analytical complexities (e.g., Lin and Ariaratnam 1982, and Li and Lin 1995). It will be shown in the paper that the omission of the quadratic noise term results in overestimation of the stabilization effect due to turbulence.

The lateral-torsional bridge motion is modelled by a linear, two degrees of freedom oscillator. To consider the effect of turbulence, a random fluctuation in the wind velocity is introduced in the form of a periodic excitation with random phase modulation. The stability condition refers to the asymptotic sample stability, for which the necessary and sufficient condition is that the largest Lyapunov exponent be negative. The Lyapunov exponent is essentially a measure of the average exponential growth rate of the amplitude process corresponding to a given system over a long period of time, and is known to be a deterministic number with probability one.

2 Linear Bridge Vibration

The bridge structure is modelled as a two degree-of-freedom oscillator representing the first torsional and vertical bending modes of vibration of the bridge. The torsional and bending modes are uncoupled structurally, but coupled aerodynamically. The equations of motion are given as:

$$\ddot{\alpha} + 2\zeta_\alpha \omega_\alpha \dot{\alpha} + \omega_\alpha^2 \alpha = \frac{M(t)}{I} \tag{1}$$

$$\ddot{y} + 2\zeta_y \omega_y \dot{y} + \omega_y^2 y = \frac{L(t)}{m} \tag{2}$$

where α and y are generalized torsional and vertical coordinates, respectively. ω_α and ω_y are natural circular frequencies, ζ_α and ζ_y are modal damping ratios, m is modal mass, I is modal moment of inertia, $M(t)$ and $L(t)$ are modal aerodynamic moment and force, respectively. The analysis considers self-excited loads $L(t)$ and $M(t)$ only, and ignores buffeting terms which do not affect the stability condition. The aerodynamic moment, for example, can be expressed in terms of convolution integrals as follows

$$M(t) = M_\alpha(t) + M_y(t) = \int_{-\infty}^{t} f_1(t-\tau)\alpha(\tau)d\tau + \int_{-\infty}^{t} f_2(t-\tau)y(\tau)d\tau \tag{3}$$

where $f_1(t)$ and $f_2(t)$ are impulse response functions. Approximation of experimentally obtained indicial functions by exponential functions leads to the following expression for $f_1(t)$:

$$f_1(t) = \rho\, B^2 u^2 \left[c_1\alpha(t) + \frac{c_2 B}{u}\dot{\alpha}(t) + \sum_{k=3}^{4} \int_{-\infty}^{t} c_k \exp\{-d_k u(t-\tau)/B\}\dot{\alpha}(\tau)d\tau \right] \tag{4}$$

in which ρ = air density, B = half-width of bridge deck, u = mean wind velocity, and c_i are constants. Mathematical expression for other impulse response functions can be obtained in a similar manner (Bucher and Lin 1988). Two convolution integrals encountered in eqn(4) can be treated as state variables, $M_{\alpha 3}(t)$ and $M_{\alpha 4}(t)$, for ease of mathematical analysis (Lin and Ariaratnam 1980):

$$M_{\alpha i}(t) = \frac{\rho B^2 u^2}{I} \left[\sum_{k=3}^{4} \int_{-\infty}^{t} c_k \exp\{-d_k u(t-\tau)/B\}\dot{\alpha}(\tau)d\tau \right], \quad i=3,4 \tag{5}$$

Eqn(4) then becomes

$$f_1(t) = \rho\, B^2 u^2 \left[c_1\alpha(t) + \frac{c_2 B}{u}\dot{\alpha}(t) + \right] + I \sum_{k=3}^{4} M_{\alpha k}(t) \tag{6}$$

The aerodynamic load has two components corresponding to torsion and bending, namely

$$L(t) = L_\alpha(t) + L_y(t) \tag{7}$$

The treatment of the remaining self-excited forces, M_y(t), $L_\alpha(t)$ and $L_y(t)$ is similar to that for $M_\alpha(t)$, given by eqns(3) to (6).

The first-order system of equations of motion involves 12 state variables denoted by vector $\mathbf{x}$:

$$\mathbf{x} = \{\alpha, \dot{\alpha}\ y, \dot{y}, M_{\alpha 3}(\mathrm{t}), M_{\alpha 4}(\mathrm{t}), M_{y3}(\mathrm{t}), M_{y4}(\mathrm{t}), L_{\alpha 3}(\mathrm{t}), L_{\alpha 4}(\mathrm{t}), L_{y3}(\mathrm{t}), L_{y4}(\mathrm{t})\}^T \tag{8}$$

Thus, the equations of motion can be written in a concise form

$$\dot{x}_i = A_{ij}\, x_j \quad (i,j = 1, 2, \ldots 12) \tag{9}$$

where elements of the system matrix, $\mathbf{A}$, are defined in the Appendix.

The wind velocity, $u(t)$, is randomized as

$$u(t) = u[1 + f(t)] \quad \text{and } u^2(t) = u^2[1 + f(t)]^2 \tag{10}$$

where u is the mean wind velocity and $f(t)$ is a non-dimensional random process with zero mean and a finite mean square value, modelled as

$$f(t) = \varepsilon \sin(\nu t + \sigma W(t) + \chi) \tag{11}$$

where ε = turbulence intensity, ν = harmonic frequency, σ = bandwidth factor, χ=a random variable uniformly distributed in $[0, 2\pi)$, and $W(t)$ = unit Weiner process. The mean square value of $f(t)$ is $\varepsilon^2/2$. The shape of the spectral density is determined by ν and σ. $f(t)$ becomes a narrow band process when $(\nu/\sigma^2) >> 1$. For large values of σ, $f(t)$ tends to approach a white noise excitation. σ is referred to as the bandwidth factor, as it controls the bandwidth of the spectral peak in the case of narrowband excitation. In Figure 1, spectral density functions for different values of ν and σ are plotted using the analytical expression derived by Lin and Li (1993).

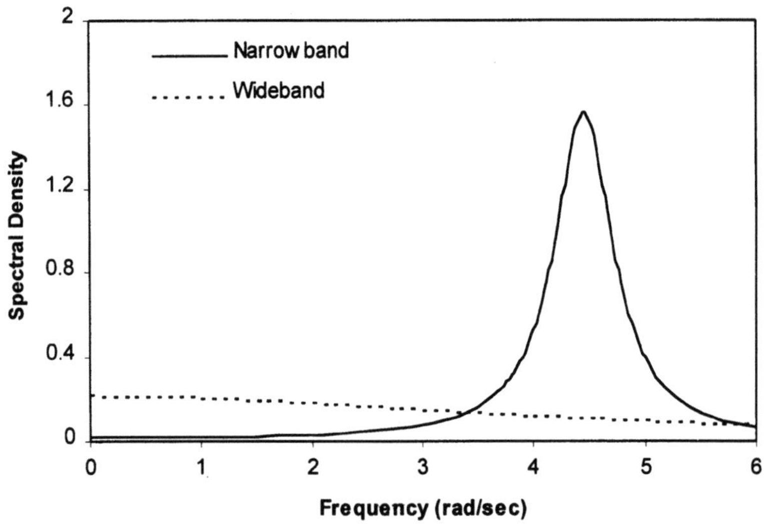

Figure 1: Possible spectra for the wind turbulence (narrowband: σ=0.8, ν = 4.45; wideband: σ = 3, ν = 0.2)

The equations of motion may be expressed as a system of first-order stochastic differential equations

$$\dot{\mathbf{x}} = \mathbf{A}\,\mathbf{x} + \mathbf{G}\,\mathbf{x}\,[\varepsilon\, sin\theta(t)] + \mathbf{H}\,\mathbf{x}\,[\varepsilon\, sin\theta(t)]^2 \tag{12}$$

where **G** and **H** are (12×12) matrices described in the Appendix. It should be noted that Li and Lin (1995) approximated $u^2(t)$ by $u^2[1 + 2 f(t)]$ for small values of ε, which is equivalent to ignoring the last term of eqn(12).

Realizing that $\sin^2\theta = (1\text{-}\cos 2\theta)/2$, eqn(12) can be rephrased as

$$\dot{\mathbf{x}} = \overline{\mathbf{A}}\,\mathbf{x} + \mathbf{G}\,\mathbf{x}\,[\varepsilon\, sin\theta(t)] - \mathbf{H}\,\mathbf{x}\,[\frac{\varepsilon^2}{2}\, cos2\theta(t)] \qquad (13)$$

where $\overline{\mathbf{A}}$ is the modified system matrix, defined as $\overline{\mathbf{A}} = [\mathbf{A} + (\varepsilon^2/2)\,\mathbf{H}]$. Since the contribution of a term of the order of ε^2 to the stochastic response is of the order of ε^4, the significance of the last term of eqn.(13) in the stability analysis is very limited for $\varepsilon < 1$. However, the $u^2(t)$ process has a non-zero mean which modifies the original system matrix **A** to $[\mathbf{A} + (\varepsilon^2/2)\,\mathbf{H}]$. This modification must be accounted for in the stability analysis to derive accurate solutions. In this respect, the stochastic averaging approximations that ignore the quadratic noise term are conceptually inaccurate. It will be shown that the quadratic wind turbulence term amplifies the destabilizing effect of wind velocity by effectively reducing the stiffness. Therefore, the omission of the quadratic term tends to overestimate the stabilization effect of turbulence (Pandey and Ariaratnam 1998).

The system of equations may be written in the eigenspace using the transformation

$$\mathbf{x} = \mathbf{T}\,\mathbf{z} \qquad (14)$$

where the transformation matrix, **T**, is constructed from the eigenvectors of $\overline{\mathbf{A}}$ (Lin and Li 1993). The transformation leads to

$$\dot{\mathbf{z}} = \mathbf{Q}\,\mathbf{z} + \mathbf{D}\,\mathbf{z}\,[\varepsilon\, sin\theta(t)] - \mathbf{E}\,\mathbf{z}\,[\frac{\varepsilon^2}{2}\, cos2\theta(t)] \qquad (15)$$

where

$$\mathbf{Q} = \mathbf{T}^{-1}\overline{\mathbf{A}}\mathbf{T}, \quad \mathbf{D} = \mathbf{T}^{-1}\mathbf{G}\mathbf{T}, \quad \text{and} \quad \mathbf{E} = \mathbf{T}^{-1}\mathbf{H}\mathbf{T} \qquad (16)$$

In summary, the lateral-torsional motion of the bridge under random wind excitation, represented by $\mathbf{z}(t)$, is modelled as a 12-dimensional stochastic process governed by a system of linear stochastic differential equations (15).

3 Stochastic Stability

The asymptotic sample stability of the torsional vibration of the bridge can be investigated in terms of the Lyapunov exponent defined as

$$\lambda = \lim_{t\to\infty} \frac{1}{t} \log\|\mathbf{z}(t)\| \qquad (17)$$

where $\|\cdot\|$ denotes the Euclidian norm. Under certain mild ergodicity conditions on the random process $f(t)$, the Lyapunov exponent obtained in this manner is a deterministic number and has been shown to converge to the largest Lyapunov exponent with probability one (Oseledec 1968). The necessary and sufficient condition for sample stability is given as $\lambda \le 0$, whereas $\lambda > 0$ implies the bifurcation of $\mathbf{z}(t)$ into a vector with unbounded exponential growth with time.

3.1 EVALUATION OF TOP LYAPUNOV EXPONENT

To develop a numerical algorithm for evaluating the largest Lyapunov exponent, the state variables, **z**, are mapped on to the surface of a 12-dimensional hyper-sphere by introducing spherical coordinates

$$s_k = \frac{z_k}{a}, \quad (|s_k| \le 1, k = 1, 2, \ldots, 12), \quad \text{and} \quad a = \|\mathbf{z}\| = \sqrt{\sum_{k=1}^{12} (z_k^2)} \tag{18}$$

The resulting process, **s**(t), is now a 11-dimensional stochastic process, as **s**(t) must satisfy the unit hyper-sphere condition $\sum_{k=1}^{12} s_k^2 = 1$. The stochastic differential equations (15) are now transformed to

$$\dot{s}_i = \left[(Q_{ij} - \beta\delta_{ij}) + (D_{ij} - \gamma\delta_{ij})f(t) + (E_{ij} - \eta\delta_{ij})f_2(t)\right]s_j \tag{19}$$

where $f(t) = \varepsilon \sin\theta(t)$, $f_2(t) = -\varepsilon^2 \cos 2\theta(t)/2$, δ_{ij} is the Kronecker delta and summation is implied over repeated subscripts. β and η are defined as

$$\beta = Q_{jk}\, s_j\, s_k\,, \quad \gamma = D_{jk}\, s_j\, s_k \quad \text{and} \quad \eta = E_{jk}\, s_j\, s_k \tag{20}$$

The differential equation for the amplitude, a, is of the form:

$$\dot{a} = [\beta + \gamma f(t) + \eta f_2(t)]\, a \tag{21}$$

which leads to the following equation for $\rho = \log a$:

$$\dot{\rho} = [\beta + \gamma f(t) + \eta f_2(t)] \tag{22}$$

It can be written in an integrated form

$$\rho(t) = \rho(0) + \int_0^t [\beta(\tau) + \gamma(\tau) f(\tau) + \eta(\tau) f_2(\tau)] d\tau \tag{23}$$

Dividing both sides by t, taking the limit as $t\to\infty$, and recalling eqn(17), lead to the following expression for the Lyapunov exponent:

$$\lambda = \lim_{t\to\infty} \frac{1}{t} \log a = \lim_{t\to\infty} \frac{1}{t}\, \rho = \lim_{t\to\infty} \frac{1}{t} \int_0^t [\beta(\tau) + \gamma(\tau) f(\tau) + \eta(\tau) f_2(\tau)] d\tau \tag{24}$$

The convergence of expression (23) to the largest Lyapunov exponent is based on the assumption that **s**(t) is an ergodic process. In essence, the Lyapunov exponent is a measure of the average exponential growth rate of the amplitude process $a(t)$ for large values of t, and is a deterministic number with probability one.

3.2 SIMULATION

To compute λ according to expression (24), the stochastic differential equations, (19), are discretized using Euler's scheme and written in an iterative form (Wedig 1996)

$$s_i^{(n+1)} = s_i^{(n)} + \left\{ \left[Q_{ij} - \beta^{(n)}\delta_{ij} + (D_{ij} - \gamma^{(n)}\delta_{ij}) f^{(n)} + (E_{ij} - \eta^{(n)}\delta_{ij}) f_2^{(n)} \right] s_j^{(n)} \right\} \Delta t$$

$$(i, j = 1 \text{ to } 12, \text{ and } n = 1 \text{ to } N) \tag{25}$$

where n is the number of the iteration cycle, Δt is the time step used in the discretization, and N is the total number of iterations in the simulation. Note that $\beta^{(n)}$, $\gamma^{(n)}$ and $\eta^{(n)}$ are calculated from eqn(20) using $s_j^{(n)}$. The stochastic wind turbulence, $f(t)$, is discretized in a similar manner

$$f^{(n+1)} = \varepsilon \sin\{\theta^{(n+1)}\} \tag{26}$$

where

$$\theta^{(n+1)} = \theta^{(n)} + [\nu\, \Delta t + \sigma\, \Delta W^{(n)}] \tag{26}$$

$\Delta W^{(n)}$ represents an increment of the unit Wiener process, simulated as

$$\Delta W^{(n)} = R_n \sqrt{\Delta t} \tag{28}$$

where R_n is the standard normal random variable with zero mean and unit standard deviation. Finally, the largest Lyapunov exponent is evaluated as

$$\lambda = \frac{1}{N} \sum_{n=1}^{N} [\beta^{(n)} + \gamma^{(n)} f^{(n)} + \eta^{(n)} f_2^{(n)}] \tag{29}$$

4 Numerical Results

Numerical calculations were carried out using structural and aerodynamic data for a bridge studied by Li and Lin (1995), which are as follows:

$\omega_\alpha = 2.32$ rad/s, $\zeta_\alpha = 0.01$, $\omega_y = 1.005$ rad/s, $\zeta_y = 0.01$
$B = 16.4$ m, $I = 8.94 \times 10^5$ kg-m, $m = 1.36 \times 10^4$ kg/m, and $\rho = 1.226$ kg/m^3.

Other aerodynamic constants for the bridge are presented in the Appendix. The deterministic flutter wind velocity (u_c) and the corresponding frequency (ω) were found to be 49.2 m/s, and 2.25 rad/s, respectively.

Three different turbulence conditions were considered: (1) narrowband turbulence nearly tuned to $2\omega_1$ (ν=4.49, σ=0.2), (2) detuned narrowband turbulence (ν=2.5, σ=0.2), and (3) wideband turbulence (ν=0.2, σ=10). In all cases, numerical simulation involved N = 10^7 samples, and a time step of $\Delta t = 10^{-3}$ *sec* was used in Euler's integration scheme.

Firstly, the quadratic noise term was ignored from the stability analysis which amounts to neglecting the **H x** $[\varepsilon \sin\theta(t)]^2$ from eqn. (12). The effect of turbulence intensity ε on the sample stability for different values of the mean wind velocity is shown in Figure 2. The mean wind velocity plotted on the vertical axis was normalized by the deterministic flutter velocity (u_c) of 49.2 m/s.

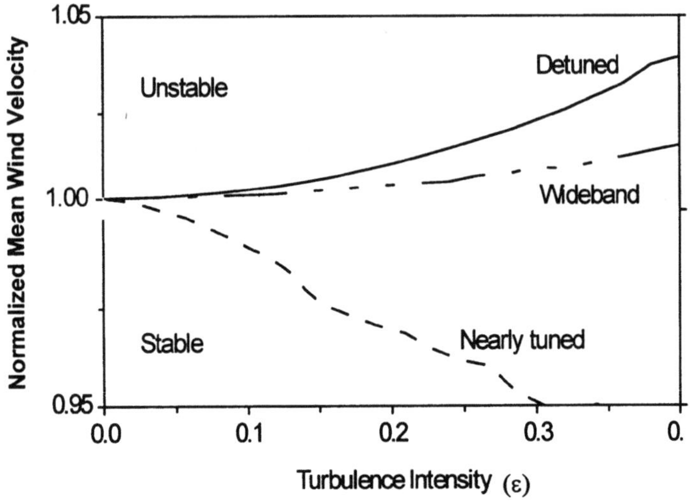

Figure 2: Stability boundaries for bridge motion without considering the quadratic noise term

In the first case of narrowband nearly tuned excitation, the turbulence intensity has a destabilizing effect in the sense that an increase in the intensity decreases the critical mean wind velocity below the flutter velocity. However, the modest stabilizing effect of increasing turbulence intensity is seen in the cases of narrowband detuned and wideband excitations. Similar to earlier analytical studies (e.g., Li and Lin 1995), the simulation results confirm qualitatively the stabilization of the bridge motion above the flutter velocity as a result of an increase in the turbulence intensity.

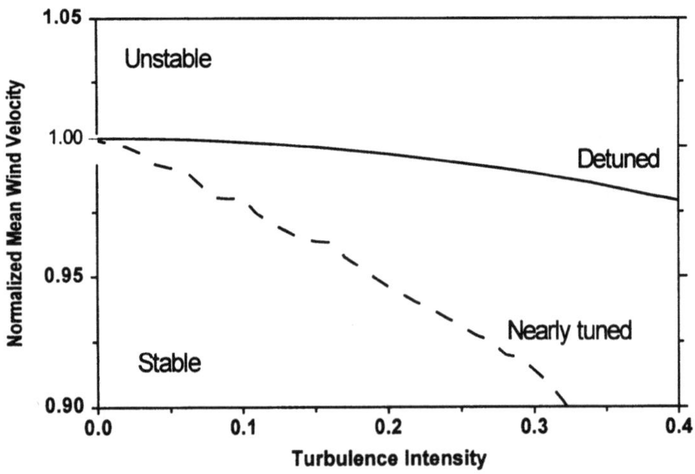

Figure 3: Stability boundaries for bridge motion considering the quadratic noise term

In the second stage, the quadratic noise term was retained in the analysis. In Figure 3, stability boundaries for detuned and nearly tuned excitations are plotted. It is interesting to note that the stabilization effect of turbulence intensity is now absent.

Moreover, the destabilization of nearly-tuned response, case (1), is further amplified by the quadratic noise term. In this respect, analytical approximations that ignore the quadratic noise term lead to qualitatively different results.

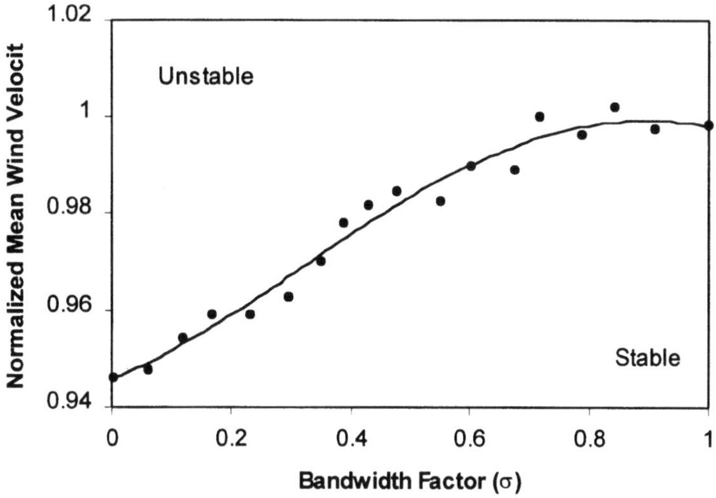

Figure 4: Effect of turbulence bandwidth on the bridge stability: Nearly tuned excitation (ε= 0.2, ν=4.49)

The effect of the bandwidth factor, σ, on the stability of bridge motion under nearly tuned narrowband excitation is displayed in Figure 4. An increase in σ also increases the critical mean wind velocity, and in this sense stabilizes the bridge motion as compared to the purely harmonic excitation.

5 Concluding Remarks

A simulation-based numerical approach is presented to investigate the effect of stochastic wind turbulence on the stability of lateral-torsional bridge motion, modelled by a two degrees of freedom oscillator. In the analysis, the random fluctuation of wind velocity is represented by a versatile mathematical model that involves a periodic excitation with random phase modulation. The stability condition refers to the asymptotic sample stability, which is evaluated in terms of the largest Lyapunov exponent.

The lateral-torsional instability under three different turbulence conditions was examined: (1) narrowband turbulence nearly tuned to parametric resonance frequency, (2) detuned narrowband turbulence, and (3) wideband turbulence.

In the case of detuned excitation, increasing values of turbulence intensity resulted in a decreasing trend for the critical mean wind velocity in comparison to the deterministic flutter velocity. The stabilization effect of turbulence, as indicated by stochastic averaging approximations reported in the literature, is seen only when the quadratic noise term is ignored from the simulation. In the case of narrowband nearly tuned excitation, the turbulence intensity has a destabilizing effect, which is overestimated in absence of the quadratic noise term from the analysis.

In summary, the possible stabilizing effect of detuned narrowband turbulence, due perhaps to aerodynamic coupling, is not verified when the quadratic noise term was incorporated in the simulation-based analysis . It should however be noted that this conclusion is valid within the construct of the mathematical model and may not reflect the true physical mechanism of bridge instability, since the analysis ignores various nonlinear effects, spanwise correlation of wind excitation, and changes in flutter derivatives due to variable flow pattern.

6 References

Bucher, C.G. and Lin, Y.K. (1988). Stochastic stability of bridges considering coupled modes. *J. Engineering Mechanics*, ASCE, 114(12), 2055-2071.

Huston, D.R., Bosch, H.R., and Scanlan, R.H. (1988). The effect of fairings and of turbulence on the flutter derivatives of a notably unstable bridge deck. *J. Wind Engineering and Industrial Aerodynamics*, 29, 339-349.

Li, Q.C. and Lin, Y.K. (1995). New stochastic theory for bridge stability in turbulent flow-II. *J. Engineering Mechanics*, ASCE, 121(1), 102-116.

Lin, Y.K. and Li, Q.C. (1993). New stochastic theory for bridge stability in turbulent flow. *J. Engineering Mechanics*, ASCE, 119(1), 113-127.

Lin, Y.K. and Ariaratnam, S.T. (1980). Stability of bridge motion in turbulent winds. *J. Structural Mechanics*, 8(1), 1-15.

Oseledec, Y.I. (1968). "A Multiplicative ergodic theorem: Lyapunov characteristic numbers for dynamical systems". *Trans. Moscow Math. Society*, 19, 197-231.

Pandey, M.D. and Ariaratnam, S.T. (1998). Stability analysis of wind-induced torsional motion of slender bridges. Accepted for Publication in *J. Structural Safety*.

Scanlan, R.H., Beliveau, J.G., and Budlong, K.S. (1974). Indicial aerodynamic functions for bridge decks. *J. Engineering Mechanics Div.*, ASCE, 100(4), 657-672.

Wedig, W. (1996). Iterative schemes for stability problems with non-singular Fokker-Planck Equations. *Int. J. Nonlinear Mechanics*, 31(5), 701-715.

Acknowledgements

The authors gratefully acknowledge the financial support for this study provided by the Natural Sciences and Engineering Research Council of Canada.

Appendix : Elements of Matrices [A], [G] and [H]

Introducing constants $p = \rho/m$ and $q = \rho B/I$, non-zero elements of the matrices **A** and **G** (eqn 12) are defined as

A(1,2) = A(3,4) = A(2,5) = A(2,6) = A(2,7) = A(2,8) = 1.0
A(4,9) = A(4,10) = A(4,11) = A(4,12) = 1.0

$G(2,1) = 2qB\, C_1(1)\, u^2$, $G(2,2) = qB^2\, C_2(1)\, u$

$A(2,1) = -\,\omega_\alpha^2 + G(2,1)/2$, $A(2,2) = G(2,2) - 2\,\omega_\alpha\,\zeta_\alpha$, $A(2,3) = G(2,3)/2 = q\, C_1(2)\, u^2$

$A(2,4) = G(2,4) = qB\, C_2(2)\, u$; $A(4,1) = G(4,1)/2 = pB\, C_1(3)\, u^2$; $A(4,3) = p\, C_1(3)\, u^2 - \omega_h^2$

$A(4,2) = G(4,2) = pB^2\, C_2(3)\, u$; $G(4,3) = = 2\, p\, C_1(3)\, u^2$, $G(4,4) = pB\, C_2(3)\, u$

$A(4,4) = G(4,4) - 2\,\zeta_y\,\omega_y$; $A(5,5) = G(5,5) = -D_3(1)\, u/B$; $A(6,6) = G(6,6) = -D_4(1)\, u/B$

$A(7,7) = G(7,7) = -D_3(2)\, u/B$; $A(8,8) = G(8,8) = -D_4(2)\, u/B$

$A(9,9) = G(9,9) = -\, D_3(3)\, u/B$; $A(10,10) = G(10,10) = -D_4(3)\, u/B$

$A(11,11) = G(11,11) = -\, D_3(4)\, u/B$; $A(12,12) = G(12,12) = -D_4(4)\, u/B$

$A(5,2) = G(5,2)/2 = qB\, C_3(1)\, u^2$; $A(6,2) = G(6,2)/2 = qB\, C_4(1)\, u^2$

$A(7,4) = G(7,4)/2 = q\, C_3(2)\, u^2$; $A(8,4) = G(8,4)/2 = q\, C_4(2)\, u^2$

$A(9,2) = G(9,2)/2 = pB\, C_3(3)\, u^2$; $A(10,2) = G(10,2)/2 = pB\, C_4(3)\, u^2$

$A(11,4) = G(11,4)/2 = p\, C_3(4)\, u^2$; $A(12,4) = G(12,4)/2 = p\, C_4(4)\, u^2$

The non-zero elements of matrix **H** of eqn.(12) are defined as

H(2,1) = G(2,1)/2,	H(2,3) = A(2,3),	H(4,1) = A(4,1)
H(4,3) = G(4,3)/2 ,	H(5,2) = A(5,2),	H(6,2) = A(6,2)
H(7,4) = A(7,4),	H(8,4) = A(8,4),	H(9,2) = A(9,2)
H(10,2) = A(10,2),	H(11,4) = A(11,4),	H(12,4) = A(12,4)

Note that $C_i(j)$ and $D_k(j)$, i,j=1 to 4 and k = 3,4, are aerodynamic constants. Numerical values of these constants for a bridge considered in the paper are given in the following Table 1.

Table 1: Aerodynamic Data

Components		Aerodynamic Constants					
No	Type	C_1	C_2	C_3	C_4	D_3	D_4
1	M_α	0.043	- 0.33	0.22	0.84	0.18	2.2
2	M_y	0.0019	1.5	29	-34	95	74
3	L_α	-2.6	- 1.1	- 2.2	5.2	2.2	1
4	L_y	0	- 2.6	0.035	- 0.91	0.72	13

INFLUENCE OF STOCHASTIC EFFECTS ON FLOW INDUCED VIBRATIONS IN TUBE BUNDLES

K. Popp and O. Romberg

Institute of Mechanics, University of Hannover
Appelstr. 11, 30167 Hannover, Germany

1. Introduction

There are different excitation mechanisms that cause fatal damages in heat exchanger tube bundles subjected to cross-flow, cp. Weaver & Fitzpatrick [27]. In the present paper the two most important mechanisms are considered, i.e. *fluidelastic instability* and *turbulent buffeting*. These mechanisms which can be investigated separately influence each other. Particularly the intensity of the turbulent buffeting can have a significant influence on the onset of instability. Figure 1 illustrates the experimental determination of the stability threshold for a tube within a bundle. Here, the amplitude $\hat{a}$ related to the tube diameter d is plotted in dependence of the reduced gap velocity V_r (Figure 1a). A reliable determination of stability boundaries requires a stability criterion, cf. Chen [11]. In this study the amplitude criterion used also by Austermann & Popp [8] has been applied. A sudden change of small amplitudes due to turbulent buffeting to large amplitudes caused by fluidelastic instability characterises the stability point that depends on the adjusted damping, described by the mass-damping parameter δ_r(Figure 1b).

1.1. Fluidelastic instability

Fluidelastic instability is a very dangerous excitation mechanism. Because of the resulting large tube amplitudes this phenomenon must be avoided in any case. To represent the stability behaviour the Connors diagram (cf. Figure 1b) which shows the stability boundaries as a function of the mass-damping parameter $\delta_r = \mu\delta/(\rho d^2)$ and the reduced gap velocity $V_r = u/(f_1 d)$, where μ is the mass per unit length, δ is the logarithmic decrement of damping, ρ is the fluid density, u is the velocity in the gap between the tubes and f_1 is the first natural frequency of the tube. This standard stability diagram gives a general idea of the stability behaviour of tube arrays in cross-flow which is essential for heat exchanger design. Since a reliable theoretical model is missing an experimental investigation of the stability boundaries is required.

S. Narayanan and R.N. Iyengar (eds.),
IUTAM Symposium on Nonlinearity and Stochastic Structural Dynamics, 197–208.

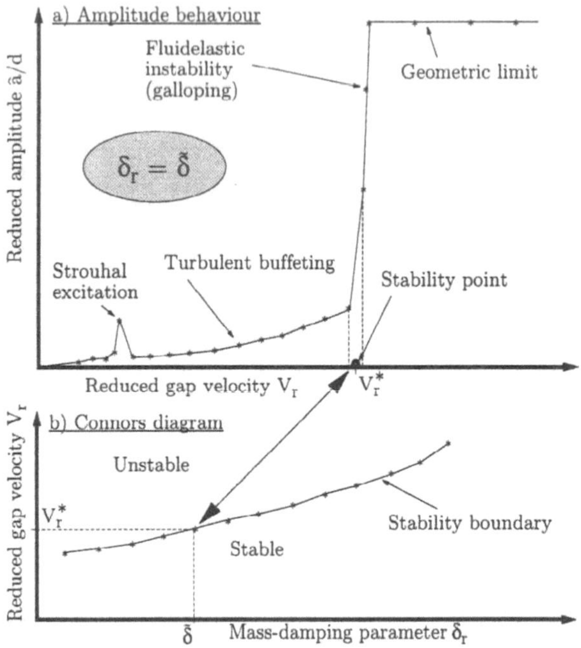

Figure 1: Experimental determination of the stability boundary of a flexible tube in a bundle subjected to cross-flow, a) amplitude behaviour, b) Connors diagram

Fluidelastic instability appears in two different mechanisms: as fluid-damping-controlled (galloping) and as fluidelastic-stiffness-controlled instability. The galloping mechanism leads generally to tube vibrations in cross-flow direction. For the tube excitation those components of the fluid force are important which are proportional to the tube velocity. In case of the fluidelastic-stiffness-controlled mechanism the instability results from coupling effects of several tubes in an array by the fluid. Here, the stability investigations requires a fully flexible bundle of tubes.

Which of these two mechanisms is dominant depends on the configuration of the tube array as well as on the fluid density. It is also possible that a combination of both effects occurs. In case of galloping instability the equilibrium position of *one single flexibly mounted tube* in an otherwise fixed array becomes unstable. This mechanism can be *isolated* using such a single tube. Moreover, for certain array configurations, where the galloping mechanism is dominant, there is a coincidence of the stability boundaries with that of a fully flexible bundle. Thus, with restrictions, the stability of a fully flexible bundle can be checked by the stability of a single flexibly mounted tube, cf. Lever & Weaver [18] and Austermann & Popp [8].

A theoretical model that describes the stability boundaries with respect to the galloping mechanism has been introduced by Lever & Weaver [17]. This analytical model is based on fully developed non stationary flow channels between the tubes which cause instability. Therefore, the aim of one part of the recent studies is to *disturb* these flow channels by tripwires and turbulence grids without changing the flow capacity and to investigate the influence on the stability behaviour. The observed damage in the first few rows leads also to the question of the influence of *increased turbulence* at the inlet of the bundle on the stability behaviour of the tubes.

1.2. Turbulent buffeting

Turbulent buffeting occurs in tube bundles subjected to cross-flow because the first tube rows of a bundle are working as turbulence generators. Thus, within the first few rows inside a tube array turbulence intensity increases downstream. The resulting excitation mechanism causes small vibration amplitudes and represents the reason for long-term wear and fatigue in heat exchanger tubes. Here, for the design of heat exchangers as well as for the calculation of lifecycles the tube amplitudes are important quantities. Half-empirical formulations for the calculation of the amplitude behaviour as a function of the gap velocity are founf in the literature, cp. Romberg [24]. To calculate the amplitudes due to turbulent buffeting and to get informations about fluidelastic instability it is useful to determine experimentally the fluid forces acting on the tubes. There are two different possibilities to investigate these fluid forces, either they can be measured as integral values from the reaction forces or they can be gained integrating the measured pressure distribution on the surface of the tube. The latter has been done in the present investigation.

2. Experimental apparatus

The experimental investigations have been carried out at extended set-ups used by Andjelić & Popp [3], Austermann & Popp [8] as well as Romberg & Popp [22]. The arrays consist of aluminium tubes with an outside diameter of d=80mm and a length of l=800mm. The arrays are mounted within the test section of a wind tunnel. The fixings in the side walls of the test section allow to realize the tube patterns which are usually found in heat exchangers, cf. Table 1, where P is the distance of the tube midpoints. The expressions 'triangle' and 'square' mean the geometric unit cell of a configuration, 'rotated' and 'normal' shows the orientation to the cross-flow, cf. Table 1 and Figure 2. For most of the array configurations one single flexibly mounted tube in an otherwise fixed bundle has been investigated. This tube is equipped with a linear iso-viscoelastic mounting, cf. [7]. The tube is free to vibrate in streamwise (x) and cross-stream (y) direction and it is possible to vary row number, equilibrium position and damping of the tube using adjustable eddy current damper. As a matter of fact the stability boundaries for a fully flexible normal triangular bundle are not in accordance with that of only one flexibly mounted tube. Therefore, the investigations for the normal triangular array with P/d = 1.25 are carried out at a fully flexible bundle, where 9 tubes have variable damping and linear iso-viscoelastic mountings. The tube motions in both apparatusses have been determined by a set of strain gauges.

To determine the exciting fluid forces the flexible tube inside the fixed array is equipped with 44 pressure sensors. The sensors are placed directly under the surface inside the test tube to avoid cavity. The acting fluid force is determined by

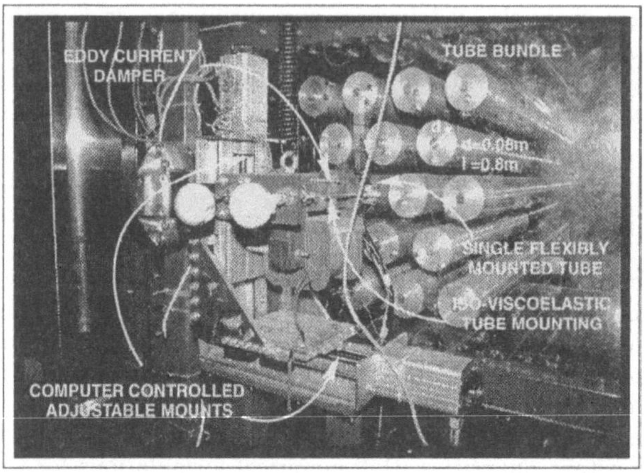

Figure 2: One of the experimental set-ups, with a rotated triangular array

Table 1: Tube bundle configuration investigated here

Symbol	Configuration	P/d	One single flexible tube in a fixed array	fully flexible bundle
△ :	rotated triangle	1.375	×	
△ :	rotated triangle	1.25	×	
◁ :	normal triangle	1.25	×	
◁ :	normal triangle	1.25		×
□ :	in-line square	1.25	×	
◇ :	normal square	1.25	×	

integrating over the whole tube surface, cf. Romberg & Popp [24]. The tube motions in $x-$ and $y-$direction can be measured together with the acting fluid forces.

3. The influence of turbulence

The stability behaviour for several tube configurations visualized by means of Connors diagrams can be found in Andjelić [2] and Austermann [7] together with attempts of a theoretical prediction on the basis of the model of Lever & Weaver [17], [18]. Figure 3 shows a photo of the excited test tube within the flow scanned out of a video film, where the flow field is visualized by a laser light sheet procedure, cf. Romberg & Popp [22]. The single flexible tube oscillates exclusively in cross-flow (lift) direction due to the fluid-damping-controlled (galloping) mechanism.

In the following the influence of a disturbed flow due to thin tripwires as well as an increased turbulence due to grids on this common behaviour is described for one configuration.

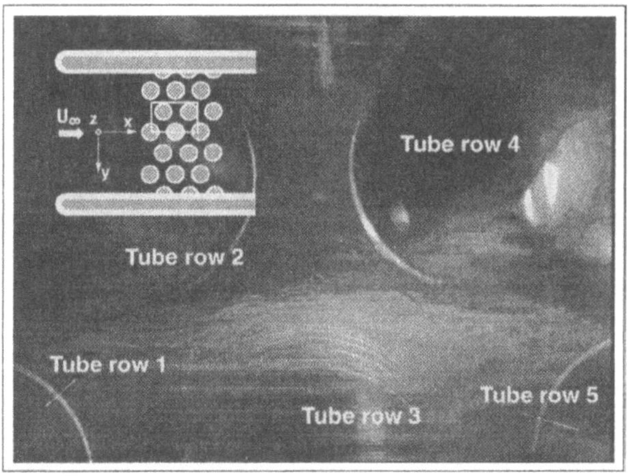

Figure 3: Oscillating pressure test tube within the visualized flow field in the third row of a rotated triangular array with $P/d = 1.375$

3.1. Influence of disturbed flow due to tripwires

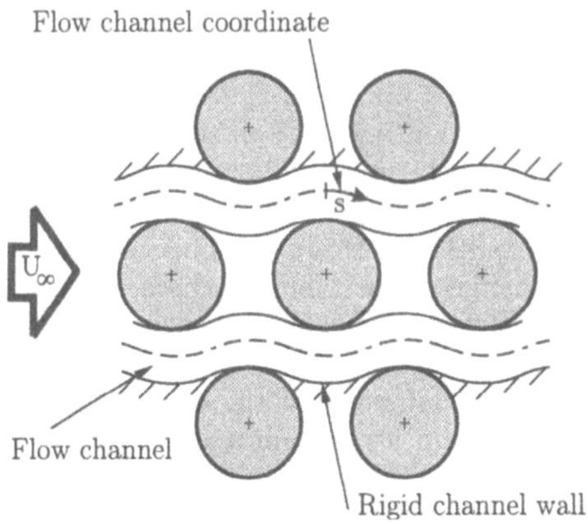

Figure 4: Flow pattern in the 'Wavy wall rigid channel model', by Weaver & Yetisir [28].

Figure 4 shows the fully developed non stationary flow channels for a rotated triangular tube configuration assumed by Weaver & Yetisir [28]. These channels are essential for the 'Wavy wall rigid channel model' introduced by Lever & Weaver [17]. This simple linear analytical model provides the stability boundaries of a flexibly mounted tube in an otherwise fixed array in the form of

$$\delta_r = F(V_r). \tag{1}$$

The model assumes a non stationary cross-section $A(s,t)$ of an one-dimensional flow channel. The non stationary continuity equation yields the flow velocity $u(s,t)$ and the non stationary Bernoulli equation provides the pressure distribution $p(s,t)$.

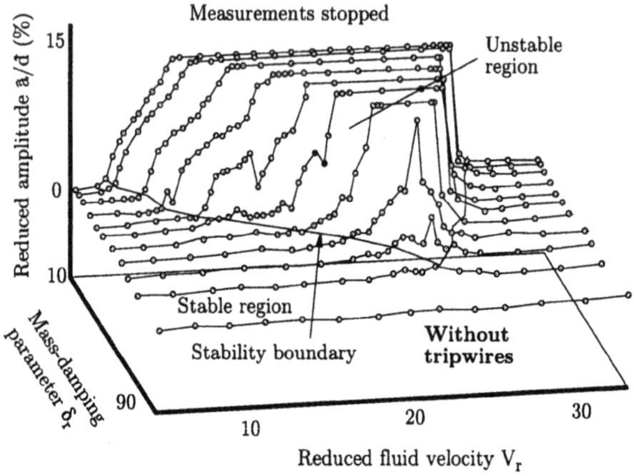

Figure 5: Measured reduced amplitudes as a function of V_r and δ_r for one single flexible mounted tube in the critical third row of an otherwise fixed array with ideal geometry, without tripwires.

The integration over the pressure distribution $p(s,t)$ on the tube surface delivers the fluid force for the equation of motion of the tube. Here, the locations of flow-attachement and flow-separation are decisive parameters. If the term of the total damping disappears the stability boundary of the tube is reached. This results in eq. (1).

To influence the flow, very thin wires (1.25% of the tube diameter d) are pasted on the top and the bottom of each tube of a bundle parallel to the tube axis.

The first documented attempt to disturb the flow field around rigid bodies by using tripwires is due to Prandtl [21]. He put a thin wire on the surface of a ball and found a tremendous change of the flow detachement due to disturbances of the boundary layer. Further investigations with tripwires can be found in [1], [12], [14], [15], [16] and [20].

In the following, the significant influence of the disturbed flow channels on the stability behaviour is shown again for the rotated triangular array with $P/d = 1.375$. Figure 5 shows the measured reduced amplitudes for undisturbed flow as a function of the reduced gap velocity V_r and the mass-damping parameter δ_r for the same tube and ideal geometry. In the stable region the tube moves with small amplitudes in cross- and in-flow direction (turbulent buffeting). In the unstable mode the tube executes galloping motions exclusively in cross-flow direction.

To protect the experimental set-up the measurements have been stopped when the tube reaches a reduced amplitude of about 12%. After the application of the thin Prandtl-tripwires at the same experimental set-up with the same array configuration and parameters, the system shows the amplitude behaviour presented in Figure 6. It can be seen easily that the unstable mode is fully suppressed. Small distortions of the tripwire position due to tube rotations lead to the same stabilization effect.

However, the tripwires were pasted solely on the single flexible tube, this leads to a slight shift of the stability boundary to higher reduced gap velocities only but does not change the stability behaviour. The tripwires pasted on the tubes which lie

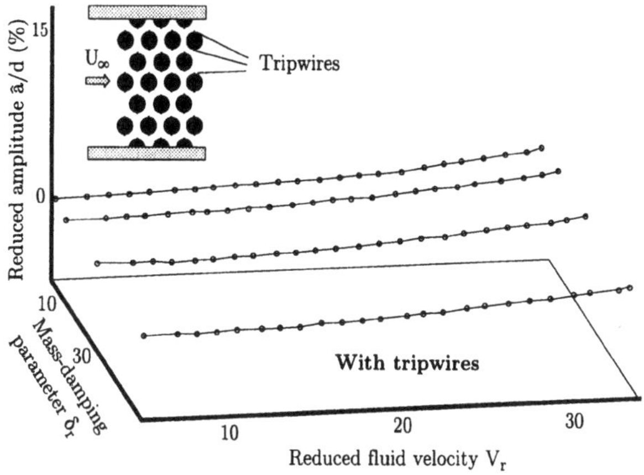

Figure 6: Measured reduced amplitudes as a function of V_r and δ_r for one single flexible mounted tube in the critical third row of an otherwise fixed array with ideal geometry, with tripwires.

upstream to the test tube have much more influence than tripwires on the test tube. A calculation using the Lever & Weaver-model with changed geometry parameters due to tripwires does *not* show the trend of a stabilization. The investigations were carried out for several bundle configurations (also for a fully flexible bundle) as well as several shifts of the equilibrium position of the test tube. The different results are summerized as Connors diagrams in Romberg & Popp [22].
The influence on the stability behaviour of a flow disturbance *inside* a tube bundle leads to the question of the influence of increased turbulence of the upstream flow.

3.2. Influence of increased upstream turbulence

To vary the turbulence field in the test section of a wind tunnel, turbulence grids are generally used. Here, a grid is placed in front of the tube array at a distinct but variable distance X, cf. Figure 7. The grid yields a broadband turbulence distribution. The turbulence has been measured using hot wire anemometers. In this study the turbulence is characterized by two parameters, the turbulence intensity Tu (velocity-fluctuation related to the mean velocity) and the scale length L_x (averaged length of a turbulence eddy). Details about the theoretical description of the turbulence field can be found in Romberg & Popp [23].
The turbulence behaviour of the flow behind a grid has been investigated by many authors, e. g. Baines & Peterson [9], Soper [25] and Southworth & Zdravkovich [26]. With increasing grid distance X the turbulence intensity Tu decreases and the scale length L_x grows. The experimental investigations in [23] show a good agreement with the theoretical consideration of the decay of turbulence by Frenkiel [13]. Depending on the used grid and the investigated bundle configuration homogeneous isotropic turbulence from $Tu = 3.5\% - 23\%$ have been generated. The $Tu-$values are largely independent from the fluid velocity U_∞ in the considered parameter region. However, the grid distance X, the mesh size s, the bar width a, the geometry of the bars and the tube bundle itself have an influence on the turbu-

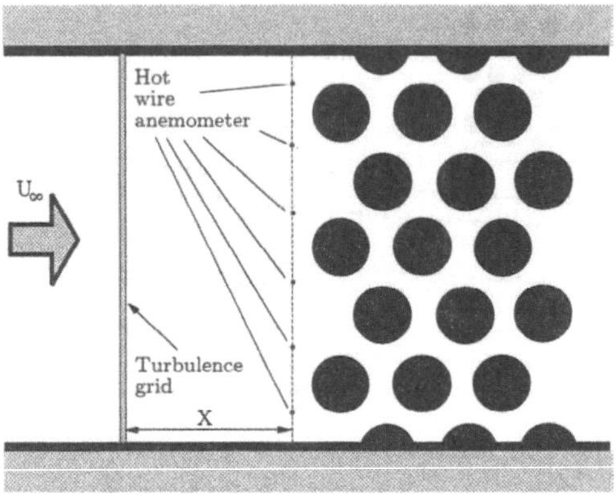

Figure 7: Experimental set-up for increasing the turbulence at the inlet of the tube bundle.

lence intensity Tu and the scale length L_x. Figure 8 shows the corresponding three dimensional stability diagram for increased turbulence again for the configuration described above (third row, rotated triangular array, $P/d = 1.375$). Here also a

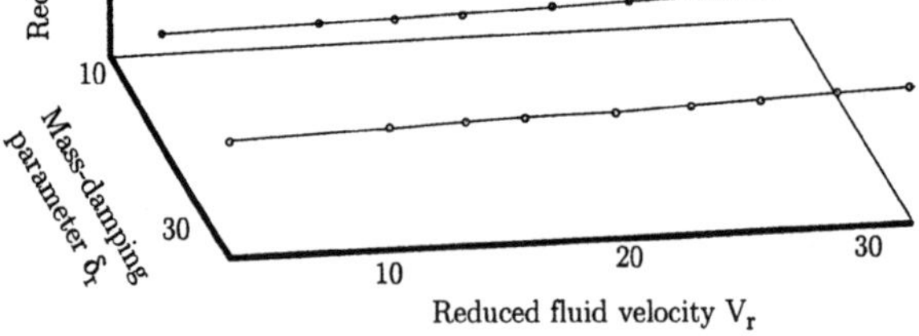

Figure 8: Measured reduced amplitudes as a function of V_r and δ_r for one single flexibly mounted tube in the third row of an otherwise fixed rotated triangular array with increased turbulence

significant stabilization due to increased turbulence can be seen. Figure 9 shows the transition of the reduced amplitudes depending on the reduced gap velocities V_r for constant damping ($\delta_r = 15$) when the distance X and, thus, the turbulence intensity Tu as well as the scale Length L_x is changed. This transition is similar to the case when for undisturbed flow ($X/a = \infty$) the damping is changed, cp. Figure 5. The summarized results concerning increased turbulence for other tube rows as well as for a fully flexible bundle can be found in Romberg & Popp [23].

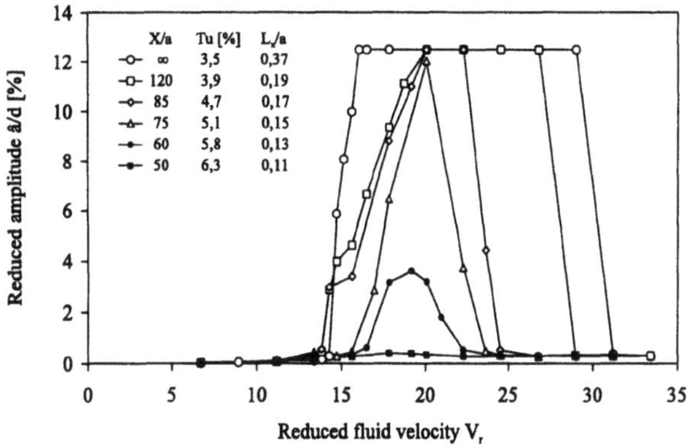

Figure 9: Transition of the reduced amplitude for changing grid distances X/a and corresponding turbulence parameters Tu and L_x (grid: $s = 25mm, a = 3.4mm$)

To find an explanation for the stabilizing effects of tripwires and increased turbulence, in the following we take a look at the fluid forces acting on the tubes and the damping.

4. Investigation of fluid forces and tube damping

With the pressure test tube it is possible to determine the non-stationary fluid force vector $\mathbf{f}(t)$ as a force per unit length in tube midplane as well as the correlation length L_c for each angle φ at the circumference. The fluid force vector $\mathbf{f}(t)$ is calculated by integrating the simultaneously measured pressure distribution $p(z = 0, \varphi, t)$ over the whole circumference. The correlation length L_c is coming from a statistical derivation, where the cross-correlation functions between the pressure signals $p(z, \varphi_i, t)$ somewhere on the tube surface and a reference signal $p(z = 0, \varphi_i, t)$ in tube mid plane are determined. Therefore, the fluid force $\mathbf{F}(t)$ acting on the whole tube can be considered as an excitation. Details of the theoretical description of the fluid forces can be found in Romberg [24].
The experimental results show that for the *stable* parameter region the spatial correlations are nearly identical for a flexibly and a rigid mounted test tube. Moreover, it has been found that these values are largely independent from the reduced velocity V_r. Beyond the stability boundary, where fluidelastic instability due to galloping appears, suddenly the correlation lengths increase to very large values, cf. Figure 10. This holds for all investigated tube configurations where galloping can be found. Tests in still air show clearly that this is *not* due to inertia effects inside the pressure transducers or the cross-flow velocity of the tube. The dynamic pressure due to the tube velocity is 2 to 4 orders of magnitude lower than the measured pressure fluctuation inside the flow field. If we consider the 'route to instability', it can be found that the correlation increases already if *small* stable limit cycles occur, cf. Figure 10. The fluctuation for the smallest amplitude is due to the difficulty of adjusting such a small stable limit cycle for the whole recording time. The measured fluid forces show harmonic character and lead to a classical resonance behaviour of the tube. Finally, it has been found that beyond the stabil-

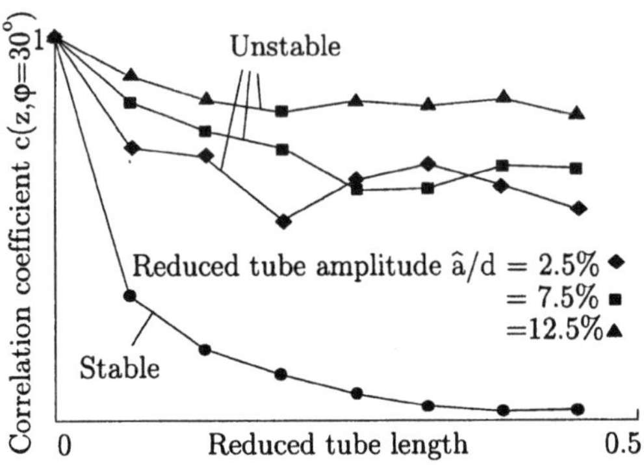

Figure 10: Local correlation coefficient c (at $\varphi = 30^o$)at the stability boundary for several stable limit cycles

ity boundary, where the fluid-damping-controlled mechanism (galloping) occurs, the correlation length of the acting fluid forces is in the order of magnitude of the tube length.

A possible explanation for the stabilizing effect due to flow disturbance and increased turbulence is a reduction of the correlation length. Recent studies shows that tripwires seem to destroy the large correlation length while galloping is dominant. By now it is not clear whether the suppression of the correlation is due to the presence of turbulence or just a consequence of the absence of the large tube amplitudes.

In the literature the stabilizing effect of turbulence has been reported e.g. by Bucher an Lin [10] for bridge structures under turbulent wind. An intuitive explanation has been given which says that turbulence may help to feed energy from the least stable mode to the more stable modes. Thus, coupled modes have to be considered in the analysis as done by Namachchivaya and Lin [19]. Also in the mathematical literature stabilizing by noise is an important topic, see Arnold et. al. [4], [5], [6]. For linear systems it has been found, that stabilization by white noise is possible if and only if the trace of the system matrix, i. e. the sum of the eigenvalues, is negative. This supports the above mentioned intuitive explanation. Regarding these results, in our case we have performed damping measurements for tube motions in flow direction (x-direction) and compared the results with those for motions across the flow direction (y-direction). In fact, it has been found that the tube damping for the highly damped in-flow motion decreases slightly with incrasing turbulence, that also supports the above mentioned explanation. However, further experiments are needed.

5. Conclusions

Experimental investigations of the stability behaviour of tube bundles subjected to cross-flow have been carried out. The experimental set-up allows to realise several bundle configurations with one single flexibly mounted tube in an otherwise fixed array as well as with a fully flexible bundle for a certain array configuration.

The presented results show clearly that a flow disturbance inside the bundle due

to tripwires on the tube surfaces as well as an increased turbulence intensity at the inlet of the bundle using turbulence grids can suppress the dangerous fluid-damping controlled (galloping) instability of a flexible tube in a bundle. These effects can be of interest for designers and operators of heat exchangers.
For the explanation of the measured effects further investigations with respect to the fluid forces as well as the damping behaviour for different tube modes are necessary.

Acknowledgement

The authors gratefully acknowledge the financial support of this research work by the DFG (Deutsche Forschungsgemeinschaft), contract number Po 136/17/1+2).

6. References

[1] AIBA, S.: Heat Transfer Around a Circular Cylinder with a Single Tripwire, *Heat Transfer-Japanese Research* **21** (5) (1992), 466-482.

[2] ANDJELIĆ, M.: Stabilitätsverhalten querangeströmter Rohrbündel mit versetzter Dreieckteilung. Ph. D. Thesis, Hannover University, Inst. f. Mechanik, 1989, in German.

[3] ANDJELIĆ, M.; POPP, K.: Stability Effects in a Normal Triangular Cylinder Array, *Journal of Fluids and Structures* **3** (1989), 165-185.

[4] ARNOLD, L.;CRAUEL, H.; WIHSTUTZ, V.: Stabilization of Linear Systems by Noise, *SIAM J. Control and Optimation* **21** (1983), 451-461.

[5] ARNOLD, L.: Stabilization by Noise Revisited, *Z. Angew. Math. Mech.* **70** (1990), 235-246.

[6] ARNOLD, EIZENBERG, A.; WIHSTUTZ, V.: Large Noise Asymptotics of Invariant Measures, with Applications to Lyapunov Exponents, *Stochastics and Stochastics Reports* **59** (1996), 71-142.

[7] AUSTERMANN, R.: Zum Schwingungsverhalten querangeströmter Rohrbündel bei Anregung vom Galloping-Typ. Ph. D. Thesis, Hannover University, Inst. f. Mechanik, 1993, in German.

[8] AUSTERMANN, R.; POPP, K.: Stability Behaviour of a Single Flexible Cylinder in Rigid Tube Arrays of Different Geometry Subjected to Cross-Flow, *Journal of Fluids and Structures* **9** (1995), 303-322.

[9] BAINES, W. D.; PETERSON, E. G.: An Investigation of Flow Through Screens, *Transactions of the ASME* **73** (1951), 467-480.

[10] BUCHER, C.G.; LIN, Y.K.: Stochastic Stability of Bridges Considering Coupled Modes,*J. Eng. Mechanics, ASCE* **114** (1988), 2055-2071.

[11] CHEN, S. S.: Some Issues Concerning Fluidelastic Instability of a Group of Circular Cylinders in Cross-Flow, *Proceedings of 1988 International Symposium on Flow Induced Vibrations and Noise in Cylinder Arrays.* Vol. 3: Flow Induced Vibrations and Noise in Cylinder Arrays, (ed. by M. P. Païdoussis, S. S. Chen & M. D. Bernstein), (1988), 1-24.

[12] FAGE, A.; WARSAP, J. H.: The Effects of Turbulence and Surface Roughness on the Drag of a Circular Cylinder, *Aeronautic Research Council R&M No. 1283*, (1929).

[13] FRENKIEL, F. N.: The Decay of Isotropic Turbulence, *Transactions of the ASME* **70** (1948), 311-321.

[14] HEINECKE, E. P.: Fluidelastic Vibrations in Heat Exchangers with Tubes in Cross-Flow, *Keswick Conf. U.K. Session 2*: Flow Induced Vibrations in Heat Exchangers (1978).

[15] JAMES, D.; TRUONG, Q.-S.: Wind Load on Cylinder with Spanwise Protrusion, *Journal of the Engineering Mechanics Division* **1** (1972), 72-81.

[16] KRAEMER, K.: Umschlagmessungen hinter Stolperdrähten. In: *Bericht 59/A/21, Aerodynamische Versuchsanstalt Göttingen*, Göttingen, Germany, 1959, in German.

[17] LEVER, J. H.; WEAVER, D. S.: A Theoretical Model for Fluidelastic Instability in Heat Exchanger Tube Bundles, *Journal of Pressure Vessel Technology* **104** (1982), 147-158.

[18] LEVER, J. H.; WEAVER, D. S.: On the Stability of Heat Exchanger Tube Bundles, Part I: Modified Theoretical Model, *Journal of Sound and Vibration* **107** (1986), 375-392.

[19] NAMACHCHIVAYA, N. SRI; LIN, Y. K.: Stochastic Averaging for Nonlinear Dynamical Systems with High Damping, *Probab. Engng. Mechanics* **3** (1988), 159-167.

[20] PEARCEY, H.; CASH, R.; SALTER, I: Flow Past Circular Cylinders: Simulation of Full-Scale Flows at Model Scale, *NMI R 131; OT-R-8208*, (1982).

[21] PRANDTL, L.: Der Luftwiderstand von Kugeln, *Nachr. Ges. Wiss. Goettingen*, (1914), 177-190, in German.

[22] ROMBERG, O.; POPP, K.: The Influence of Tripwires on the Stability of Tube Bundles Subjected to Cross-Flow, *ASME Pressure Vessels and Piping Conference*, Montreal, Canada **328** (1996), 3-10.

[23] ROMBERG, O.; POPP, K.: The Influence of Upstream Turbulence on the Stability of Tube Bundles Subjected to Cross-Flow, *ASME Pressure Vessels and Piping Conference*, Montreal, Canada **328** (1996), 11-18.

[24] ROMBERG, O.: Zum Turbulenzeinfluß auf das Schwingungsverhalten querangeströmter Rohrbündel. Ph. D. Thesis, Hannover University, Inst. f. Mechanik, 1998, in German.

[25] SOPER, B. M. H.: The Effect of Grid Generated Turbulence on the Fluidelastic Instability of Tube Bundles in Cross-Flow, *Taborek, J.; Hewitt, G. F.; Afgan, N.: Heat Exchangers: Theory and Praxis* (1982), 325-337.

[26] SOUTHWORTH, P. J.; ZDRAVKOVICH, M. M: Effect of Grid Turbulence on the Fluidelastic Vibrations of In-Line Tube Banks in Cross-Flow, *Journal of Sound and Vibration* **39** (4) (1975), 461-469.

[27] WEAVER, D. S. AND FITZPATRICK, J.A.: A Review of Flow Induced Vibrations in Heat Exchangers, *Proceedings BHRA Conference on Flow-Induced Vibrations* (ed. by R. King), Bowness-on-Windermere, UK, pp. 1-17.

[28] WEAVER, D. S.; YETISIR, M: On an Unsteady Theory for Fluidelastic Instability of Heat Exchanger Tube Arrays, *Int. Symp. on Flow-Induced Vibration and Noise*, Chicago **3** (1988), 181-195.

A PASSIVE HOPPER WITH LOSSLESS COLLISIONS

C. K. REDDY and R. PRATAP
Department of Mechanical Engineering,
Indian Institute of Science, Bangalore, India - 560012.

Abstract

A passive vertical hopping robot is here highly idealised as two vertically arranged masses acted on by gravity and coupled by a linear spring. The lower mass makes dead ($e = 0$) collisions with the rigid ground. The equations of motion can be reduced to a one dimensional map. Fixed points of the map are found in which case the robot hops incessantly. For these conservative solutions the lower mass collides with the ground with zero impact velocity. The interval of attraction for these conservative fixed points depends on system parameters

1. Introduction

There have been successful attempts [1] at making one-legged and multi-legged driven hopping robots that exhibit stable legged locomotion. These robots have to be driven not only to provide stability of motion but also to compensate for the energy lost during collisions. Because of the phased thrust input to these robots one can assume the impacts of the foot with the ground to be an elastic impact [2,3]. However the collisions in reality are always inelastic. The study of driven robots usually clouds the impact of energy loss during collisions on the motion of the system. On the other hand, the motion of passive(uncontrolled) robots heavily depends on these impacts and the consequential energy losses.

The question of energy efficiency in the gait of humans and in simple uncontrolled models of walking robots that mimic human walking is fundamental to the understanding of human walking and has been studied by several researhers [4-6]. Critical to the issue of energy efficiency in walking is the loss of energy during impacts of the foot with the ground. In passive robots, the energy loss in the plastic collisions of the robot with the ground is usually compensated by gravity by making the robot walk on a slope [6].

Passive hopping robots are representatives of a larger class of mechanical systems called impact oscillators. Impact oscillators model a number of physical systems [8,9] . All such systems have usually periodic motion that involves some kind of impact(that leads to non-smooth dynamical systems). The central interest in studying these systems is around stability of such motions, possible bifurcations and chaotic motions. In all the systems represented by impact oscillators, the inelastic collisions produce large stresses in the material which may result in

S. Narayanan and R.N. Iyengar (eds.),
IUTAM Symposium on Nonlinearity and Stochastic Structural Dynamics, 209–220.

premature failure of the different parts of the system. That apart, the energy loss during the collision is crucial when one studies the efficiency of these machines.

Passive hopping robots have not received much attention probably because of the trivial dynamics expected. But a thorough understanding of the dynamics of a passive hopping robot would not only lead to better understanding of the driven hopping robot but would also lead to their efficient performance. Garcia et al. [6] have studied a simple model of a one dimensional walking robot which is based on McGeer's work [7]. The authors show that steady walking on smaller slopes is possible only at smaller speeds so that the energy loss is minimised. So the only way to make a passive robot walk on a level ground with nonzero speed is to design the robot such that the heel strikes the ground gently, ideally with zero speed, which would eliminate the energy lost in the plastic collisions. Schiehlen et al. [10] have studied nonsurge motion of a passive one dimensional hopping robot. The possibility that collisions could be avoided with a totally passive mechanism was suggested by Anindya Chatterjee – private communication. His suggestion motivated this study. In the present paper we study the vertical motion of a one dimensional passive robot modelled as a simple spring mass system and suggest a model that eliminates the loss of energy during impacts completely. It is observed that there are some interesting aspects of the dynamics of this robot around the special solutions obtained for incessant hopping.

2. Mathematical Model and Assumptions

The one dimensional simplified hopping robot appears in Fig. 1. The entire mass of the robot is concentrated in the two rigid masses. The spring is a linear spring and assumed to be massless. The hopping robot is passive. This means that there is no external energy input. The impacts of the lower mass with the rigid ground are perfectly plastic. This means that the velocity of the lower mass becomes zero when it strikes the ground while the velocity of the upper mass is undisturbed. All energy dissipating processes other than the plastic collisions are neglected. The two masses themselves can impact against each other. Relaxed length of the spring is assumed large enough to avoid mid-air collisions.[1]

3. Problem Formulation

3.1. EQUATIONS OF MOTION

The forces acting on both the masses are shown in Fig. 1. The equations of motion for both the masses are written down as,

$$m_1\ddot{x}_1 = -k(x_1 - x_2) - m_1 g \tag{1}$$

$$m_2\ddot{x}_2 = k(x_1 - x_2) - m_2 g \tag{2}$$

[1]The value of the minimum relaxed length to avoid mid-air collisions can be easily found out for a given set of initial conditions and system parameters.

where,

$$\begin{aligned} m_1 &= \text{mass of the upper mass.} \\ m_2 &= \text{mass of the lower mass.} \\ x_1 &= \text{displacement of the upper mass.} \\ x_2 &= \text{displacement of the lower mass.} \\ k &= \text{spring constant.} \\ g &= \text{acceleration due to gravity.} \\ L &= \text{relaxed length of the spring.} \end{aligned}$$

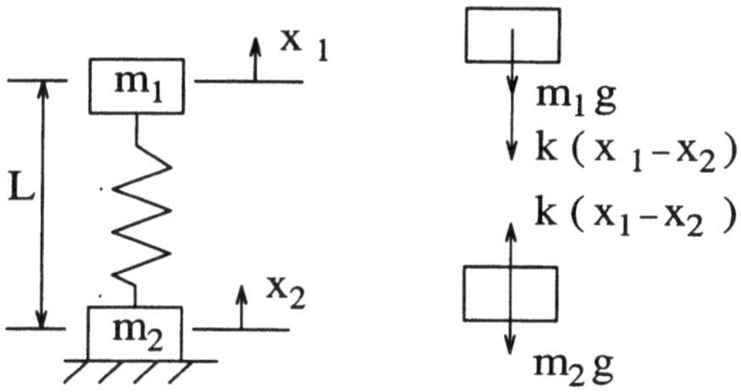

Figure 1. The spring-mass model of the hopping robot.

Eliminating g from Eq. (1) and Eq. (2), we get

$$\ddot{x}_1 - \ddot{x}_2 = \left(\frac{-k(m_1 + m_2)}{m_1 m_2}\right)(x_1 - x_2)$$

Let $x_1 - x_2 = \theta$. Hence,

$$\ddot{\theta} + \omega^2 \theta = 0 \tag{3}$$

where,

$$\omega = \left(\frac{k(1 + M)}{m_2}\right)^{1/2}; \; M = \frac{m_2}{m_1} = \text{mass ratio}$$

3.2. SOLUTION

Equation (3) can be solved to get

$$\theta = \frac{\dot{\theta}(o)}{\omega} \sin \omega t + \theta(o) \cos \omega t; \; \dot{\theta} = \dot{\theta}(o) \cos \omega t - \theta(o) \omega \sin \omega t \tag{4}$$

where, $\theta(o)$ and $\dot{\theta}(o)$ are the initial conditions.

$$\theta(o) = x_1(o) - x_2(o); \; \dot{\theta}(o) = \dot{x}_1(o) - \dot{x}_2(o)$$

3.3. CONDITION TO MAKE THE ROBOT HOP FOREVER

The only loss of energy in the model is due to the collisions of the lower mass with the ground. From basic collision mechanics [12], we know that if the velocity of the lower mass becomes zero when it hits the ground, then there would not be any loss of energy. We shall use the initial conditions when the lower mass is about to move, which would mean

$$x_2(o) = \dot{x}_2(o) = 0; \ x_1(o) = \frac{m_2 g}{k} = x_o$$

Let $\dot{x}_1(o) = V_o$ be initial velocity of the upper mass. Equations in (4) become

$$\theta = \frac{V_o}{\omega} \sin \omega t + x_o \cos \omega t; \ \dot{\theta} = V_o \cos \omega t - x_o \omega \sin \omega t$$

The above equations can be written in terms of dimensionless quantities α and τ as

$$\frac{\theta}{x_o} = \alpha \sin \tau + \cos \tau; \ \frac{\dot{\theta}}{x_o \omega} = \alpha \cos \tau - \sin \tau \tag{5}$$

where,

$$\alpha = \frac{V_o}{x_o \omega} \text{ and } \tau = \omega t$$

We require that when $x_2 = 0$, $\dot{x}_2 = 0$. A second condition is $\ddot{x}_2 = 0$, so that the lower mass remains at rest as the upper mass moves down. The above two conditions imply

$$\theta|_{x_2=0} = x_o \text{ and } \dot{\theta}|_{x_2=0} = -V_o$$

Equations in (5) become

$$1 = \alpha \sin \tau + \cos \tau; \ \ -\alpha = \alpha \cos \tau - \sin \tau \tag{6}$$

Let us determine the time of flight t. Since gravity is the only external force acting, the equation of the centre of mass becomes

$$\ddot{x}_{cm} = -g$$

where,

$$x_{cm} = \frac{m_1(x_1 + L) + m_2 x_2}{m_1 + m_2}; \ \dot{x}_{cm} = \frac{m_1 \dot{x}_1 + m_2 \dot{x}_2}{m_1 + m_2}.$$

Now,

$$V_{cm} = U_{cm} - gt \tag{7}$$

Lct,

$$U_{cm} = \frac{m_1 V_o}{m_1 + m_2} = \bar{V}_{cm}$$

At the end of the flight,

$$V_{cm} = -\bar{V}_{cm}$$

Hence substituting the values in Eq. (7)

$$t = \frac{2\bar{V}_{cm}}{g} = \frac{2m_1 V_o}{g(m_1 + m_2)}$$

which means,

$$\tau = \frac{2\omega m_1 V_o}{g(m_1 + m_2)} = 2\alpha$$

Hence Equations in (6) reduce to,

$$1 = \alpha \sin 2\alpha + \cos 2\alpha; \quad -\alpha = \alpha \cos 2\alpha - \sin 2\alpha \tag{8}$$

Equations in (8) can be solved to get,

$$\alpha = \tan \alpha \tag{9}$$

This condition for conservative passive motions was apparently first found out by Andy Ruina – private communication. Equation (9) can be solved to get α. It has infinitely many solutions. See Fig. 2.

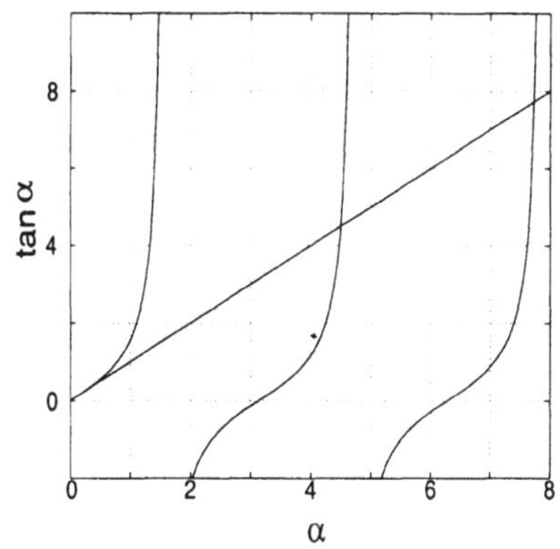

Figure 2. Solutions for $\alpha = \tan \alpha$.

The solutions are $\alpha = 4.493409458,\ 7.725251837,\ 10.9041216...$ We have,

$$\alpha = \frac{V_o}{g\sqrt{m_2(1+M)/k}}$$

The values of the parameters can be suitably chosen to get V_o, for a given α. We shall use $\alpha = 7.725251837$ and $\alpha = 4.493409458$. The solution obtained can be checked by solving Eq. (1) and Eq. (2) numerically. The value of g is assumed to be equal to 1 m/s^2 for all simulations. See Fig. 3. It can be seen that the velocity and the acceleration of the lower mass become zero when it touches the ground.

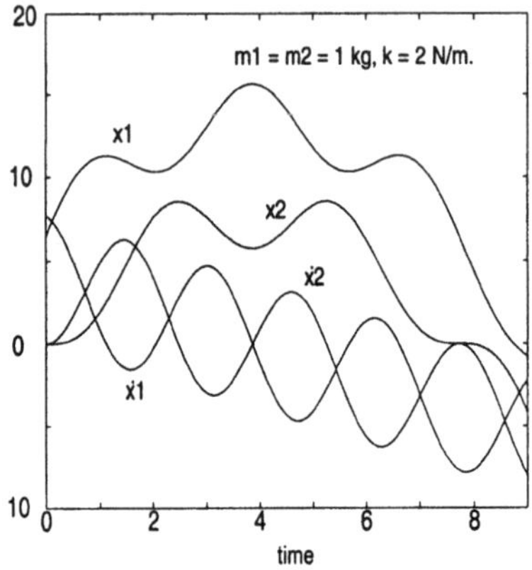

Figure 3. Time histories of displacement and velocity of both the masses for the special solution $\alpha = 7.72525$.

4. Basin of Attraction

We have,

$$V_o = g\sqrt{\frac{m_2(1+M)}{k}}\alpha = \beta\alpha$$

where

$$\beta = g\sqrt{\frac{m_2(1+M)}{k}}$$

V_o would be large if β is large and small if β is small for a given α. Let $\Delta\alpha$ be a small change in α such that $\Delta\alpha \to 0$ after a number of cycles.
Now,

$$\Delta\alpha = \frac{\Delta V_o}{\beta}$$

For large β, we would have a corresponding large ΔV_o and vice-versa. But when β is large the separation between the different solutions is large. Hence the size of ΔV_o as compared to the separation between the solutions remains the same. Let β =1, so that $V_o = \alpha$.
Let,
V_o+$\Delta V_o = V$ = velocity of the upper mass when the lower mass is about to move.
V_c = velocity of the lower mass when it hits the ground.
V_1 = velocity of the upper mass when the lower mass is about to move again.
Energy lost during the impact is $\frac{1}{2}m_2V_c^2$. See Fig. 4. Making an energy balance we get,

$$V_1 = \sqrt{V^2 - MV_c^2} \tag{10}$$

That is,

$$V_1 = \sqrt{V_o^2 + 2V_o\Delta V_o + \Delta V_o^2 - MV_c^2} \tag{11}$$

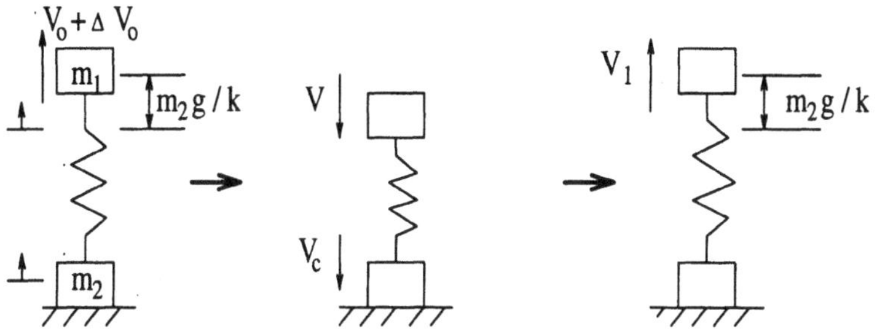

Figure 4. Different phases during the robot hop.

Now if the quantity MV_c^2 is small, so that $2V_o\Delta V_o + \Delta V_o^2 > MV_c^2$, then we could have $V_1 > V_o$. Similarly after the next impact we would have V_2, V_3,.. , such that $\Delta V_o \to 0$, which means $V_n \to V_o$, after a few impacts, so that the robot hops forever. Hence we can think of a one dimensional map that maps V to V_1, V_1 to V_2 and so on.[2] Hence V_o becomes the fixed point of the map. The form of the map near the fixed point is shown in Fig. 5.

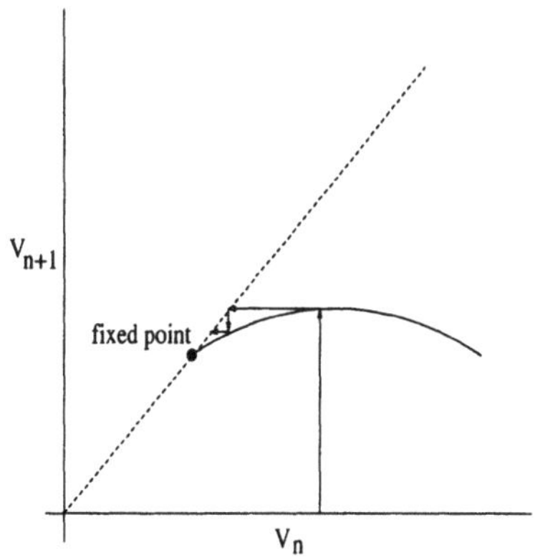

Figure 5. Likely one dimensional map near the special solutions.

[2]It should be noted that the extension of the spring can be more than m_2g/k when the lower mass hits the ground so that the lower mass moves up after it hits the ground. We here consider the case when the extension is less than m_2g/k so that the lower mass rests for sometime as the extension becomes equal to m_2g/k as the upper mass moves. The velocity of the upper mass at this instant is the mapped quantity. This means that in between mappings, we can have oscillations ensuing when the extension is greater than m_2g/k when the lower mass hits the ground.

The basin of attraction would be the interval ΔV_o such that $V_n \to V_o$ for large n. For any initial condition $V > V_o + \Delta V_o$, the energy lost would be large to make $V_1 < V_o$. Hence there would not be sustained oscillations. The dynamics would generally settle down to periodic oscillations of the upper mass with the lower mass at rest.

4.1. ONE DIMENSIONAL MAP

We could find ΔV_o by noting that,

$$2V_o\Delta V_o + \Delta V_o^2 - MV_c^2 \geq 0$$

But this requires V_c as a function of ΔV_o. Hence one would need the time of flight, which requires numerical methods because of the transcendental nature of the equation to be solved. We can find successive velocities of the upper mass, V_1, V_2,.., when the lower mass is about to move. The one dimensional map can obtained by plotting V_{n+1} against V_n [11]. Let $M = 1$ and $\beta = 1$. Using different initial conditions ranging from 0 m/s to 8 m/s, the map was plotted. See Fig. 6.

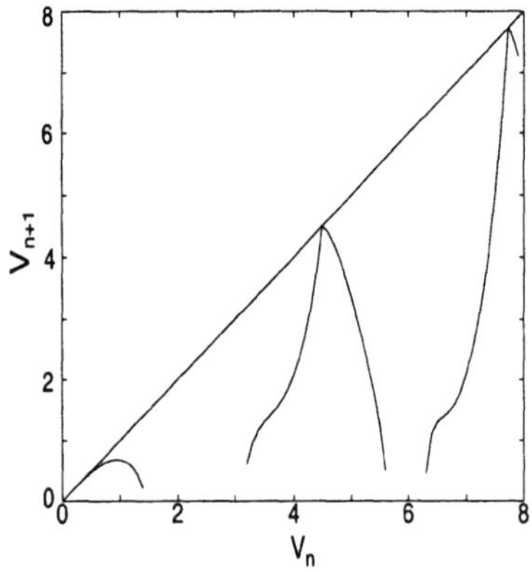

Figure 6. The one dimensional map with $\beta = 1$ and $M = 1$.

It can be seen that $V_o = 4.4934094$ and $V_o = 7.7253518$ are the fixed points of the map as expected. The disconnected region of the map gives the interval of initial conditions where we do not have a second mapping. This means that for the case of the spring extension being less than m_2g/k when the lower mass strikes the ground, the velocity of the upper mass is not large enough to lift the lower mass and we only have oscillations of the upper mass with the lower mass at rest. The nature of the map around the fixed point $V_o = 4.4934094$ is shown in Fig. 7. It can be easily concluded from the nature of the map that the fixed point is an attracting fixed point from one side. This means that there is a small interval of initial conditions around the fixed point(basin of attraction), where after a few impacts and energy loss the dynamics settles down to sustained oscillations. It can be seen from Fig. 7 that the interval of attraction is till $V = 4.509$ for the fixed point $V_o = 4.493409$.

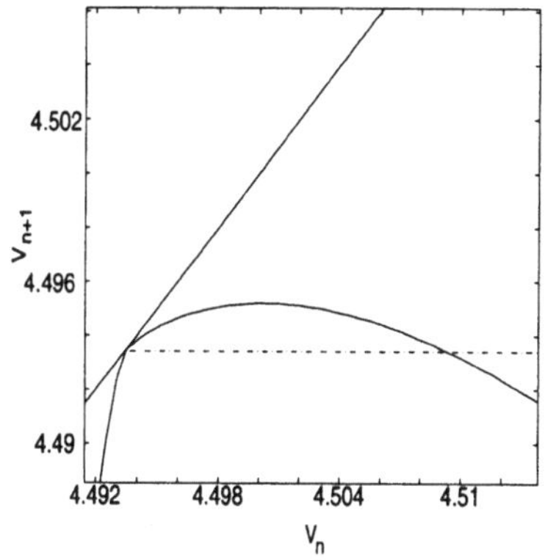

Figure 7. One dimensional map around $V_o = 4.4934$ with $\beta = 1$ and $M = 1$.

4.2. DEPENDANCE OF THE INTERVAL OF ATTRACTION ON THE SYSTEM PARAMETERS

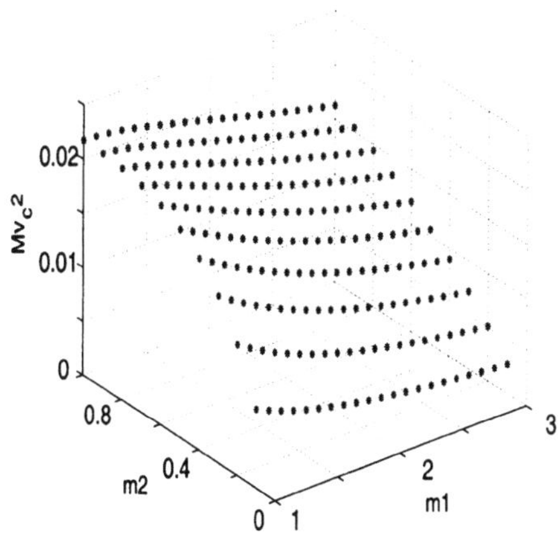

Figure 8. Variation of MV_c^2 against m_1 and m_2.

To see how the interval of attraction depends upon the system parameters let $\alpha = 7.7252518$. We know from previous analysis that smaller the quantity MV_c^2, less is the energy lost and hence it is plausible that the interval of attraction would increase with decreasing values of MV_c^2. The quantity MV_c^2 can be found numerically for different values of m_1 and m_2 and the results can be compared. See Fig. 8. $V = 7.727$ is used as the initial condition. Note that $V = 7.727$ lies in the interval of attraction for $M = 1$.
It can be seen from Fig. 8, that smaller the value of M, smaller is the quantity MV_c^2. This should suggest that the interval of attraction can be increased by keeping M low. This can be checked by plotting the one dimensional map for various parameter values, keeping $\beta = 1$. Figures 9-10 show that as the value of M is decreased, the interval of attraction increases(with $\beta = 1$).

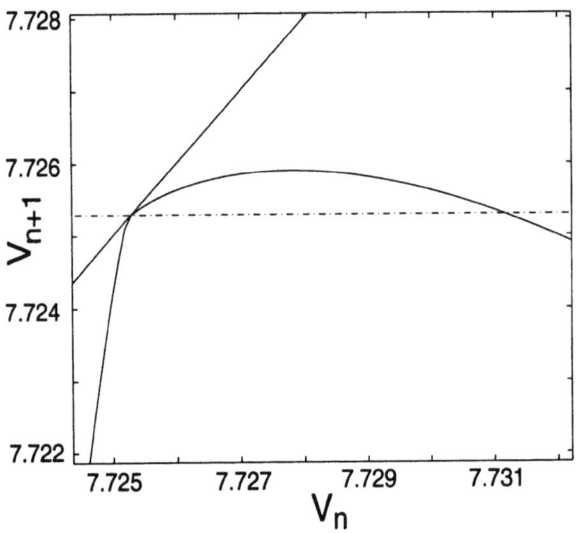

Figure 9. Interval of attraction around $V_o = 7.7253$ with $M = 0.40$.

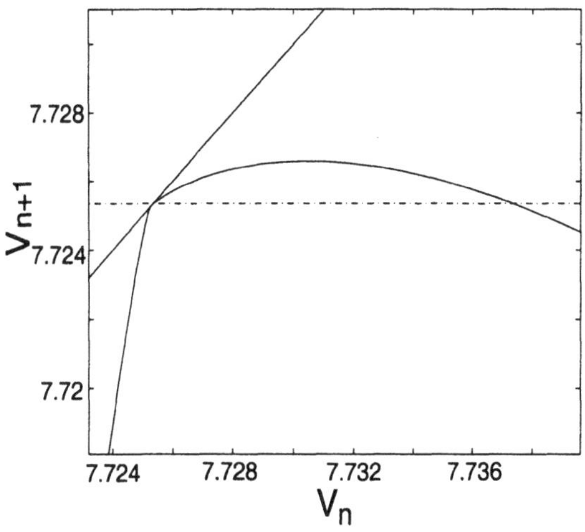

Figure 10. Interval of attraction around $V_o = 7.7253$ with $M = 0.25$.

5. Conclusions

We have presented a simple two degree of freedom spring mass model for a one-dimensional(vertical) hopping robot. We have constructed special solutions for this model that make it hop forever without any loss of energy during "plastic" collisions with the ground. These solutions have been obtained by solving for initial conditions that lead to such motions. We have shown that, theoretically there are infinite number of such initial conditions. The incessant hopping motion has one-way stability; that is, these motions are stable only if perturbed on the positive side. We have also constructed the basins of attraction of these motions and shown that the basins can be enlarged by changing the system parameters.

References

[1] Raibert, M.H. (1986) *Legged Robots that Balance.* The MIT Press, Cambridge, Massuchussets.

[2] M'Closkey, R.T., Burdick, J.W. (1993) Periodic motions of a hopping with vertical and forward motion, *International Journal of Robotic Research* **12(6)**:197-218.

[3] Vakakis, A.F., Burdick, J.W. (1991) An "interesting" strange attractor in the dynamics of a hopping robot, *International Journal of Robotic Research* **10(6)**:606-618.

[4] Alexander, R.M. (1991) Energy-saving mechanisms in walking and running, *J. Exp. Biol* **160**:55-69.

[5] Garcia, M., Ruina, A., Chatterjee, A. (1997) Speed, efficiency and stability of small slope 2D passive-dynamic bipedal walking, *International Conference on Robotics and Automation.*

[6] Garcia, M., Ruina, A., Chatterjee, A. (1998) Passive dynamic walking: Small slope implies low speed. (to be published).

[7] McGeer, T. (1990) Passive dynamic walking, *International Journal of Robotic Research* **9**:62-82.

[8] Thomson, J.M.T., Ghaffari, R. (1982) Chaos after period doubling bifurcations in the resonance of an impact oscillator, *Physics Letters* **91A**:5-8.

[9] Senator, M. (1970) Existence and stability of periodic motions of a harmonically forced impacting oscillator, *Journal of the Acoustical Society of America* **47**:1390-1397.

[10] Schiehlen, W., Gao, J. (1989) Simulation des stobfreien hüpfens, *ZAMM* **69**:303-305.

[11] Devaney, R. (1987) *An Introduction to Chaotic Dynamical Systems.* New York, Addison-Wesley.

[12] Pratap, R., Ruina, A. (1998) *Introduction to Dynamics.* Oxford University Press(under publication).

RESPONSE OF NON-LINEAR OSCILLATORS TO NON-WHITE RANDOM EXCITATION USING AN ENERGY BASED METHOD

J.B. ROBERTS and M. VASTA
School of Engineering, University of Sussex
Falmer, Brighton , East Sussex, BN1 9QT, UK

Abstract

It is shown that, for a non-linear oscillator responding to non-white random excitation, the energy envelope of the response can be modelled approximately as a one-dimensional Markov process. Using a self-consistent asymptotic analysis, for small damping, simple expressions for the drift and diffusion coefficients of this Markov process are derived: these involve Fourier coefficients derived from the solution for free, undamped oscillation. The theory is validated through comparisons with some digital simulation results.

1. Introduction

The method of stochastic averaging, due to Stratonovitch (1963), is a well known technique for analysing the response of oscillators with non-linear damping and linear stiffness to random excitation. Based on the Stratonovitch-Khasminskii (SK) limit theorem, it leads to a Markov model for the amplitude envelope, $a(t)$, of the response process from which many statistics of the response can be calculated (e.g. see Roberts and Spanos, 1986). Results obtained by this means are asymptotically correct as the damping approaches zero.

If the stiffness is strongly non-linear then it is possible to generalise the normal stochastic averaging method through the introduction of the energy envelope, $E(t)$, in place of $a(t)$. This was first achieved, for the case of white noise excitation only, by Stratonovitch (1963). He showed that $E(t)$ is approximately a one-dimensional Markov process and hence found an expression for the probability distribution of the response. This agreed with an exact result obtained earlier by Caughey (1971), for a specific form of damping of an energy dependent type.

This analysis has been extended by the first author to the case where the excitation is non-white (Roberts, 1982). Through a consideration of $E(t)$, a companion phase process, $\Phi(t)$, and a perturbation analysis based on the solution for free, undamped oscillations, expressions for the drift and diffusion coefficients for a Markov model of $E(t)$ were obtained. It was shown that these coefficients are dependent on the shape of the excitation spectrum and on Fourier coefficients relating to the free, undamped

S. Narayanan and R.N. Iyengar (eds.),
IUTAM Symposium on Nonlinearity and Stochastic Structural Dynamics, 221–231.

solution. In the case of a Duffing oscillator a reasonably good agreement with some simulation results has been obtained (Roberts, 1983). Subsequently this approach has been used to obtain results for oscillators with parametric excitation (Cai, 1995) and with hysteretic damping (Cai and Lin, 1996). Some comparisons with a simpler, heuristic approach have also been made (Dimentberg et al., 1995).

A difficulty with the earlier treatment, acknowledged at the time (Roberts, 1982), was that, in order to apply the SK limit theorem, it was necessary to introduce an approximation with respect to the phase process. This approximation is not consistent with other approximations arising from the perturbation analysis based on the free, undamped response. In the present paper it is shown that this inconsistency can be eliminated without introducing significant additional complexity. The analysis shows that the expression for the diffusion coefficient, obtained earlier, is correct but that a modification to the corresponding expression for the drift coefficient is necessary. Through a comparison with simulation results, for the case of a Duffing oscillator, an assessment is made of the effect of this correction.

2. The Equations of Motion

Consider the following non-linear oscillator:

$$\ddot{x}+\varepsilon^2 f(x,\dot{x})\dot{x}+g(x)=\varepsilon y(t) \tag{1}$$

Here x is the displacement response of the oscillator, $f(\dot{x},x)$ (in conjunction with $\dot{x}$) represents arbitrary, non-linear damping and $g(x)$ is an arbitrary, anti-symmetric restoring force. ε is a scaling parameter which, it will be assumed at the outset, is *small*: i.e. it is assumed that the rate of energy dissipation is low. The scaling of the excitation in equation (1) ensures that the standard deviation of the response is of order ε^0 and does not imply that the excitation is weak in an absolute sense.

The excitation process, $y(t)$, is assumed here to be representable as a stationary random process, with zero mean, and a power spectrum, $S(\omega)$, defined by the relation

$$S(\omega)=\frac{1}{2\pi}\int_{-\infty}^{\infty} w(\tau)\cos\omega\tau\, d\tau, \quad \text{where} \quad w(\tau)=E\{y(t)y(t+\tau)\} \tag{2,3}$$

is the covariance function of the excitation and $E\{\,\}$ is the expectation, or ensemble averaging, operator. Since the restoring force is taken to be anti-symmetric it follows that the response also has a zero-mean.

The total energy envelope process, $E(t)$, associated with the response is

$$E(t)=\frac{\dot{x}^2}{2}+V(x) \quad \text{where} \quad V(x)=\int_0^x g(\xi)d\xi \tag{4,5}$$

Clearly $\dot{x}^2/2$ represents the kinetic energy of the oscillator and $V(x)$ is the potential energy function.

In parallel with $E(t)$ it is convenient to introduce a phase process, Φ, such that

$$\text{sgn}(x)\sqrt{V(x)}=\sqrt{E}\cos\Phi, \qquad \dot{x}=-\sqrt{2E}\sin\Phi \tag{6,7}$$

Written in terms of E and Φ, equation (1) becomes

$$\dot{E}=-\varepsilon^2\alpha_1(E,\Phi)+\varepsilon\beta_1(E,\Phi)y(t),\quad \dot{\Phi}=-\varepsilon^2\alpha_2(E,\Phi)+\varepsilon\beta_2(E,\Phi)y(t)+\gamma(E,\Phi) \quad (8,9)$$

where

$$\alpha_1=2E\sin^2\Phi\, f(E,\Phi) \qquad \alpha_2=-\sin\Phi\cos\Phi\, f(E,\Phi) \quad (10,11)$$

$$\beta_1=-\sqrt{2E}\sin\Phi, \qquad \beta_2=-\frac{\cos\Phi}{\sqrt{2E}}, \qquad \gamma=\frac{g(E,\Phi)}{\sqrt{2E}\cos\Phi} \quad (12\text{-}14)$$

and

$$f(E,\Phi)=F(\dot{x},x)\dot{x} \qquad g(E,\Phi)=g(x) \quad (15,16)$$

3. Stochastic Averaging

3.1 MARKOV MODELLING

An interval of time, Δt, will now be considered such that $\Delta t \gg t_{cor}$, where t_{cor} is the correlation time scale of the excitation. If $\Delta E = E(t+\Delta t)-E(t)$ is the change in $E(t)$ during Δt then the drift coefficient, $m(E)$, and the diffusion coefficient, are defined by

$$m(E)=\lim\frac{1}{\Delta t}E\{(\Delta E)\} \qquad D(E)=\lim\frac{1}{\Delta t}E\{(\Delta E)^2\} \quad (17,18)$$

where the limit corresponds to $\varepsilon^2\Delta t\to 0$. If higher order coefficients of this type are zero then $E(t)$ is a one-dimensional Markov process (Gardiner, 1985). In the following section the conditions under which such conditions apply, at least in an approximate sense, will be investigated.

3.2 AVERAGING THE DISSIPATION TERMS

Let s be some time during an interval Δt. Then one can write

$$E(s)=E_0(s)+e(s) \qquad \Phi(s)=\Phi_0(s)+6(s) \quad (19,20)$$

where E_0 and Φ_0 are the solutions of equation (1) for the case where $\varepsilon=0$: i.e. free, undamped solutions. From equations (8) and (9), for $\varepsilon=0$,

$$\dot{E}(t)=0 \qquad \dot{\Phi}_0(t)=\gamma(E_0,\Phi_0) \quad (21,22)$$

Hence

$$E_0(t)=\text{a constant} \qquad \Phi_0(t)=h(E_0,t) \quad (23,24)$$

where $h(E,t)$ is a cyclic function of time, with period $T(E)$ corresponding to the period of free undamped oscillations, with energy level E. From equation (1) one finds that

$$T(E)=\frac{2\pi}{\omega(E)}=2\sqrt{2}\int_0^b\frac{dx}{\sqrt{[E-V(x)]}} \quad (25)$$

where b is such that $V(b)=E$ and $\omega(E)$ is the corresponding frequency. Evidently both $e(t)$ and $\theta(t)$ are small and are of order ε. Thus one can approximate the dissipation terms, α_1 and α_2, in equations (8) and (9) by averaging over one period of oscillation, $T(E)$, treating E as constant during this period and setting $\Phi(t)=\Phi_0(t)$. Hence equations (8) and (9) may be replaced by the approximations

$$\dot{E}=-\varepsilon^2 A_1(E)+\varepsilon\beta_1(E,\Phi)y(t), \quad \dot{\Phi}=-\varepsilon^2 A_2(E)+\varepsilon\beta_2(E,\Phi)y(t)+\gamma(E,\Phi) \qquad (26,27)$$

where

$$A_1(E)=\frac{2E}{T(E)}\oint \sin^2\Phi_0 f(E,\Phi_0)dt \qquad (28)$$

$$A_2(E)=\frac{1}{T(E)}\oint \sin\Phi_0 \cos\Phi_0 f(E,\Phi_0)dt \qquad (29)$$

3.3 EVALUATION OF THE DRIFT COEFFICIENT

It will be supposed that the term $\gamma(E,\Phi)$ can be expanded about its "deterministic" evolution, corresponding to $\varepsilon=0$, denoted $\gamma^0(t)=\gamma(E_0,\Phi_0)$. Thus, correct to order ε,

$$\gamma(t)=\gamma^0(t)+\gamma_E^0(t)e(t)+\gamma_\Phi^0(t)\theta(t) \qquad (30)$$

where

$$\gamma_E^0(t)=\left[\frac{\partial\gamma}{\partial E}\right]_t^0, \quad \gamma_\Phi^0(t)=\left[\frac{\partial\gamma}{\partial\Phi}\right]_t^0 \qquad (31,32)$$

are time-dependent coefficients. It follows from equations (20), (24), (27) and (30) that $\theta(t)$ is governed by

$$\dot{\theta}-\gamma_\Phi^0(t)\theta=-\varepsilon^2 A_2(E)+\varepsilon\beta_2(E,\Phi)y(t)+\gamma_E^0(t)e(t) \qquad (33)$$

It is now possible to evaluate the drift coefficient, $m(E)$. As a first step equation (26) can be integrated over the interval Δt, to find ΔE : thus

$$\Delta E=-\varepsilon^2 A_1(E)\Delta t+\varepsilon\int_t^{t+\Delta t}\beta_1(u)y(u)du \qquad (34)$$

Now β_1 can be expanded about its deterministic evolution, corresponding to $\varepsilon=0$, in a similar manner to that previously used for γ (see equation (30)). Thus, correct to order ε,

$$\beta_1(u)=\beta_1^0(u)+\left[\frac{\partial\beta_1}{\partial E}\right]_u^0 e(u)+\left[\frac{\partial\beta_1}{\partial\Phi}\right]_u^0 \theta(u) \qquad (35)$$

where $t<u<t+\Delta t$. Since $\Delta t >> \tau_{cor}$ the first term in this expansion will not contribute to the mean of E. Moreover, correct to order ε, the contribution to $e(u)$ from the excitation is given by (from equations (26) and (35))

$$e(u)=\varepsilon\int_t^u \beta_1^0(v)y(v)dv \tag{36}$$

Similarly, from equation (33), the contribution to $\theta(u)$ from the excitation, correct to order ε, is given by

$$\theta(u)=\varepsilon\int_t^u \beta_2^0(v)y(v)h(u,v)dv+\int_t^u \gamma_E^0(v)e(v)h(u,v)dv \tag{37}$$

Here $h(u,v)$ is the impulse response function for the filter described by the left hand side of equation (33) and $\beta_2^0(v)$ is the "deterministic" evolution of β_2, corresponding to $\varepsilon=0$. Since equation (33) is a simple, time-dependent first order equation for θ $h(u,v)$ can be expressed in terms of an integrating factor

$$g(t)=\exp\left(-\int \gamma_\Phi^0(u)du\right) \tag{38}$$

Thus

$$h(u,v)=g(v)/g(u) \tag{39}$$

Here $g(t)$ is a periodic function, with period $T(E)$.

From equations (36) and (37) one has

$$\theta(u)=\varepsilon\int_t^u \beta_2^0(v)y(v)h(u,v)dv+\varepsilon\int_t^u\int_t^v \gamma_E^0(v)h(u,v)\beta_1^0(\xi)y(\xi)d\xi dv \tag{40}$$

On combining equations (34) to (40) an expression for ΔE may be obtained. On taking expectations of this, and using equation (17), the following expression for the drift coefficient, correct to order ε^2, may be obtained (Roberts and Vasta, 1999).

$$m(E)=-\varepsilon^2 A_1(E)+\frac{\varepsilon^2\pi}{2}\sum_{n=1}^{\infty}\left(s_n^2+c_n^t c_n^b\right)S[n\omega(E)] \tag{41}$$

Here s_n c_n^t and c_n^b are Fourier coefficients defined by

$$\sin\Phi_0(t)=\sum_{n=1}^{\infty} s_n \sin n\omega(E)t \tag{42}$$

$$\cos\Phi_0(t)g(t)=\sum_{n=1}^{\infty} c_n^t \cos n\omega(E)t\ , \qquad \frac{\cos\Phi_0(t)}{g(t)}=\sum_{n=1}^{\infty} c_n^b \cos n\omega(E)t \tag{43,44}$$

It is noted that, if the terms γ_Φ^0 and γ_E^0 are neglected then $g(t)=1$ and

$$c_n^t=c_n^b\equiv c_n \tag{45}$$

where

$$\cos\Phi_0(t)=\sum_{n=1}^{\infty} c_n \cos n\omega(E)t \tag{46}$$

then equation (41) reduces to

$$m(E) = -\varepsilon^2 A_1(E) + \frac{\varepsilon^2 \pi}{2} \sum_{n=1}^{\infty} \left(s_n^2 + c_n^2\right) S[n\omega(E)] \tag{47}$$

This is a result obtained previously by the first author (Roberts, 1982))

3.4 EVALUATION OF THE DIFFUSION COEFFICIENT

The diffusion coefficient is much easier to calculate than the drift coefficient. From equation (26) one finds immediately that, correct to order ε^2,

$$(\Delta E)^2 = \varepsilon^2 \int_t^{t+\Delta t} \int_t^{t+\Delta t} \beta_1^0(u) \beta_1^0(v) X(u) X(v) du dv \tag{48}$$

Using the definition of β_1 (see equation (12)), equation (48) may be rewritten as

$$(\Delta E)^2 = 2\varepsilon^2 E \int_t^{t+\Delta t} \int_t^{t+\Delta t} \sin \Phi_0(u) \sin \Phi_0(v) X(u) X(v) du dv \tag{49}$$

Taking expectations, using the fact that $t_{cor} << \Delta t$ again, and averaging over time, one finds that the drift coefficient $D(E)$ (see equation (18)) is given by

$$D(E) = \varepsilon^2 2\pi E \sum_{n=1}^{\infty} s_n^2 S[n\omega(E)] \tag{50}$$

3.5 THE FPK EQUATION AND THE STATIONARY RESPONSE

Using an analysis similar to that given in the previous section one finds that the higher order coefficients referred to in section 3.1 all approach zero in the limit as $\varepsilon^2 \Delta t \to 0$. Thus $E(t)$ is approximately a Markov process, running on time $\varepsilon^2 t$, as $\varepsilon \to 0$. From the Fokker-Planck-Kolmogorov (FPK) equation corresponding to this process one obtains the expression

$$p(E) = \frac{C}{D(E)} \exp\left\{ 2 \int_0^E \frac{m(\xi)}{D(\xi)} d\xi \right\} \tag{51}$$

4. Application to the Duffing Oscillator

To illustrate the method a Duffing oscillator is now considered with linear damping. A non-dimensionalised form of the equation of motion is as follows: i.e.

$$\ddot{x} + \alpha \dot{x} + x + x^3 = Ay(t) \tag{52}$$

where x is the displacement response, y is the excitation, α is the linear damping coefficient and A is an input scaling factor.

The excitation process considered here can be obtained by filtering white noise, $n(t)$ through two first order linear filters, in series. The output, $\eta(t)$, is obtained from

$$\dot{\eta} + \beta\eta = \beta\lambda(t)\,, \qquad \dot{\lambda} + \beta\lambda = \beta n(t) \tag{53}$$

If two such filtering operations are applied to independent white noises the resulting outputs, $\eta_1(t)$ and $\eta_2(t)$, may be combined to form the input process

$$y(t) = h\lfloor \eta_1(t)\cos\omega_p t + \eta_2(t)\sin\omega_p t \rfloor \tag{54}$$

where h is an amplitude scaling parameter. The spectrum of $y(t)$, so defined, is given by

$$S(\omega) = k\left[\left\{\frac{1}{\beta^2 + (\omega-\omega_p)^2}\right\}^2 + \left\{\frac{1}{\beta^2 + (\omega+\omega_p)^2}\right\}^2\right] \tag{55}$$

where k is a scaling constant.

For small β the spectrum has a single peak in the neighbourhood of the frequency ω_p. There are thus three useful parameters for this spectrum:

(a) the band-width of the peak , measured by β ,

(b) the position of the peak relative to the linear, undamped natural frequency ($\omega = 1$),

(c) the level of the excitation, measured by A.

4.1 ANALYTICAL EVALUATION OF $P(E)$

As an initial step in the application of the general theory the equation of motion for free undamped oscillations, i.e.

$$\ddot{x} + x + x^3 = 0 \tag{56}$$

may be solved. The solution can be expressed as

$$x = x_o cn(\Omega t | m) \tag{57}$$

where cn is the Jacobian elliptic function, x_o is the amplitude of oscillation and

$$\Omega = \left(1 + x_o^2\right)^{1/2}, \qquad m = \frac{x_o^2}{2\Omega^2} \tag{58,59}$$

The period of oscillation, $T(E)$, is found to be

$$T(E) = \frac{4K(m)}{\Omega} \tag{60}$$

where $K(m)$ is the complete elliptic integral.

To evaluate the drift and diffusion terms in the FPK equation it is also necessary to evaluate $\sin\Phi_o(t)$ and $\cos\Phi_o(t)$. These can also be expressed in term of elliptic functions (Roberts, 1983) . Hence the Fourier coefficients can be evaluated, through application of equations (43) and (44).

The term $A_1(E)$ is also required in the calculation of $P(E)$ (see equation (28)). In this example $\varepsilon^2 = \alpha$ and $f(E,\Phi) = 1$; hence

$$A_1(E)=\frac{2E}{T(E)}\int_0^{T(E)}\sin^2\Phi_0(t)dt \tag{61}$$

This can be evaluated numerically by using equation (42).

4.2 COMPARISON WITH SIMULATION RESULTS

To test the proposed theory some sample functions of the energy and phase response were studied. These were obtained by generating a sample function of the excitation, with time step $\Delta t = 0.01$, using equations (56) and (57), and then integrating the equation of motion using the fourth order Runge Kutta algorithm. Figure 1(a) shows a typical sample function for the energy response, with initial value $E_0 = 3$, compared to the (constant) free oscillation result. Here $\beta = 1$, $\omega_p = 5$, and the damping is zero. The corresponding variation of the incremental energy , ΔE, with time is shown in Figure 1(b). Also shown in Figure 1(b) is the variation of ΔE calculated using the perturbation approximation given by equation (36). This shows that, for this level of excitation, the perturbation approximation is reasonably accurate.

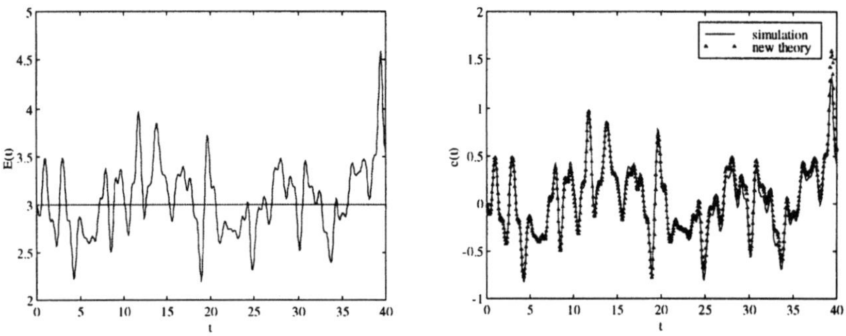

(a) (b)

Figure 1: (a) Variation of energy level, E(t), with time –exact free and forced response. (b) Variation of incremental energy with time –comparison between exact simulation and simulation using a perturbation expansion.

The corresponding time history of the phase angle, Φ, is shown in Figure 2(a). Here the variation with excitation is compared to the freeoscillation variation. The difference, 6 , between these variations is shown in Figure 2(b). Also shown in Figure 2(b) is the variation of 6 with time according to the pertubation expansion given by equation(40), using the $h(u,v)$ function given by equation (39) (new) and also using $h(u,v)$=1 (old - corresponding to $\gamma_\Phi^0 = \gamma_E^0 = 0$). This last result shows that the new theory gives an improved approximation for 6 .

Figure 3(a) shows some simulation estimates for the mean of the energy level, m_E, for various values of the linear damping coefficient: each estimate was obtained by simulating a very long sample function of the energy envelope ($\Delta t = 0.05$, 10 million samples). Here $\beta = 1$ and ω_p =2 and $A=\sqrt{1.80\alpha}$. For comparison the theoretical variation of m_E with α is shown (a constant). Also shown, for comparison purposes, is the theoretical variation corresponding to a white noise approximation, where the (constant) spectral level of the white noise is equated to the spectral level of the actual excitation at $\omega = 1$. For the parameters chosen the difference between the correction term contained in the present theory is negligibly small and, at small to moderate values of damping there is a very good agreement between the theory and the simulation estimates. The variability of the simulation estimates does, however, increase significantly as the damping reduces to very small values, as indicated by the large scatter in the various estimates obtained for $\alpha = 0.0001$.

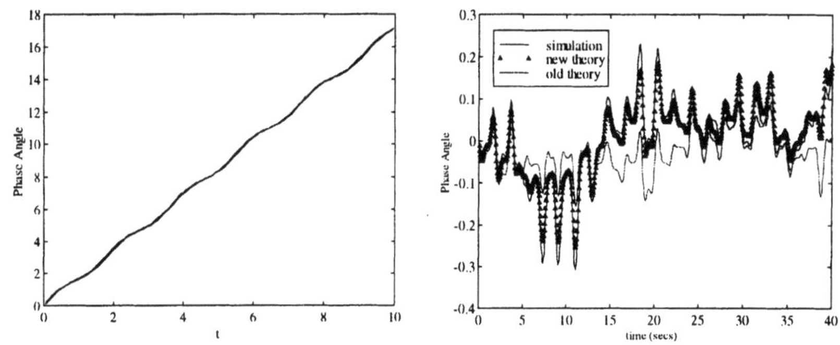

(a) *(b)*

Figure 2 : (a) Variation of phase angle $\Phi(t)$ with time –exact free and forced response (b) Variation of phase angle $\theta(t)$ with time –comparison between exact simulation and simulations using perturbation expansions (old and new).

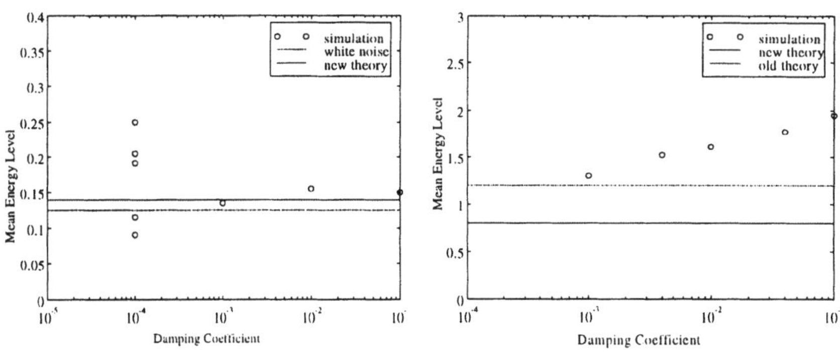

(a) *(b)*

Figure 3 : Variation of the mean energy level with damping

Figure 3(b) shows a similar comparison for a more extreme case, where $\beta = 1$ and $\omega_p = 5$ and $A = \sqrt{722\alpha}$. Here there is a significant difference between the present theory and the "old" theory. This is because the spectral values corresponding to the third harmonic in the displacement response are larger than those relating to the first harmonic. Hence, also the third harmonic Fourier coefficients are much smaller than the first harmonic coefficients, even at large energy levels, the contribution of the third harmonic components in the drift and diffusion coefficients is significant. However, in this case the simulation estimates indicate a very slow convergence to the theoretical value of m_E and, due to the high variability of the simulation estimates at very low values of damping it is not possible to validate the correction terms in the theory by this means.

In Figure 4(a) the theoretical variation of $p(E)$ with E, as computed from equation (51), is shown and compared with some corresponding digital simulation estimates. In this case the correction to the original theory is negligible. Here $\beta = 1$ and $\omega_p = 2$ and $A = \sqrt{1.80\alpha}$, as for the results in Figure 3(a) and $\alpha =$ 0.005. For comparison purposes the white noise theoretical result is also shown. The trend here can be predicted from simple physical reasoning. When $\omega_p = 2$ the resonant frequency for small oscillations $(\omega \approx 1)$ lies to the left of the peak in the input spectrum. Now increasing amplitude in the response is associated with increasing frequency of oscillation, which in turn is associated with higher levels in the input spectrum. Thus one can expect that the response is more likely to reach high amplitude levels than the white noise approximation predicts. Figure 4(b) shows a similar comparison for the case where $\beta = 1$, $\omega_p = 5$ and $A = \sqrt{722\alpha}$ and $\alpha = 0.005$.

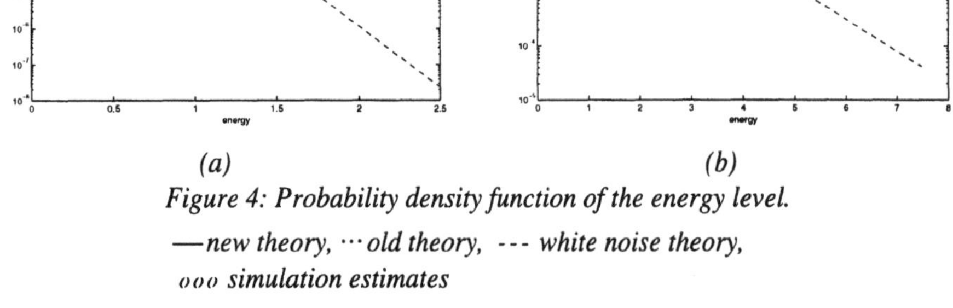

(a) *(b)*

Figure 4: Probability density function of the energy level.
—new theory, ··· old theory, --- white noise theory,
ooo simulation estimates

As indicated by the results for the mean energy level (see Figure 3(b)) in this more extreme case the correction to the original theory is significant and the agreement between theory and simulation is not so close. Here the accuracy of the present theory will be poor unless the damping level is an extremely low level, and below a value for which reliable simulation estimates can be obtained.

5. Conclusions

The principal conclusions are summarised as follows:

1. It is possible to develop a consistent Markov model for the energy envelope of a non-linear oscillator responding to non-white excitation, based on a perturbation of the free, undamped solution.
2. An inconsistency in the theory, with respect to phase, present in earlier treatments, can be eliminated.
3. The shape of the input spectrum, and the characteristics of the free oscillation solution, influence the magnitude of the drift and diffusion coefficients.
4. Numerical studies, for the case of a Duffing oscillator, show that the theory gives results which agree well with simulation when the band-width of the excitation is large and/or the damping is small.

6. References

Cai, G.Q. (1995) Random vibration of nonlinear systems under nonwhite excitations, *Journal of Engineering Mechanics* **121**, 633-639.

Cai, G.Q. and Lin, Y.K., (1996) Response of a hysteretic system under non-stationary earthquake excitations, in A. Naess and S. Krenk (eds.), *IUTAM Symposium on Advances in Nonlinear Stochastic Mechanics*, Kluwer Academic Publishers, Dordrecht, pp. 99-108.

Caughey, T. K., (1971) Nonlinear theory of random vibrations, in C.S. Yih (ed.) *Advances in Applied Mechanics*, Vol. 11, Academic Press, New York, pp. 209-253.

Dimentberg, M., Cai, G.Q. and Lin, Y.K. (1995) Application of quasi-conservative averaging to a non-linear system under non-white excitation, *International Journal of Non-Linear Mechanics* **30**, 677-685.

Gardiner, C.W. (1985) *Handbook of Stochastic Methods*, Springer-Verlag, Berlin.

Roberts, J.B. (1982) A stochastic theory for nonlinear ship rolling in irregular seas, *Journal of Ship Research* **26**, 229-245.

Roberts, J.B. (1983) Energy methods for non-linear systems with non-white excitation, in K. Hennig (ed.), *IUTAM Symposium on Random Vibrations and Reliability*, Akademie-Verlag, Berlin, pp. 285-294.

Roberts, J.B. and Spanos, P.D. (1986) Stochastic averaging: an approximate method of solving random vibration problems, *International Journal of Non-Linear Mechanics* **21**, 111-134.

Roberts, J.B. and Vasta, M. (1999) A Markov model for the response of non-linear systems to non-white random excitation (to be published).

Stratonovich, R.L. (1963) *Topics in the Theory of Random Noise, Vols. 1 & 2*, Gordon and Breach, New York.

APPLICATION OF BIFURCATION THEORY TO THE IDENTIFICATION OF THE CONSTITUTIVE RELATIONS FOR SLIDING FRICTION

A.R. SAVKOOR
Vehicle Research Laboratory

AND

J.P. MEIJAARD
Laboratory for Engineering Mechanics,
Delft University of Technology,
Mekelweg 2, NL-2628 CD Delft, The Netherlands

Abstract. This paper considers the consequences of some assumptions concerning the constitutive relations of friction on the non-linear dynamic behaviour of a simple tribometer device. The identification of the constitutive relations for stationary and non-stationary friction is based on the characteristic features of dynamic behaviour such as friction induced vibrations including stick-slip. Stability and bifurcations are investigated, mostly with numerical methods. The tribometer device is represented by one- and two-degree-of-freedom models with lumped parameters. The main focus of the study is the law of dry friction of elastomers sliding on smooth surfaces of hard solids, for which friction depends markedly on the speed of sliding. Based on results for stationary sliding friction, the first friction model considered is one where friction depends only on the instantaneous speed of sliding. A more complex law of friction, considered next, takes into account the influence of the history of motion through an internal state variable. Finally, a third friction model is included that displays the typical, more or less random dependence of friction on the distance slid. From a comparison of predictions of the different theoretical models with experimental results, it is concluded that a friction law based only on the dependence of friction on the sliding velocity is inadequate to explain the observed variation of the friction force in the stick-slip regime. The trend observed is qualitatively similar to that according to the friction model including a state variable.

S. Narayanan and R.N. Iyengar (eds.),
IUTAM Symposium on Nonlinearity and Stochastic Structural Dynamics, 233–248.

1. Introduction

Dry friction between solids is undoubtedly one of the most important phenomena in every day life, because it both enables and hinders control of the relative motion that is desired for proper functioning of several human activities. The scientific study of friction has been pursued along two different but complementary paths. The first path is directed to understanding and quantifying the basic mechanisms of friction from the physicist's viewpoint that can lead to understanding and control of physical factors that are its root cause. The second path followed in the field of engineering mechanics is a more pragmatic one in that not the physical mechanism but the ability to characterize friction is the main concern. The aim is to define a constitutive relation between frictional force and relative motion of solids using a minimum number of coordinates and parameters that are consistent with experimental observations. One of the greatest difficulties in modelling friction is that the modelling of the contact interface between solids requires the knowledge of many details of both geometrical and physical nature. Material points within the solid bodies that are convenient to observe and remote from the contact interface serve as coordinates to define the state variables of motion of the solids. The choice of the coordinates of observation and measurements filter out the physical details of the interfacial contact as far as possible without affecting the essential features of the relative motion.

This paper concerns with the friction of elastomer in contact with hard solids. The main purpose of the present investigation is to throw some light on the constitutive law of quasi-isothermal friction. It is a well-known fact that both the sliding speed and temperature have a strong influence on the kinetic friction of elastomer on hard and smooth solids. Since the experimental work of Grosch [1], it is known that this influence is strongly correlated to the corresponding influence on the visco-elastic properties of elastomers. In experimental studies, the effects of varying speed and the temperature on friction can be studied separately at a constant ambient temperature by restricting the maximum speed of sliding to the order of 10^{-2} m/s. At sufficiently low speed of sliding, friction may be regarded as a quasi-isothermal process, because the thermal field associated with frictional heating is then insignificant relative to the uniform ambient temperature of the solids. The time-temperature equivalence principle applies for thermo-rheologically simple and linear visco-elastic elastomers. For such materials a master curve for isothermal friction over a very wide range of speed may be constructed by shifting appropriately the friction-speed curves measured at different temperatures over a limited speed.

At first sight, it appears a simple matter to conduct experiments at constant sliding velocity and establish the law for stationary friction experimentally. The standard experimental method is to apply a constant normal force to the test specimen and measure the force of friction of an elastomer slider in contact with a hard solid with the aid of a tribometer for different values of the relative sliding speed. However, it turns out that it is not always possible to enforce uniform sliding motion even if the imposed pulling speed of the tribometer is maintained at a constant value. This is due to the occurrence of significant frictional vibrations. In certain operating ranges of normal pressure and temperature, sliding occurs with large variations of sliding speed and frictional force that is typical of stick-slip. Following the seminal work of Blok [2], the instability of the stationary motion is generally attributed to the negative slope of the stationary friction characteristic, that is where the stationary friction force decreases with increasing speed of sliding. More recently, a large number of publications have appeared regarding this problem, which propose various mechanisms and dynamic models for the onset of instability of the uniform sliding motion. Some of these models ascribe the excitation of frictional oscillations to the normal vibrations of the slider, which are assumed to be coupled to the tangential motion by some mechanism. While it is well known that normal vibrations influence the frictional behaviour of solids, this explanation is too specialized to be true more generally in tribometer experiments conducted at low speeds.

Irrespective of the origin of stick-slip motion, the difficulty remains that the stationary friction characteristic in the speed range of decreasing friction cannot be determined properly unless the frictional vibrations are suppressed. However, the technique used to suppress frictional vibrations requires, in turn, the knowledge of the constitutive relation governing friction. On the other hand, the frictional vibrations observed in the tribometer experiments can provide useful information for the identification of the non-stationary frictional behaviour. The identification of the friction law for relative motion of solids requires proper modelling of the two solids and their interfacial coupling.

The present paper describes and analyses one- and two-degree-of-freedom models with friction with or without an internal state variable and also friction that depends on the sliding distance, and some comparisons with experimental observations are made. Stability and bifurcations are investigated mostly numerically, as described in other places [3, 4].

2. Experimental set-up

The tribometer test set-up shown in Figure 1 is installed in a chamber with controlled ambient temperature which can be set within the range from

Figure 1. Tribometer test set-up. The upper photograph shows the climate chamber with the driving mechanism outside of it, the lower photograph shows the slider assembly.

−60°C to +70°C. The test specimen of elastomer mounted on the slider assembly is pulled across a plate of smooth and hard material by a steel wire wound on a drum. The drum is driven by a speed-controlled direct current motor through a gear-box which has a number of discrete reduction ratios.

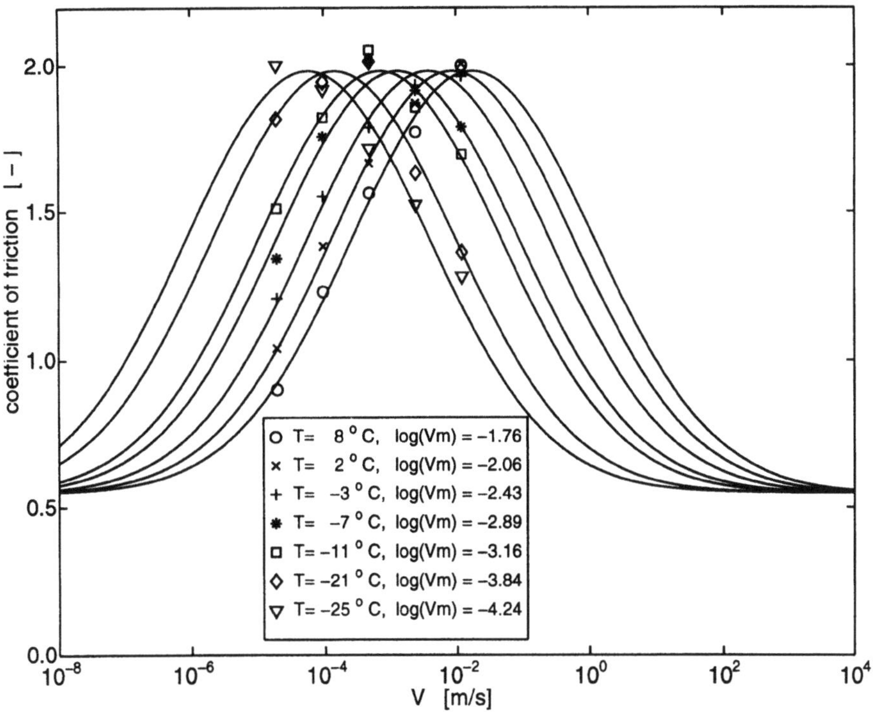

Figure 2. Stationary isothermal friction as a function of the sliding speed for several temperatures. The different markers identify different ambient temperatures at which friction was measured at five different driving speeds V ($V = 1.2 \cdot 10^{-2}$, $2.4 \cdot 10^{-3}$, $5.0 \cdot 10^{-4}$, $1.0 \cdot 10^{-4}$, and $2.0 \cdot 10^{-5}$ m/s).

TABLE 1. Amplitudes and frequencies of vibrations; for v_m see formula (3).

T °C	V mm/s	v_m mm/s	Amplitude (top-top)	Frequency Hz
-7	2.4	1.288	0.20	4.7
-11	2.4	0.692	0.20	4.7
-21	2.4	0.145	0.20	4.7
	0.5		1.15	0.6
-25	2.4	0.058	0.20	4.2
	0.5		1.12	0.7
	0.1		1.53	0.1

The results of stationary isothermal friction conducted with this test set-up are summarized in Figure 2 and Table 1. Experimental data points away from the region of peak friction have been omitted for the sake of clarity. The experiments were performed on an oil-extended SBR copolymer compound with 23% bound styrene level sliding on a plate of smooth stainless steel under dry conditions. In the region of very low driving speed of the tribometer, a uniform sliding motion is observed and the stationary force of friction is found to increase with increasing sliding speed until a point where frictional vibrations start to appear. Table 1 gives the top-top value of the normalized force variation and the frequency of some clearly identified vibrations. With sufficient damping, it appeared that the decreasing part of the friction-speed curve could be determined without oscillations.

The isothermal friction curves including the measured data have been scaled using suitable factors in order to compensate for the surface changes due to changes in temperature and relative humidity (surface energy or the limit of static friction) at these temperatures. In the region of the negative slope of the friction speed curve where stick-slip may occur, the marked points represent the peak values of the measured tangential force. The shift of the isothermal friction curves corresponding to the various temperatures T, measured in Kelvin or degrees centigrade, is given by the well-known W.L.F. (Williams-Landel-Ferry) transform [1] $\log_{10} a_T = -8.86(T - T_s)/(101.5 + T - T_s)$, where T_s is the characteristic temperature for the elastomer (generally $T_s \approx T_g + 50°C$, where T_g is the glass transition temperature). The estimated value of T_s for the elastomer is found to be roughly $-16°C$.

The low-speed isothermal friction curves in the range of higher temperatures can be determined easily, because uniform sliding motion occurs corresponding to the prescribed driving speed of the tribometer. The speed at which frictional vibrations start to appear becomes lower as the temperature is reduced. At low ambient temperature, for driving speeds slightly exceeding the speed corresponding to the point of maximum friction the relative motion becomes strongly oscillatory. It resembles stick-slip and the tangential force exhibits a strong variation with a large peak to peak value. At still higher driving speed the force variations are considerably reduced and in some cases uniform sliding motion reappears.

3. One-degree-of-freedom models

Most textbooks dealing with dynamic behaviour of systems involving sliding friction, for instance [5], treat the case of a rigid body of mass m which is pressed against a rigid moving belt with a constant normal force N. The moving belt is assumed to translate at a constant velocity V. The rigid

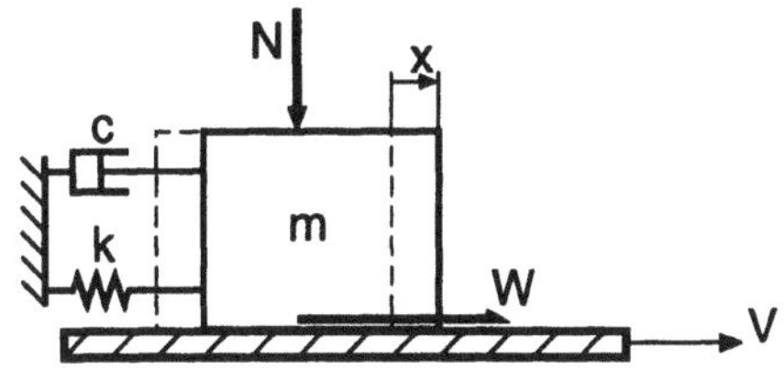

Figure 3. A basic one-degree-of-freedom model of a simple tribometer device.

body is retained by a linear spring-damper combination (with spring rate k and damper rate c), which is attached to a rigid frame. The basic system with one degree of freedom is shown in Figure 3.

The friction force between the body and the driven surface, which is denoted by W, is given by some friction law, which may have a simple or more complicated form. This law may show dependence on the relative sliding displacement and velocity and their histories. Furthermore, there can be a dependence on ambient circumstances such as the temperature, chemistry of the surfaces, and the composition of the atmosphere.

If the tangential displacement of the body is denoted by x, the equation of motion for the system becomes

$$m\ddot{x} + c\dot{x} + kx = W, \tag{1}$$

where the overdot denotes a derivative with respect to the time t. Usually the coefficient of friction is introduced as the ratio of the absolute value of the friction force and the normal force, or the maximal available friction force divided by the normal force if there is no relative motion, which is assumed to be independent of the normal force.

3.1. FRICTION THAT DEPENDS ONLY ON THE SLIDING VELOCITY

If the friction force depends only on the relative sliding velocity $v_s = V - \dot{x}$, the equation of motion can be written in a non-dimensional way as

$$\ddot{x} + 2\zeta\dot{x} + x = W(V - \dot{x}) = W(v_s). \tag{2}$$

Here the time is scaled by the reciprocal natural frequency of the system, $\sqrt{(m/k)}$, the distances are scaled by the deflection of the spring caused by the maximal static friction force, $\max\{W(0)\}/k$, and the forces are scaled by the same maximal friction. Scaled quantities are denoted by the same symbols as the unscaled quantities. The parameters in this model are the relative damping $\zeta = (c/2)/\sqrt{(mk)}$ and the function $W(v_s)$ which may depend on additional parameters to be defined later. It is assumed that this function is odd. It is multivalued at zero sliding velocity, where the value of the friction force is determined by the requirement that this velocity remains zero or, if the friction force needed for holding this condition is not in the allowed range of values between, and including, its minimum and maximum, it is at one of its extremes, opposing the incipient sliding motion. A continuous transition from static to dynamic friction is assumed, $W(0^+) = \max\{W(0)\}$.

We assume that the friction function has a bell-like shape as in Figure 2, where it is increasing at low sliding velocities and decreasing at high sliding velocities, as has been observed in the sliding of elastomer on a smooth stiffer material [1, 6, 7]. This shape can be approximated by the Gaussian function on a logarithmic scale [6],

$$W(v_s)\text{sgn}(v_s) = 1 + (W_m - 1)\exp\left[-\frac{h^2}{2}\ln^2\left(\frac{|v_s|}{v_m}\right)\right]. \tag{3}$$

In this expression, $W_m > 1$ is the maximal friction force that occurs for $|v_s| = v_m$ and h is a dimensionless parameter. The friction force increases with increasing sliding speeds if $|v_s| < v_m$ and it decreases if $|v_s| > v_m$.

For every prescribed positive driving velocity of the surface, there is a unique stationary solution given by $x = W(V)$. Linearization of the equation of motion in this solution yields the equation for small perturbations,

$$\Delta\ddot{x} + [2\zeta + \frac{\partial W}{\partial v_s}(V)]\Delta\dot{x} + \Delta x = 0, \tag{4}$$

where the prefix Δ denotes a perturbation of a variable with respect to its nominal value. The linearized equation shows that the stationary solution is stable if the sum of the damping and the slope of the friction function is positive, while it is unstable if this sum is negative, which can occur in the part where the friction function decreases with increasing sliding velocity. If the sum is exactly zero, a Hopf bifurcation occurs and the stability is determined by the sign of the third derivative of the friction function: if this derivative is negative, the solution is unstable and a subcritical, or hard, bifurcation occurs, while if it is positive, the solution is stable and a supercritical, or soft, bifurcation occurs. From the Hopf bifurcation point periodic solutions, limit cycles, originate [8]. Apparently the stationary solution can be stabilized by adding a sufficiently large damping, while other measures, such as modifying the slider mass or the stiffness, give no solace. By decreasing the damping from a sufficiently large value, a Hopf bifurcation is found if the driving velocity is in the decreasing part of the friction function. As the third derivative is negative for most of the decreasing part of the friction-velocity curve, this bifurcation is often of the hard type and a large-amplitude limit cycle, mostly of the stick-slip kind, is found.

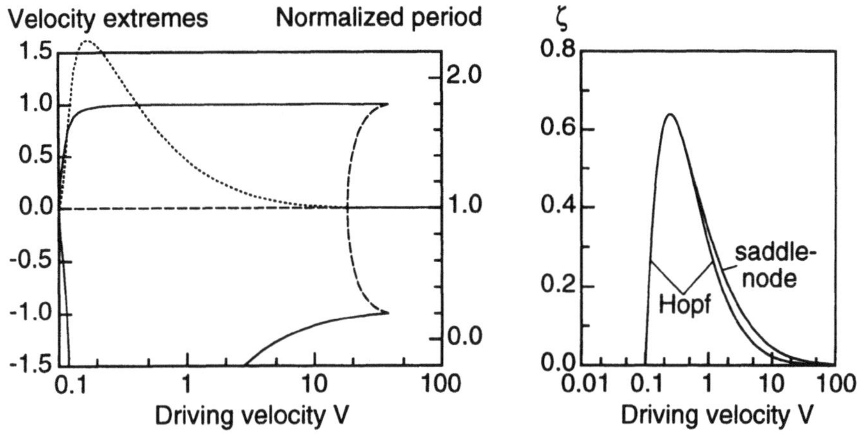

Figure 4. Left: velocity extremes, normalized with respect to the speed of the driven surface, and the period of the limit cycles, divided by 2π, for $\zeta = 0.01$. Drawn lines denote stable branches, dashed lines denote unstable branches and the dotted line denotes the period. Right: bifurcation values for Hopf bifurcations of the stationary solutions and saddle-node bifurcations for limit cycles if the speed V as well as the damping ζ are varied. Note that the scale along the abscissa is logarithmic.

As an example, we choose a friction function described by the formula (3) with $W_m = 4$, $v_m = 0.1$, $h = 0.35$ and $\zeta = 0.01$, while the speed V is the main bifurcation parameter. The stationary solution, as seen in Figure 4, is unstable in the interval $0.1005 < V < 18.203$. As the third derivative of the friction function is positive at the low-velocity boundary, the corresponding Hopf bifurcation is supercritical and a stable limit cycle is found. The amplitude of the limit cycle grows rapidly with increasing driving velocity. At the upper boundary the third derivative is negative and an unstable limit cycle originates. At $V = 37.84$, the unstable and stable limit cycles come together and disappear as the velocity is increased further. Figure 4 (left) summarizes the results for the velocity extremes of the limit cycles and their periods. For the part where the maximal velocity is approximately equal to the velocity of the driven surface, we can have situations in which this maximal velocity is either a little smaller, or exactly equal, or a little larger than the driving velocity. In all of these cases, a small slip occurs during a part of the motion, which was observed experimentally by Papenhuyzen [9].

Furthermore, if the damping ζ is considered as a second variable parameter, the bifurcation values are shown in Figure 4 (right): the Hopf bifurcations for the stationary solutions as well as the saddle-node bifurcations for the limit cycles are shown. It is seen that for sufficiently large damping, the stationary solution is globally stable for all driving velocities.

The line of saddle-node bifurcations ends at the driving velocity where the third derivative of the friction function is zero.

3.2. INTERNAL STATE VARIABLE MODEL FOR FRICTION

If the friction model includes internal state variables, as has been proposed in the friction literature, the model of the previous subsection is only valid if the evolution of motion is so slow that faster transient aspects of friction can be neglected and the friction force can be approximated by its stationary value. For motions in which rapid variation of velocity may occur, various authors suggest that it is necessary to include the dynamics of the internal state variables. A model based on the ideas of Rabinowicz [10, 11] and Ruina [12] was proposed by one of the authors [13, 14]. In this model, for any sliding velocity the stationary value of friction force for steady sliding with velocity v_s, is contributed only partly by a component $W_i(v_s)$, which is due to the instantaneous frictional response at the current value of the sliding velocity v_s. The rest of the contribution to friction comes from the effect of the internal state that reflects the past of the sliding motion and evolves continuously. For non-uniform sliding the internal state is defined with the help of a state variable μ that satisfies the following first-order differential equation,

$$\dot{\mu} = (1/\tau_s + |v_s|/d_k)(\mu_k(v_s) - \mu). \tag{5}$$

For constant v_s, the state variable μ approaches the value $\mu_k(v_s)$ asymptotically with a time constant $\tau_s d_k/(d_k + |v_s|\tau_s)$. In the sticking phase, this time constant equals τ_s, while in the sliding phase, it equals half of the harmonic mean of τ_s and $d_k/|v_s|$, a characteristic sliding distance divided by the sliding speed. For large sliding velocities, this equation gives an approximation by a model with one state variable of the mean value of the stationary friction corresponding to the sliding velocity over a past slid distance d_k, as was proposed by Rabinowicz. In non-stationary sliding the instantaneous value of the friction force W is given by

$$W(t) = W(v_s, \mu) = \mu \mathrm{sgn}(v_s) + W_i(v_s). \tag{6}$$

The model contains two functions, μ_k and W_i, that have to be identified from experiments. In addition, it has the fixed parameters τ_s and d_k.

By linearizing the equations around the stationary solution with driving speed V, one obtains

$$\begin{aligned} &\Delta\ddot{x} + (2\zeta + \eta)\Delta\dot{x} + \Delta x - \Delta\mu = 0, \\ &\Delta\dot{\mu} + (1/\tau_s + V/d_k)(\mu_k'\Delta\dot{x} + \Delta\mu) = 0, \end{aligned} \tag{7}$$

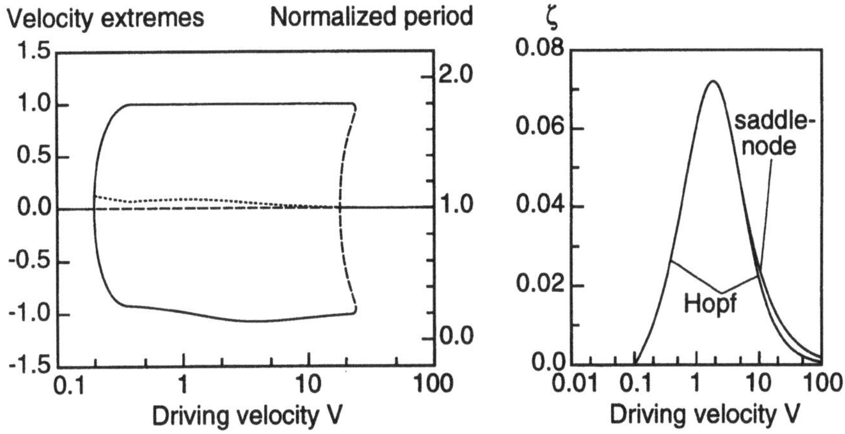

Figure 5. Left: normalized velocity extremes and period for a friction model with state variable. The meaning of the lines is the same as in Figure 4. Right: Bifurcation values for Hopf bifurcations of the stationary solutions and saddle-node bifurcations for limit cycles if the speed V as well as the damping ζ are varied. Note that the scale along the abscissa is logarithmic.

where μ_k' is the derivative of the friction function μ_k and $\eta \geq 0$ is the derivative of W_i at the stationary solution. From this one finds the stability condition by means of the Routh-Hurwitz criteria as

$$(2\zeta+\eta)[1+(1/\tau_s+V/d_k)(\mu_k'+\eta+2\zeta)]+(1/\tau_s+V/d_k)^2(\mu_k'+\eta+2\zeta) > 0. \quad (8)$$

From this it can be concluded that the stationary solution is stable if $\mu_k' + 2\zeta + \eta > 0$, as in the model without state variables. On the other hand, if this expression is negative and the stationary solution is unstable, this solution can be stabilized not only by increasing the damping, but also by increasing the stiffness, which flattens the dimensionless friction function and increases the dimensionless time constants. A similar argument yields that increasing the mass has a destabilizing effect.

The results for the limit cycles, corresponding to Figure 4, are shown in Figure 5 (left) for the parameter values $W_i \equiv 0$, $\tau_s = 25$, $d_k = 2.5$ and the other parameters as before. The range of speeds for which limit cycles may occur is reduced, and the sliding velocity does not reach negative values. The Hopf bifurcations occur at $V = 0.1973$ and $V = 17.966$ and the saddle-node bifurcation for the limit cycle at $V = 24.037$. Figure 5 (right) shows the bifurcation values, from which may be concluded that the amount of external damping needed for stabilizing the stationary motion for the whole range of driving speeds is significantly reduced in virtue of the non-stationary dynamics of friction.

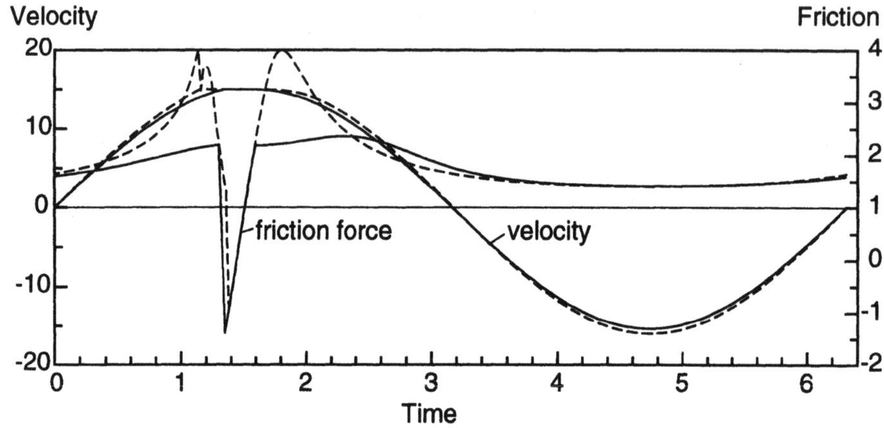

Figure 6. Velocities and friction forces for a friction model without internal state variable (dashed lines) and with internal state variable (drawn lines) at $V = 15$.

The results for the friction model with and without state variable are qualitatively similar with respect to the nature of motion and the bifurcations. However, if the magnitude of the variation of the friction force is examined, a considerable difference is found for the larger driving velocities. Figure 6 shows the friction force during a period of the limit cycle for the case $V = 15$. A large variation is noticed in the magnitude of the friction force. More in particular, this variation in magnitude does not change for changing prescribed velocities V if the friction force depends on the instantaneous sliding velocity only. This contrasts with experimental observations shown in Table 1, which is an argument for assuming that the friction force also depends on other state variables.

3.3. INFLUENCE OF RANDOM DISPLACEMENT-DEPENDENT VARIATIONS OF FRICTION

In the models for friction, one usually assumes that the sliding surfaces are homogeneous, that is, one assumes that the constitutive relation does not depend on the relative position of the contacting surfaces, which is also assumed in the models discussed above. In practice, however, the surfaces are rough on a small scale and the places where the normal and friction force are transmitted from one body to the other may change in a haphazard way at a high frequency. For the basic model shown in Figure 3, this means that the friction force not only depends on the relative sliding speed, but also on the relative position of the sliding body with respect to the contact area. If we make the simplification that the friction force depends only on the current relative displacement and velocity, the equation of motion becomes

$$\ddot{x} + 2\zeta\dot{x} + x = W(Vt - x, V - \dot{x}) = W(x_s, v_s), \tag{9}$$

where $x_s = Vt - x$ is the distance slid. If we start from a Coulomb friction model with variable coefficient of friction, W depends only on the distance slid and the sign of v_s and for positive sliding velocities, the equation of motion can be written as

$$\ddot{x} + 2\zeta\dot{x} + x = W(Vt - x) = 1 + W_1(Vt - x), \tag{10}$$

where we have split the friction function in its average value, which is normalized to one, and a variation W_1. If this equation is expanded in terms of the deviations from a nominal solution $x = 1$, one obtains

$$\Delta\ddot{x} + 2\zeta\Delta\dot{x} + (1 + W_1')\Delta x - \frac{1}{2}W_1''(\Delta x)^2 + \ldots = W_1(Vt - 1), \tag{11}$$

where the primes denote the derivatives of W_1 with respect to its argument. It can be seen that the system has a combination of parametric and external excitation, as is quite common in mechanics.

The random variation of the friction force can be approximated by a finite sum of harmonics with incommensurate frequency ratios,

$$W_1 = \sum_i a_i \sin(\kappa_i x_s - \phi_i), \tag{12}$$

with participation factors a_i, wave numbers κ_i and phases ϕ_i. Alternatively, if the variations are so large that the coefficient of friction can become negative, a multiplicative approximation with positive terms only can be used. A perturbation analysis for the response can be made, which shows that the average over time of the friction force is larger than one, and that, if all wave numbers are sufficiently large with respect to $1/V$, the excess is in first approximation equal to

$$\mathrm{E}[\Delta x] \approx \sum_i \frac{a_i^2 \kappa_i^2 V\zeta}{(1 - \kappa_i^2 V^2)^2} \approx \int_0^\infty \frac{S(\kappa)\zeta\kappa^2 V}{(1 - \kappa^2 V^2)^2} d\kappa, \tag{13}$$

where we have introduced the spectral density function $S(\kappa)$ for the irregularities in the friction force. This shows that the friction force averaged over time, which is usually taken in measurements as the observed value, depends on the sliding velocity, and typically decreases as the driving speed is increased. Furthermore, the average friction force depends on the damping and on the stiffness of the measuring apparatus, where it increases with increased damping and stiffness.

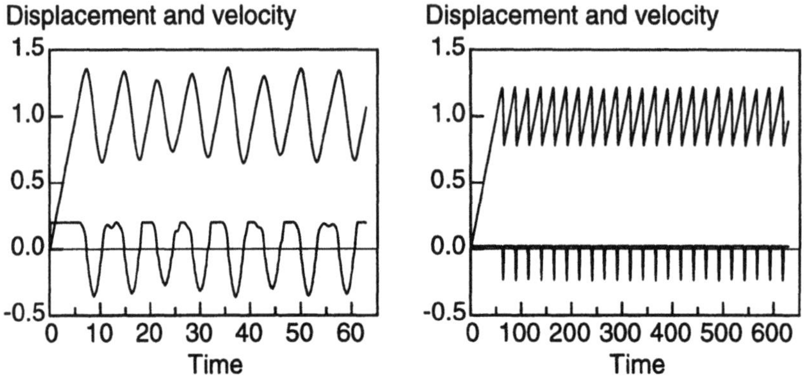

Figure 7. Examples of chaotic stick-slip motions for $V = 0.2$ (left) and $V = 0.02$ (right).

For a better understanding of the behaviour at lower velocities, the distance-dependent term is further simplified to a single harmonic. This can be a valid approximation of a realization of a narrow-band process for short time intervals. In examples we choose $a_1 = 0.2$, $\phi_1 = 0$, $\zeta = 0.01$, and $\kappa_1 = 50$ and V variable. By following the low amplitude periodic solution from high values of the driving velocity for decreasing V, then between $V = 0.06650$ and $V = 0.06420$, one finds a sequence of period-doubling bifurcations leading to chaotic motion. Because the damping in the system is small, the observed Feigenbaum number approaches the limiting value for Hamiltonian systems, $\delta_H = 8.7210\ldots$, in the first few bifurcations before ultimately approaching the value $\delta_F = 4.6692\ldots$. For a wide range of driving velocities, stick-slip-like motions may occur, which can be periodic or chaotic and may coexist with small-amplitude periodic solutions. Examples of chaotic stick-slip motions are given in Figure 7 for $V = 0.2$ and $V = 0.02$, for which a positive value of the maximal Liapunov exponent was found. No significant change of the amplitude of the variation of the friction force with the driving velocity is observed for the sinusoidal variation.

4. Two-degree-of-freedom models

As another model for a simple tribometer system, a two-degree-of-freedom model as shown in Figure 8 is considered. This model consists of two rigid bodies that can move in a horizontal direction. The first body with mass m_1 is connected to a fixed frame by a spring with spring constant k_1 and a damper with constant c_1. This first body represents the loading mass on the tested specimen. The second body with mass m_2 is connected to the first by means of a similar spring and damper with constants k_2 and c_2 and represents the specimen of the tested material, which is much smaller.

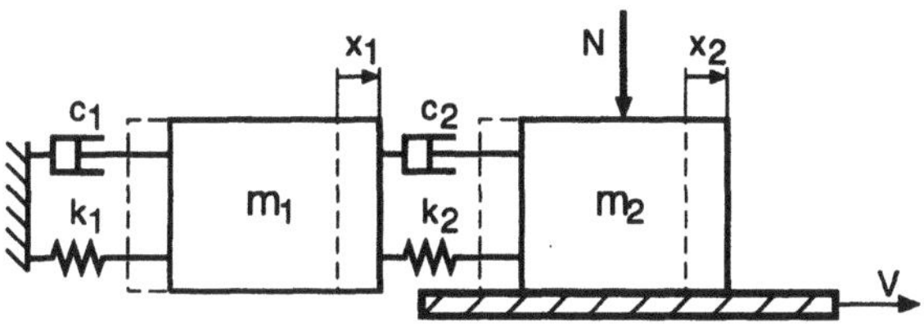

Figure 8. A two-degree-of-freedom model of a tribometer system.

This second body rests on a surface driven at a constant velocity V. The displacements of the two bodies are measured by x_1 and x_2, respectively.

For a friction model in which the friction force depends only on the relative sliding velocity, the linearized equations of motion near a stationary solution for a given driving velocity V, $x_1 = W(V)/k_1$, $x_2 = W(V)(1/k_1 + 1/k_2)$, read

$$\begin{aligned} &m_1\Delta\ddot{x}_1 + (c_1 + c_2)\Delta\dot{x}_1 - c_2\Delta\dot{x}_2 + (k_1 + k_2)\Delta x_1 - k_2\Delta x_2 = 0 \\ &m_2\Delta\ddot{x}_2 - c_2\Delta\dot{x}_1 + (c_2 + c_s)\Delta\dot{x}_2 - k_2\Delta x_1 + k_2\Delta x_2 = 0 \end{aligned} \tag{14}$$

Here, c_s is the derivative of the friction force with respect to the sliding velocity. A stability analysis with the aid of the Routh-Hurwitz criteria shows that if $c_2 + c_s$ is negative, the stationary solution becomes unstable if m_2 is sufficiently small [14], regardless of the values of the external stiffness and damping k_1 and c_1. However, if a friction model with an internal state variable as in equations (5) and (6) is used with $W_i \equiv 0$, the linearized equations become

$$\begin{aligned} &m_1\Delta\ddot{x}_1 + (c_1 + c_2)\Delta\dot{x}_1 - c_2\Delta\dot{x}_2 + (k_1 + k_2)\Delta x_1 - k_2\Delta x_2 = 0 \\ &m_2\Delta\ddot{x}_2 - c_2\Delta\dot{x}_1 + c_2\Delta\dot{x}_2 - k_2\Delta x_1 + k_2\Delta x_2 - \Delta\mu = 0, \\ &\Delta\dot{\mu} + (1/\tau_s + V/d_k)(\mu_k'\Delta\dot{x}_2 + \Delta\mu) = 0. \end{aligned} \tag{15}$$

Stable motion may still be possible if $c_2 + \mu_k'$ is negative, even if the mass m_2 approaches a zero value, provided k_2 is sufficiently large. As experiments have shown that stable stationary motion can be achieved in the decreasing part of the friction-speed curve as in Figure 2 for a small test specimen [7], this is an additional argument for assuming the presence of an internal state variable in the model.

5. Concluding remarks

The present study has shown that the bifurcation behaviour of stationary solutions and the resulting limit cycles can be used for the identification of the law of friction. In particular, the insufficiency of a law for which the friction depends only on the relative sliding velocity has been demonstrated by its inability to explain the decreasing amplitude of the variation of friction force at higher driving velocities and the suppression of instability in a two-degree-of-freedom model. For models with an internal state, only a small damping is needed to stabilize stationary sliding, while an increase in stiffness has a similar effect. Full agreement with observed data has not yet been achieved. The importance of considering position dependent friction has been shown in that it may lead to the apparent dependence on speed and the measurement system of stationary friction, the stick-slip phenomenon and chaotic response even for deterministic sinusoidal variations of the friction.

References

1. K. A. Grosch, "The relation between the friction and visco-elastic properties of rubber," *Proceedings of the Royal Society of London A* **274** (1963), pp. 21–39.
2. H. Blok, "Fundamental mechanical aspects of boundary lubrication," *S.A.E. Journal* **46**(2) (1940), pp. 54–68.
3. J. P. Meijaard, "Efficient numerical integration of the equations of motion of non-smooth mechanical systems," *Zeitschrift für angewandte Mathematik und Mechanik* **77** (1997), pp. 419–427.
4. J. P. Meijaard, "Continuation of stationary and periodic solutions and their bifurcations," Report No 1179, Laboratory for Engineering Mechanics, Delft University of Technology, Delft, September 1998.
5. J. P. den Hartog, *Mechanical Vibrations* (fourth edition), McGraw-Hill, New York, 1956.
6. A. R. Savkoor, "Some aspects of friction and wear of tyres arising from deformations, slip and stresses at the ground contact," *Wear* **9** (1966), pp. 66–78.
7. A. R. Savkoor, *Dry Adhesive Friction of Elastomers, A Study of the Fundamental Mechanical Aspects*, Dissertation, Delft University of Technology, Delft, 1987.
8. B. D. Hassard, N. D. Kazarinoff and Y.-H. Wan, *Theory and Applications of Hopf Bifurcation*, Cambridge University Press, Cambridge, 1981.
9. P. J. Papenhuyzen, "Wrijvingsproeven in verband met het slippen van autobanden," *De Ingenieur* **53** (1938), pp. V.75–V.81.
10. E. Rabinowicz, "The nature of the static and kinetic coefficients of friction," *Journal of Applied Physics* **22** (1951), pp. 1373–1379.
11. E. Rabinowicz, "The intrinsic variables affecting the stick-slip process," *Proceedings of the Physical Society, London* **71** (1958), pp. 668–675.
12. A. L. Ruina, "Slip instability and state variable friction laws," *Journal of Geophysical Research* **88** (1983), pp. 10359–10370.
13. J. P. Meijaard, "Friction modelling and dynamic computation," to be published, 1999.
14. J. P. Meijaard and A. R. Savkoor, "Modelling and suppression of stick-slip in dry friction," in *Proceedings of the Asia-Pacific Vibration Conference '97*, The Korean Society of Mechanical Engineers, Seoul, 1997, pp. 27–32.

Bounds on Sample Paths of Stochastic Nonlinear Systems – A Lyapunov Function Approach

Klaus Reiner Schenk–Hoppé *
Department of Economics
University of Bielefeld
P.O. Box 100 131
33501 Bielefeld, Germany
(schenk@wiwi.uni-bielefeld.de)

Abstract. This paper presents an analytic technique to derive time-dependent bounds on solutions of nonlinear continuous-time systems which are perturbed by real noise processes. The technique provides a tool to analyze the transient behavior of such systems. Moreover, it can also be used to study the quantitative behavior of sample paths of stochastic nonlinear systems. For instance, one can obtain lower bounds on first hitting times, estimate first exit times, and prove boundedness of sample paths. Our approach, originating from the theory of random attractors, is similar to Lyapunov's direct method.

1. Introduction

A thorough understanding of the quantitative behavior of stochastic nonlinear systems is mandatory when one is concerned with safety and reliability investigations of engineering structures with motion uncertainty or with robustness properties of stochastic economic systems. In particular, it is often important to have information on the transient behavior such systems.

Our goal in this paper is to present a technique to analytically derive time-dependent bounds on solutions of random differential equations without solving the equation under consideration. This Lyapunov function technique can be used to estimate first hitting and exit times and to prove uniform boundedness of sample paths of such stochastic nonlinear systems. Since these bounds are obtained in explicit form, one gathers reliable information on the quantitative behavior of these systems. In particular, information can be derived on the short-run and transient behavior of a system with this method. The approach presented here is particularly appropriate for perturbed nonlinear differential equations in which the perturbations are real noise processes, i.e. ergodic (station-

* Part of this work has been done while the author was with the Institute for Dynamical Systems, University of Bremen.

S. Narayanan and R.N. Iyengar (eds.),
IUTAM Symposium on Nonlinearity and Stochastic Structural Dynamics, 249–260.

ary) processes. In the examples provided here, we consider parametric and additive perturbations.

Our work is remotely related to an approach pursued by Schiehlen's group Hu et al. (1996), Schiehlen et al. (1996) who also analyze boundedness of sample paths in stochastic systems, but their primary concern is directed toward a numerical study of statistical properties of upper bounds on sample paths on finite time intervals. Lin and Cai (1995), Chapter 8, investigate first-excursion times of stochastic nonlinear systems in which the excitation is gaussian white noise, see also the references given there.

We introduce the framework used. The reader is referred to Arnold (1998), Chapter 2.2 and Appendix A, for further information. In this paper we are concerned with random differential equations of the form $\dot{x}_t = f(x_t, \xi(\theta_t\omega))$, where $f : \mathbb{R}^d \times \mathbb{R}^m \to \mathbb{R}^d$ and ξ is a real noise process. These equations have a pathwise meaning, and $\varphi(t, \omega, x)$ is called a local solution of the above random differential equation if $\varphi(t, \omega, x) = x + \int_0^t f(\varphi(s, \omega, x), \xi(\theta_s\omega))\, ds$ for all t in some open interval containing zero (the maximal interval of existence $]\tau^-(\omega, x), \tau^+(\omega, x)[$ where the left- resp. right-hand boundary is the explosion time of the backward resp. forward solution). The solution is forward resp. backward complete, if the integral equation is fulfilled for all $t \geq 0$ resp. $t \leq 0$ (equivalently, $\tau^+(\omega, x) = \infty$ resp. $\tau^-(\omega, x) = -\infty$). The solution is called global, if it is forward and backward complete.

Here a real noise process is an integrable measurable function $\xi : \Omega \to \mathbb{R}^m$ over an ergodic dynamical system $((\Omega, \mathcal{F}, \mathbb{P}), (\theta_t)_{t \in \mathbb{R}})$. That is, $(\Omega, \mathcal{F}, \mathbb{P})$ is a probability space and $\theta_t : \Omega \to \Omega$ is a flow of measurable (in (t, ω)), measure preserving ($\theta_t\mathbb{P} = \mathbb{P}\,\forall t$), and ergodic (for all $A \in \mathcal{F} : \theta_t^{-1}A = A\,\forall t \Rightarrow \mathbb{P}(A) \in \{0, 1\}$) mappings. Any integrable ergodic càdlàg process has such a representation, Arnold (1998), Appendix A.2. Thus $\xi(\theta_t\omega)$, $t \in \mathbb{R}$ is an ergodic (stationary) process with $\mathbb{E}|\xi(\omega)| = \int_\Omega |\xi(\omega)|\, d\mathbb{P} < \infty$ and distribution $\xi\mathbb{P} = \xi(\theta_t\cdot)\mathbb{P}$ for all $t \in \mathbb{R}$. Fixing any $\omega \in \Omega$, the function $t \mapsto \xi(\theta_t\omega)$ is called a sample path of the perturbation. In particular $t \mapsto \xi(\theta_t\omega)$ is locally integrable.

For example the Ornstein-Uhlenbeck process is an unbounded real noise processes: It is generated by the stochastic differential equation $dx = -\alpha x dt + \sigma dW_t$, $\alpha > 0$, $\sigma \in \mathbb{R}$ and possesses a stationary probability measure which is given by the normal distribution with expected value 0 and variance $\sigma^2/(2\alpha)$. Any measurable periodic function on $\mathbb{R}$, which is locally integrable, is also a real noise processes: Let T denote the period of such a function f, and define $\Omega = [0, T]$, $\mathbb{P}$ the uniform distribution on Ω, and $\theta_t\omega := \omega + t \bmod T$. Then $f(\theta_t\omega)$ is a real noise processes. For example the periodic excitation $\cos(\alpha(t + \omega) + \beta)$ with ω uniformly distributed on $[0, 2\pi/\alpha]$ is a bounded real noise processes.

The rest of this paper is organized in the following fashion. A simple instructive example is considered in Section 2 to motivate the Lyapunov function technique introduced in Section 3. We conclude with the study of some well-known two-dimensional stochastic nonlinear systems in Section 4.

2. A simple example

We start with an instructive one-dimensional example. Consider the nonlinear random differential equation

$$\dot{x}_t = \xi(\theta_t\omega)\, x_t - \eta(\theta_t\omega)\, x_t^2 \operatorname{sgn} x_t + \zeta(\theta_t\omega) \tag{1}$$

on $\mathbb{R}$ with $\eta \geq 0$ and $\mathbb{E}\eta > 0$. Its solution is forward complete, i.e. the solution is well-defined for all $t \geq 0$, $x \in \mathbb{R}$, and $\omega \in \Omega$ (this follows from Proposition 3.4 as will be seen later).

Letting $V(x) = x^2/2$, we obtain that

$$\begin{aligned}
\dot{V}(x_t) &= \xi(\theta_t\omega)x_t^2 - \eta(\theta_t\omega)x_t^3 \operatorname{sgn} x_t + \zeta(\theta_t\omega)x_t & (2)\\
&\leq \xi(\theta_t\omega)x_t^2 - \eta(\theta_t\omega)(a\, x_t^2 - c(a)) + |\zeta(\theta_t\omega)|(x_t^2 + 1/4) & (3)\\
&= -2(a\eta(\theta_t\omega) - \xi(\theta_t\omega) - |\zeta(\theta_t\omega)|)\, V(x_t) + c(a)\, \eta(\theta_t\omega) + \frac{|\zeta(\theta_t\omega)|}{4}
\end{aligned}$$

where $c(a) = 4/27a^3$. Moving from (2) to (3), we used that $-x^3 \operatorname{sgn} x = -|x|^3 \leq -ax^2 + c(a)$ and $x \leq x^2 + 1/4$.

Fix any $a > 0$ such that $a\mathbb{E}\eta > \mathbb{E}\xi + \mathbb{E}|\zeta|$ and define

$$\alpha(\omega) := 2(a\eta(\omega) - \xi(\omega) - |\zeta(\omega)|) \text{ and } \beta(\omega) := c(a)\eta(\omega) + 1/4|\zeta(\omega)|$$

Then the affine random differential equation

$$\dot{y}_t = -\alpha(\theta_t\omega)y_t + \beta(\theta_t\omega) \tag{4}$$

on $\mathbb{R}_+$, which is associated to (3), has the unique stable stationary solution

$$r(\omega) = \int_{-\infty}^{0} \exp\left(-\int_{t}^{0} \alpha(\theta_s\omega)\, ds\right) \beta(\theta_t\omega)\, dt \tag{5}$$

i.e. denoting the solution of (4) by $\psi(t, \omega, y)$, one has $\psi(t, \omega, r(\omega)) = r(\theta_t\omega)$ almost surely for all $t \in \mathbb{R}$. Stability of the random variable $r(\omega)$ means that for each $y \in \mathbb{R}_+$, $\lim_{t\to\infty} \psi(t, \theta_{-t}\omega, y) = r(\omega)$ almost surely, which implies lim in prob $_{t\to\infty} |\psi(t, \omega, y) - r(\theta_t\omega)| = 0$ by stationarity of θ_t. Further, $r(\omega)$ depends only on the past of the noise processes, i.e. on $\alpha(\theta_t\omega), \beta(\theta_t\omega)$ for $t \leq 0$ only.

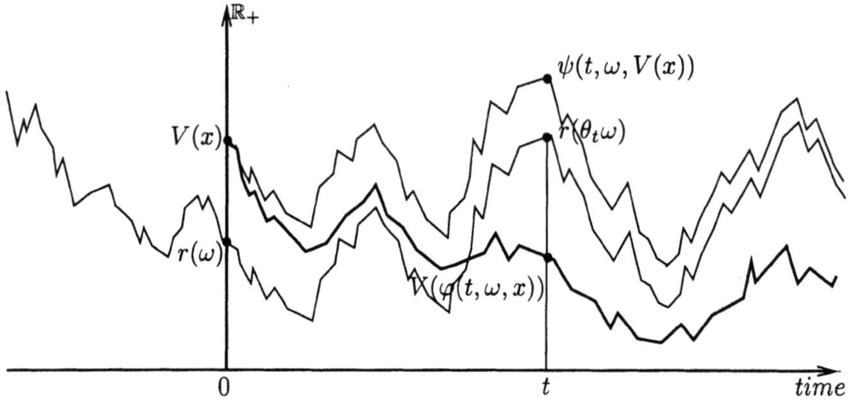

Figure 1. Sample path and stationary solution of the affine equation (4), and sample path of (2). As pointed out in the text, $\psi(t,\omega,V(x)) \geq V(\varphi(t,\omega,x))$ and $|\psi(t,\omega,V(x)) - r(\theta_t\omega)|$ converges to zero in probability as $t \to \infty$. Note that it might happen that $V(\varphi(t,\omega,x)) > r(\theta_t\omega)$ for all $t \geq 0$. However, $V(\varphi(t,\omega,x)) < (1+\varepsilon)r(\theta_t\omega)$ for all sufficiently large t.

It can be checked (see the next section) that the interval $[0,r(\omega)]$ is forward invariant, i.e. $\psi(t,\omega)[0,r(\omega)] \subset [0,r(\theta_t\omega)]$ for all $t \geq 0$. Therefore, if the system (4) is started at a point $y \in [0,r(\omega)]$, the corresponding sample path $\psi(t,\omega,y) \in [0,r(\theta_t\omega)]$ for all $t \geq 0$. Hence $r(\theta_t\omega)$ is a time-dependent upper bound on the sample path $\psi(t,\omega,y)$. It is given explicitly by (5) and can be calculated without solving equation (4).

These arguments carry over to system (1), because its solution $\varphi(t,\omega,x)$ satisfies $0 \leq V(\varphi(t,\omega,x)) \leq \psi(t,\omega,V(x))$ for all $t \geq 0$. Letting $B(\omega) := V^{-1}([0,r(\omega)])$, we have that $\varphi(t,\omega)B(\omega) \subset B(\theta_t\omega)$ almost surely for all $t \geq 0$, and further that for all $x \in \mathbb{R}$, $\lim_{t\to\infty} \operatorname{dist}(\varphi(t,\theta_{-t}\omega,x),B(\omega)) = 0$ almost surely, where $\operatorname{dist}(x,C) = \inf_{y\in C} d(x,y)$. Again this implies that lim in prob $_{t\to\infty} \operatorname{dist}(\varphi(t,\omega,x),B(\theta_t\omega)) = 0$. Hence the random sets $B(\omega)$, $\omega \in \Omega$ provide time-dependent upper bounds on the sample paths of (1) in the sense that for all $x \in B(\omega)$, $\varphi(t,\omega,x) \in B(\theta_t\omega)$ for all $t \geq 0$.

As above we can remark that $B(\omega)$ is given in explicit form and can be calculated without solving (1). Further, $B(\theta_t\omega)$ is determined completely by states of the perturbations which are prior to t.

Suppose c is a critical value for (1), e.g. a safety boundary or a tolerance limit. Then it is straightforward to give an estimate from below on the first hitting time of a function of the system state $\varphi(t,\omega,x)$ to this critical value: $\inf\{t \geq 0 \mid h(B(\theta_t\omega)) = c\} \leq \inf\{t \geq 0 \mid h(\varphi(t,\omega,x)) = c\}$ for all $x \in B(\omega)$, where $h : \mathbb{R}^d \to \mathbb{R}$ is any function. The first

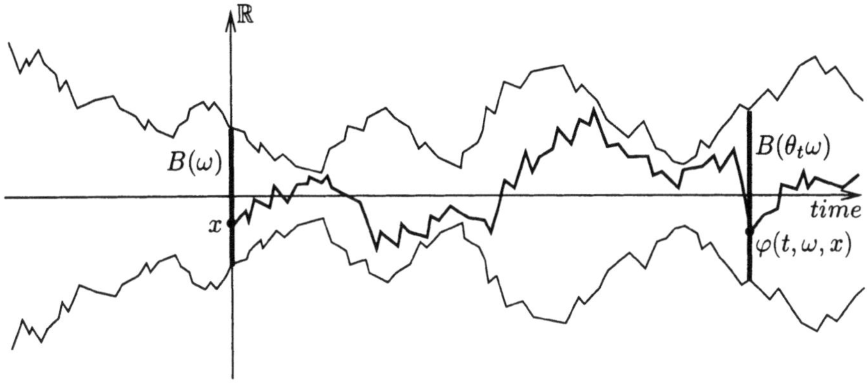

Figure 2. Sample path $\varphi(t,\omega,x)$ of (1) and compact random set $B(\omega) := V^{-1}([0,r(\omega)]) = [-\sqrt{2r(\omega)}, +\sqrt{2r(\omega)}]$. Note that $\varphi(t,\omega)B(\omega) \subset B(\theta_t\omega)$ for all $t \geq 0$, where the inclusion is proper in general.

quantity is given explicitly and can be calculated without solving (1). It is immediate that the critical value may also depend on the time and the excitation.

Note that $0 \in B(\omega)$ for all ω, where zero is a fixed point of (1) when $\zeta \equiv 0$ (purely parametric excitation). One therefore has an upper bound on the probability that $\|\varphi(t,\omega,0)\| = d(0,\varphi(t,\omega,0))$ exceeds a certain value c: $\mathbb{P}\{\|\varphi(t,\omega,0)\| \geq c\} \leq \mathbb{P}\{\sup\{\|y\| \mid y \in B(\theta_t\omega)\} \geq c\}$. This yields an upper bound on tail-ends of the distribution of $\varphi(t,\omega,0)$ for each $t \geq 0$, and one could carry out a quantitativ evaluation of the effects that the additive excitation has on the sample paths starting at zero next.

3. The technique

In this section we explain how the technique introduced in the preceding section can be applied to nonlinear random differential equations which describe higher dimensional systems.

Consider the random differential equation

$$\dot{x}_t = f(x_t, \xi(\theta_t\omega)) \tag{6}$$

on $\mathbb{R}^d$ with $f : \mathbb{R}^d \times \mathbb{R}^m \to \mathbb{R}^d$ and ξ being an $\mathbb{R}^m$-valued real noise process.

Under certain conditions (6) generates a local continuous random dynamical system φ on $\mathbb{R}^d$ and in particular possesses a unique local solution, see Arnold (1998), Theorem 2.2.1. This is true, for example, if the right-hand side of the above random differential equation has the particular form $f(x,\xi(\omega)) = g(x) + h(x)\xi(\omega)$ with locally Lipschitz continuous functions g and h. Existence of a unique local solution will be assumed from now on.

THEOREM 3.1. *Consider the random differential equation (6). Suppose there exists a function* $V : \mathbb{R}^d \to \mathbb{R}_+$ *such that*

(i) V *is continuously differentiable, surjective, and has the property that pre-images of bounded sets are bounded; and*

(ii) $LV(x,\omega) := \langle DV(x), f(x,\omega)\rangle \leq -\eta_1(\omega)V(x) + \eta_2(\omega)$ *for all* $x \in \mathbb{R}^d$, $\omega \in \Omega$, *where* η_1, η_2 *are real noise processes satisfying* $\mathbb{E}\eta_1 > 0$, *and* $\eta_2 \geq 0$.

Then the compact random set

$$B_\varepsilon(\omega) := V^{-1}\left([0, (1+\varepsilon)r(\omega)]\right) \tag{7}$$

with arbitrarily fixed $\varepsilon \geq 0$ *and*

$$r(\omega) := \int_{-\infty}^{0} \exp\left(-\int_t^0 \eta_1(\theta_s\omega)\, ds\right) \eta_2(\theta_t\omega)\, dt \tag{8}$$

has the following properties.

(i) $B_\varepsilon(\omega)$ *is forward invariant, i.e. for all* $t \geq 0$, $\varphi(t,\omega)B_\varepsilon(\omega) \subset B_\varepsilon(\theta_t\omega)$ *almost surely.*

(ii) $B_\varepsilon(\omega)$ *is attracting, i.e.* $\lim_{t\to\infty} \operatorname{dist}(\varphi(t,\theta_{-t}\omega,x), B_\varepsilon(\omega)) = 0$ *almost surely. Moreover, if* $\varepsilon > 0$, *then* $B_\varepsilon(\omega)$ *is absorbing, i.e. there is a* $T(\omega,x)$ *such that* $\varphi(t,\theta_{-t}\omega,x) \in B_\varepsilon(\omega)$ *for all* $t \geq T(\omega,x)$.

(iii) the random attractor $A(\omega) := \bigcap_{n\in\mathbb{N}} \varphi(n,\theta_{-n}\omega)B_\varepsilon(\theta_{-n}\omega)$ *is contained in* $B_\varepsilon(\omega)$.

A function V with the above properties will be referred to as a *Lyapunov function* in the following.

We point the reader to Amann (1990), §18, and Hale and Koçak (1996), Chapter 9.3, for information on Lyapunov functions of deterministic systems. It is worth to note that in the absence of the perturbation the above result is only remotely related to standard results presented there. Khasminskii (1980) has used Lyapunov-function-type methods to ensure forward completeness (and non-completeness) of solutions of stochastic differential equations.

Readers who like learn more about random attractors can consult e.g. Arnold and Schmalfuß (1996) and Schenk–Hoppé, Schenk–Hoppé (1998, 1999).

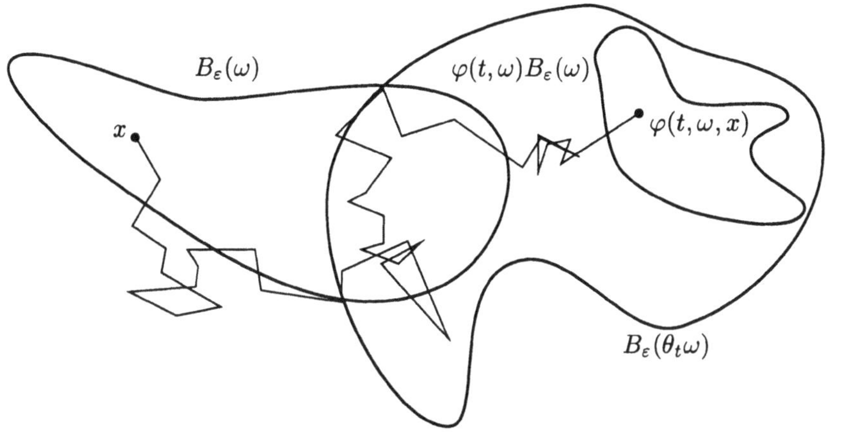

Figure 3. Illustration of a sample path $\varphi(t,\omega,x)$ and of the motion of the sets $B_\varepsilon(\theta_t\omega)$ and $\varphi(t,\omega)B_\varepsilon(\omega)$.

REMARK 3.2. *If the solution of (6), say $\varphi(t,\omega,x)$, is forward complete (i.e. exists for all $t \geq 0$), then the conclusions of Theorem 3.1 hold if V is only continuous instead of continuously differentiable, and the process $V(\varphi(t,\omega,x))$ satisfies an inequality of the type*

$$V(\varphi(t,\omega,x)) - V(x) \leq \int_0^t -\eta_1(\theta_s\omega)V(\varphi(s,\omega,x)) + \eta_2(\theta_s\omega)\,ds$$

for all $x \in \mathbb{R}^d$, $\omega \in \Omega$, and $t \geq 0$, where η_1 and η_2 are real noise processes satisfying $\mathbb{E}\eta_1 > 0$, and $\eta_2 \geq 0$. In differential form this inequality reads $\dot{V}(\varphi(t,\omega,x)) \leq -\eta_1(\theta_t\omega)V(\varphi(t,\omega,x)) + \eta_2(\theta_t\omega)$.

Sketch of the proof. The Lyapunov function V reduces the problem of finding upper bounds on the sample paths of a d-dimensional system to a tractable one-dimensional problem.

The random differential equation associated to the inequality in assumption *(ii)* is affine and can be analyzed analytically. By the variation of constants formula one finds its solution

$$\psi(t,\omega,y) = e^{-\int_0^t \eta_1(\theta_s\omega)ds}\left(y + \int_0^t e^{-\int_s^0 \eta_1(\theta_u\omega)du}\eta_2(\theta_s\omega)ds\right) \tag{9}$$

The solution of the affine equation dominates $V(\varphi(t,\omega,x))$, i.e. $0 \leq V(\varphi(t,\omega,x)) \leq \psi(t,\omega,V(x))$ for all $t \geq 0$, Walter (1970), Theorem I.5.III.

Further, ψ possesses the unique stable stationary solution $r(\omega)$ as stated in (8). We only show invariance of $r(\omega)$ here:

$$\psi(t,\omega,r(\omega)) = \int_{-\infty}^t e^{-\int_s^t \eta_1(\theta_u\omega)\,du}\eta_2(\theta_s\omega)\,ds$$

$$= \int_{-\infty}^{t} e^{-\int_{s-t}^{0} \eta_1(\theta_{t+u}\omega)\, du} \eta_2(\theta_s\omega)\, ds$$

$$= \int_{-\infty}^{0} e^{-\int_{s}^{0} \eta_1(\theta_{t+u}\omega)\, du} \eta_2(\theta_{t+s}\omega)\, ds \; = \; r(\theta_t\omega)$$

Let us next check forward invariance of $[0,(1+\varepsilon)r(\omega)]$:

$$\psi(t,\omega,r(\omega)) = \varepsilon\, r(\omega)\, e^{-\int_0^t \eta_1(\theta_s\omega)ds} + r(\theta_t\omega)$$

$$= \varepsilon \int_{-\infty}^{0} e^{-\int_s^t \eta_1(\theta_u\omega)du} \eta_2(\theta_s\omega)ds + r(\theta_t\omega) \;\le\; (1+\varepsilon)r(\theta_t\omega)$$

since $\eta_2(\omega) \ge 0$. Similar calculations show that $[0,(1+\varepsilon)r(\omega)]$ is attracting, and, if $\varepsilon > 0$, absorbing.

These properties carry over to the original system, because by surjectivity of V, $[0,(1+\varepsilon)r(\omega)] = V(B_\varepsilon(\omega))$ and hence the domination property of ψ and positivity of V imply $V(\varphi(t,\omega)B_\varepsilon(\omega)) \subset V(B_\varepsilon(\theta_t\omega))$. Finally, taking pre-images with respect to V on both sides yields the forward invariance of $B_\varepsilon(\omega)$. Analogously, one can read-off the attraction and absorbtion properties. □

Some remarks on the implications of Theorem 3.1 are in order:

REMARK 3.3. *Given a random differential equation (6) for which a Lyapunov function exists, then the compact random set B_ε, defined in Theorem 3.1, provides a time-dependent upper bound on the sample paths of this system in the following sense. If the initial state $x \in B_\varepsilon(\omega)$, then $\varphi(t,\omega,x) \in B_\varepsilon(\theta_t\omega)$ for all $t \ge 0$. Moreover, for any given realization of the perturbation ξ (i.e. $t \mapsto \xi(\theta_t\omega), t \in \mathbb{R}$, and ω fixed) and any initial value $x \in \mathbb{R}^d$, there is some $\varepsilon \ge 0$ such that $x \in B_\varepsilon(\omega)$ and hence $\varphi(t,\omega,x) \in B_\varepsilon(\theta_t\omega)$ for all $t \ge 0$.*

$B_\varepsilon(\omega)$ is given in explicit form for all ω by (7) and has been obtained without solving the random differential equation (6). Moreover, $B_\varepsilon(\theta_t\omega)$ is determined completely by states of the perturbations which are prior to t. It can be calculated numerically for all ω and all t, where the numerical approximation is, in general, simpler and more reliable than a direct numerical analysis of the solutions of (6).

The Lyapunov-function-type technique can be used to estimate first hitting times. Denote by $t_h(\omega) = \inf\{t \ge 0 \mid \operatorname{dist}(\varphi(t,\omega,x),X) = 0\}$, X a closed subset of $\mathbb{R}^d$, the first hitting time of $\varphi(t,\omega,x)$ to the set X. Then $t_h(\omega) \ge \inf\{t \ge 0 \mid B_\varepsilon(\theta_t\omega) \cap X \ne \emptyset\}$ where ε is chosen such that $x \in B_\varepsilon(\omega)$.

If $x \in X$, then one can obtain bounds on first exit times of $\varphi(t,\omega,x)$ by the above method. Suppose there is an ε such that $x \in B_\varepsilon(\omega) \subset X$, then $t_e(\omega) := \inf\{t \ge 0 \mid \varphi(t,\omega,x) \notin X\} \ge \inf\{t \ge 0 \mid B_\varepsilon(\theta_t\omega) \cap \mathbb{R}^d \backslash X \ne$

$\emptyset\}$. *In particular, if* $\cup_{t\geq 0} B_\varepsilon(\theta_t\omega)$ *is a bounded set, then sample paths starting in* $B_\varepsilon(\omega)$ *are uniformly bounded.*

It is straightforward to see that one can also estimate first hitting and exit times for time-dependent (and even noise-dependent) sets. Moreover, one can use an evaluation function, say $h : \mathbb{R}^d \to \mathbb{R}^d$, *and give estimates on first hitting and exit times of the transformed sample paths* $h(\varphi(t,\omega,x))$.

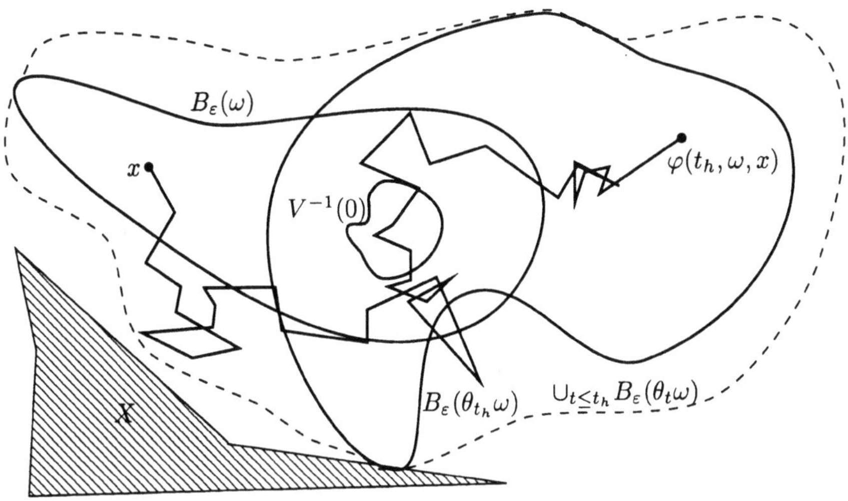

Figure 4. Lower bound on the hitting time of a sample path $\varphi(t,\omega,x)$: $\inf\{t \geq 0 \mid B_\varepsilon(\theta_t\omega) \cap X \neq \emptyset\} \leq t_h := \inf\{t \geq 0 \mid \mathrm{dist}(\varphi(t,\omega,x), X) = 0\}$.

The Lyapunov function approach introduced here also provides a tool to estimate the weight of the tail-ends of a distribution of a sample path. In particular, this method (necessarily) detects rare events such as large excursions of solutions, which often causes problems in numerical simulations of stochastic nonlinear systems.

It holds $V^{-1}(\{0\}) \subset B_0(\omega)$ for all $\omega \in \Omega$. For each $x \in V^{-1}(\{0\})$ one therefore has an upper bound on the probability that $\|\varphi(t,\omega,x)\| = d(0,\varphi(t,\omega,x))$ exceeds a certain value c: $\mathbb{P}\{\|\varphi(t,\omega,x)\| \geq c\} \leq \mathbb{P}\{\sup\{\|y\| \mid y \in B_0(\theta_t\omega)\} \geq c\}$. This yields an upper bound on tail-ends of the distribution of $\varphi(t,\omega,x)$ for each $t \geq 0$. In all examples discussed below, zero is fixed point of the systems in the absence of the additive perturbation and $0 \in V^{-1}(\{0\})$. One therefore obtains an upper bound on the probability that a solution of the additively perturbed system which starts at zero exceeds some threshold at a fixed time $t \geq 0$.

Another problem related to stochastic nonlinear systems is that of explosion of solutions. Due to the existence of nonlinearities in the generator, the standard conditions (global Lipschitz continuity) ensuring existence of a global solution do not hold in general. Hence solutions of random differential equations may explode (i.e. leave the state space in finite time) and thus are only defined locally. Existence of a Lyapunov function, however, implies that solutions are non-explosive (strictly conservative) forward in time. We refer the reader to Schenk–Hoppé (1996) for a detailed discussion and further information of this subject.

PROPOSITION 3.4. *Suppose there exists a Lyapunov function for (6). Then the solution is strictly forward complete. That is, denoting the explosion time of a solution started at* x *by* $\tau^+(\omega, x) > 0$, $\mathbb{P}\{\tau^+(\omega, x) = \infty \, \forall x \in \mathbb{R}^d\} = 1$.

Proof. The above method provides us with compact sets $B_\varepsilon(\omega)$ such that $\varphi(t,\omega)B_\varepsilon(\omega) \subset B_\varepsilon(\theta_t\omega)$ almost surely for all $t \geq 0$. The null-sets depend on t as well as on ε. Considering only times $t \in \mathbb{N}$, this reveals that $\mathbb{P}\{\tau^+(\omega, x) = \infty$ for all $x \in B_\varepsilon(\omega)\} = 1$ (here the null-set N_ε depends on ε only). Define the sequence $\varepsilon_n = n$ and let $N := \cup_{n\in\mathbb{N}} N_{\varepsilon_n}$. Since $\mathbb{R}^d = \cup_{n\in\mathbb{N}} B_{\varepsilon_n}(\omega)$, one obtains the following. Fix any $\omega \notin N$, then for each $x \in \mathbb{R}^d$ there is some ε_n such that $x \in B_{\varepsilon_n}(\omega)$ and, therefore, $\tau^+(\omega, x) = \infty$. □

Explosion of the backward solution is not excluded by this result.

4. Applications

The Lyapunov function approach introduced above will be applied to three nonlinear random differential equations. They all belong to the class of systems of the form

$$\ddot{x}_t = g(x_t, \dot{x}_t + f(x_t)) - f'(x_t)\,\dot{x}_t + h(x_t, \dot{x}_t + f(x_t))\,\xi(\theta_t\omega)$$

which we call stochastic Liénard systems, cf. e.g. Hale and Koçak (1996) and Verhulst (1996) for their (simplified) deterministic counterpart.

These systems can be represented as the system

$$\begin{aligned} \dot{x}_t &= y_t - f(x_t) \\ \dot{y}_t &= g(x_t, y_t) + h(x_t, y_t)\,\xi(\theta_t\omega) \end{aligned}$$

of first order random differential equations. If $f(x) \equiv 0$, this is simply the canonical representation.

Our first example is the stochastic nonlinear system governed by

$$\ddot{x} = (-1 + \xi(\theta_t\omega))x - \beta\dot{x} - (\beta x + \dot{x})^3 + \eta(\theta_t\omega) \tag{10}$$

with $\beta > 0$.

The Liénard representation of (10) is given by $f(x) = \beta x$, $g(x,y) = -x - y^3$, $h(x,y) = (x,1)$, and the two-dimensional real noise process $(\xi,\eta)^T$.

We show that $V(x,y) = (x^2+y^2)/2$ is a Lyapunov function, provided that $\mathbb{E}|\xi| < 2\beta$. One can check that

$$
\begin{aligned}
LV(x,y,\omega) &= -\beta x^2 - y^4 + x\,y\,\xi(\omega) + y\,\eta(\omega) \\
&\leq -(2\beta - |\xi(\omega)|)\,\frac{x^2}{2} + (|\xi(\omega)| + 2|\eta(\omega)|)\,\frac{y^2}{2} - y^4 + \frac{|\eta(\omega)|}{4} \\
&\leq -(2\beta - |\xi(\omega)|)\,V(x,y) + \frac{|\eta(\omega)| + (|\eta(\omega)| + \beta)^2}{4}
\end{aligned}
$$

using $\xi xy \leq |\xi|(x^2+y^2)/2$ and $|\eta|y \leq |\eta|(y^2+1/4)$ in the first estimation, and $|\eta|y^2 - y^4 \leq -\beta y^2 + (|\eta|+\beta)^2/4$ in the second estimation. Thus, Theorem 3.1 applies and thus all conclusions drawn in Remark 3.3 hold. Moreover, $\{(0,0)\} = V^{-1}(\{0\})$ and, as explained in the preceding section, one immediately has an estimate on the probability that a solution started at zero exceeds some threshold at a given time $t \geq 0$.

The second example is the stochastic Duffing-van der Pol oscillator which is governed by

$$\ddot{x} = (\alpha + \xi(\theta_t\omega))x + \beta\dot{x} - x^3 - x^2\dot{x} + \eta(\theta_t\omega) \tag{11}$$

with $\alpha, \beta \in \mathbb{R}$. See e.g. Lin and Cai (1995) for several engineering applications of white noise versions of (11) and the analyses provided there and in Arnold et al. (1996).

The Liénard representation of (11) is given by $f(x) = \beta x - x^3/3$, $g(x,y) = \alpha x - x^3$, $h(x,y) = (x,1)$, and the real noise process $(\xi,\eta)^T$. Using this representation, it is shown in Schenk–Hoppé (1996) that $V(x,y) = 7/24x^4 + x^2 - xy + 3/4y^2$ is a Lyapunov function for all α, β, if $7/2\mathbb{E}|\xi| + 6\mathbb{E}|\eta| < 1$.

If (11) is only perturbed parametrically ($\eta \equiv 0$), $\alpha < 0$, and $\beta > 0$ is sufficiently large such that the fixed point zero is unstable, then the Lyapunov function approach studied here can also be used to obtain a time-dependent lower bound on the distance of a solution to the steady state zero, see Schenk–Hoppé (1996).

The last example studied is the perturbed Duffing oscillator which is governed by the nonlinear equation

$$\ddot{x} = (\alpha + \xi(\theta_t\omega))x + \beta\dot{x} - x^3 + \eta(\theta_t\omega) \tag{12}$$

with $\alpha, \beta \in \mathbb{R}$. Its Liénard representation is given by $f(x) = \beta x$, $g(x,y) = \alpha x - x^3$, $h(x,y) = (x,1)$, and the real noise process $(\xi,\eta)^T$.

We check that $V(x,y) = x^4/2 + x^2 - xy + y^2$ is a Lyapunov function, provided $\beta < -1/2$ and $2\mathbb{E}(|\xi| + |\eta|) < \min\{1/2, -(2\beta+1)\}$. Fix any

$\delta < 1/2 \min\{1/2, -(2\beta+1)\}$, then

$$LV(x,y,\omega) = (2\beta+1)x^4 + (2\beta-\alpha)x^2 + (2\alpha+2-\beta)xy - y^2 + \underbrace{(2xy - x^2)\xi(\omega)}_{\leq 4V(x,y)|\xi(\omega)|} + \underbrace{(2y-x)\eta(\omega)}_{\leq (4V(x,y)+3/4)|\eta(\omega)|}$$

$$\leq -4(\delta/2 - |\xi(\omega)| - |\eta(\omega)|)\, V(x,y) + R_{\delta,\alpha,\beta}(x,y) + \frac{3|\eta(\omega)|}{4}$$

The remainder $R_{\delta,\alpha,\beta}(x,y) = (2\beta+1+\delta)x^4 + (2\beta-\alpha+2\varepsilon)x^2 + (2\alpha+2-\beta-2\delta)xy + (-1+2\delta)y^2 \leq c_{\delta,\alpha,\beta}$ is bounded by some constant, because the coefficient of x^4, $(2\beta+1+\delta) < 0$, and the coefficient of y^2, $(-1+2\delta) < 0$.

References

Amann, H. (1990) *Ordinary Differential Equations*, de Gruyter, Berlin.

Arnold, L. (1998) *Random Dynamical Systems*, Springer-Verlag, Berlin.

Arnold, L., Namachchivaya, N. Sri, and Schenk–Hoppé, K. R. (1996) Toward an understanding of stochastic Hopf bifurcation: A case study, *International Journal of Bifurcation and Chaos in Applied Science and Engineering* **6**, 1947–1975.

Arnold, L. and Schmalfuß, B. (1996) Fixed points and attractors for random dynamical systems, in A. Naess and S. Krenk (eds.), *IUTAM Symposium on Advances in Nonlinear Stochastic Mechanics*, Kluwer Academic Publishers, Dordrecht, pp. 19–28.

Hale, J. and Koçak, H. (1996) *Dynamics and Bifurcations*, 3rd printing, Springer-Verlag, Berlin.

Hu, B., Schaub, S., and Schiehlen, W. (1996) Boundedness evaluation of structural systems with randomly distributed initial conditions, *International Journal on Non-Linear Mechanics* **31**, 743–754.

Khasminskii [Has'minskiĭ], R. Z. (1980) *Stochastic Stability of Differential Equations*, Sijhoff & Noordhoff, Alphen aan den Rijen, The Netherlands.

Lin, Y. K. and Cai, G. Q. (1995) *Probabilistic Structural Dynamics*, McGraw-Hill, New York.

Schenk–Hoppé, K. R. (1996) Deterministic and stochastic Duffing-van der Pol oscillators are non-explosive, *ZAMP - Journal of Applied Mathematics and Physics* **47**, 740–759.

Schenk–Hoppé, K. R. (1998) Random attractors – General properties, existence and applications to stochastic bifurcation theory, *Discrete and Continuous Dynamical Systems* **4**, 99–130.

Schenk–Hoppé, K. R. (1999) Random Attractors – A Brief Introduction and Some Applications, *ZAMM*, 4 p., 1999. Forthcoming.

Schiehlen, W., Hu, B., and Schaub, S. (1996) Amplitude bounds of stochastic nonlinear multibody systems, in A. Naess and S. Krenk (eds.), *IUTAM Symposium on Advances in Nonlinear Stochastic Mechanics*, Kluwer Academic Publishers, Dordrecht, pp. 373–382.

Verhulst, F. (1996) *Nonlinear Differential Equations and Dynamical Systems*, 2nd printing, Springer-Verlag, Berlin.

Walter, W. (1970) *Differential and Integral Inequalities*, Springer-Verlag, Berlin.

PARAMETER IDENTIFICATION OF NONLINEAR MULTI-BODY SYSTEMS USING CORRELATION TECHNIQUES

W. SCHIEHLEN AND B. HU
Institute B of Mechanics
University of Stuttgart
Pfaffenwaldring 9
D-70550 Stuttgart
Germany

1. Introduction

Quantitative dynamical analysis of mechanical systems requires mathematical models. The multibody system method has been applied successfully for modeling of mechanical systems and symbolic equations of motion can be obtained using software packages such as NEWEUL, (Schiehlen, 1997). Some physical parameters in the equations of motion can be measured directly, other parameters, however, must be identified from experiments.

During last decades, many parameter identification methods for estimating unknown parameters in nonlinear dynamical systems have been developed. A parameter identification method is characterized by

- the systems which can be dealt with,
- the measurements which must be obtained,
- the model error which is chosen and
- the optimization technique which is used.

Different combinations of systems, measurement requirements, model errors and optimization techniques make different identification methods. The most well known parameter identification methods for nonlinear systems are the Gaussian least-square method (Gauss, 1887), the instrumental variable method (Young, 1970), and the maximum likely-hood method (Fisher, 1950). Among them, the least-square method is most commonly used due to its simplicity and wide applicability. Other methods require the statistical properties of measurement noises which are in practice very difficult to get.

S. Narayanan and R.N. Iyengar (eds.),
IUTAM Symposium on Nonlinearity and Stochastic Structural Dynamics, 261–270.

The major disadvantage of the least-square method is that it yields often biased estimates, see e.g. (Söderström and Stoica, 1989). To overcome this disadvantage, a parameter identification method for nonlinear multibody systems using correlation techniques is presented in this paper.

2. Correlation-Based Parameter Identification

For a nonlinear multibody system the equations of motion can be represented in the state space as

$$\dot{\boldsymbol{x}} = \boldsymbol{f}(\boldsymbol{\theta}, \boldsymbol{x}, \boldsymbol{u}, t) \;, \tag{1}$$

where $\boldsymbol{x}$ denotes an n-dimensional state vector composed of generalized coordinates and velocities and $\boldsymbol{\theta}$ an m-dimensional unknown parameter vector. Moreover, the vector $\boldsymbol{u}$ means an l-dimensional excitation vector which is assumed here to consist of stationary stochastic processes. The variable t stands for the time. The model errors $\boldsymbol{\varepsilon}(t)$ for parameter identification are usually chosen as the residua of the state equations (1), i.e.

$$\boldsymbol{\varepsilon}(t) = \tilde{\dot{\boldsymbol{x}}} - \boldsymbol{f}(\boldsymbol{\theta}, \tilde{\boldsymbol{x}}, \tilde{\boldsymbol{u}}, t) \,, \tag{2}$$

where the tilde means the measured signals which are noised due to measurement noises. For ideal measurements of the input and output variables

$$\tilde{\boldsymbol{u}} = \boldsymbol{u} \,, \qquad \tilde{\boldsymbol{x}} = \boldsymbol{x} \,, \qquad \tilde{\dot{\boldsymbol{x}}} = \dot{\boldsymbol{x}} \tag{3}$$

without noise, the equations (1) are well satisfied and the model parameters $\boldsymbol{\theta}$ can be found from a set of algebraic equations following from (2). However, in practice the noise in measurements is unavoidable and the model errors $\boldsymbol{\varepsilon}(t)$ become stochastic processes. Suppose measured signals are sampled from the initial time $t_0 = 0$ with equal time steps Δt, and the number of samples is N. Using classical least-square method to estimate the unknown parameters $\boldsymbol{\theta}$, it is required that the mean square of the model errors

$$J_{\mathrm{LS}}(\boldsymbol{\theta}) := \mathrm{E}[\boldsymbol{\varepsilon}^T(t)\boldsymbol{\varepsilon}(t)] \simeq \frac{1}{N}\sum_{i=0}^{N-1} \boldsymbol{\varepsilon}^T(i\Delta t)\boldsymbol{\varepsilon}(i\Delta t) \tag{4}$$

to be minimal, where E stands for the mathematical expectation operator. The solution of the least-square method is obtained from the following equations:

$$\frac{\partial J_{\mathrm{LS}}(\boldsymbol{\theta})}{\partial \boldsymbol{\theta}} = 0 \;. \tag{5}$$

For stationary processes $\boldsymbol{\varepsilon}(t)$ with zero mean values, the function $J_{\mathrm{LS}}(\boldsymbol{\theta})$ corresponds to the trace of covariance matrix of the model errors. Moreover, the trace of the correlation matrix of the model errors

$$J_\tau(\boldsymbol{\theta}) := \mathrm{E}[\boldsymbol{\varepsilon}^{\mathrm{T}}(\mathrm{t}+\tau)\boldsymbol{\varepsilon}(\mathrm{t})] \tag{6}$$

may be also considered as an objective function to be minimized. From equations

$$\frac{\partial J_\tau(\boldsymbol{\theta})}{\partial \boldsymbol{\theta}} = 0 \,, \tag{7}$$

one gets an estimation $\boldsymbol{\theta}_\tau$ for the parameter $\boldsymbol{\theta}$ for every time lag τ. It is easy to see that for the vanishing time lag $\tau = 0$ it yields

$$J_0(\boldsymbol{\theta}) = J_{\mathrm{LS}}(\boldsymbol{\theta}) \qquad \text{and} \qquad \boldsymbol{\theta}_0 = \boldsymbol{\theta}_{\mathrm{LS}} \,. \tag{8}$$

As an example, a simple linear stochastic system

$$\dot{x} + ax = u \tag{9}$$

is used to explain the idea of the correlation-based method. Here x is the state variable and u a stationary stochastic excitation. The measurement variables are u and x which are noised by two independent stationary stochastic processes ξ_x and ξ_u respectively, i.e. the measured signals read as

$$\tilde{x} = x + \xi_x \qquad \text{and} \qquad \tilde{u} = u + \xi_u \,. \tag{10}$$

It is assumed that the measurement noises ξ_x and ξ_u are not correlated with signals x and u and have zero mean values. The model error is

$$\varepsilon(t) = \dot{x}(t) + a\tilde{x}(t) - \tilde{u}(t) \tag{11}$$

and the objective function (6) reads as

$$J_\tau(a) = \mathrm{E}[\varepsilon(t+\tau)\varepsilon(t)] \,. \tag{12}$$

From (7), it follows for the estimation a_τ

$$\left.\frac{\partial J_\tau(a)}{\partial a}\right|_{a_\tau} = 0 \,. \tag{13}$$

After some mathematical manipulations and simplifications, we get the equation

$$2[R_x(\tau) + R_{\xi_x}(\tau)]a_\tau = R_{xu}(\tau) + R_{xu}(-\tau) - [R_{x\dot{x}}(\tau) + R_{x\dot{x}}(-\tau)] \tag{14}$$

where $R_\xi(\tau)$ stands for autocorrelation function of a stationary stochastic processes ξ, and $R_{\xi\eta}(\tau)$ for crosscorrelation function of two stationary stochastic processes $\xi(t)$ and $\eta(t)$. For the stationary response $x(t)$, it remains

$$\frac{d}{dt}\mathrm{E}[x(t+\tau)x(t)] = R_{x\dot{x}}(\tau) + R_{x\dot{x}}(-\tau) = 0 \,. \tag{15}$$

Substituting (15) into (14) yields

$$2[R_x(\tau) + R_{\xi_x}(\tau)]a_\tau = R_{xu}(\tau) + R_{xu}(-\tau)\,. \tag{16}$$

It is noted that for the stationary response $x(t)$ the signal $\dot{x}(t)$ is not used here and therefore need not to be measured.

On the other hand, we can prove that for the system (9) the following equation

$$2R_x(\tau)a = R_{xu}(\tau) + R_{xu}(-\tau) \tag{17}$$

holds true. Combining (16) and (17), we obtain an estimation

$$a_\tau = \frac{a}{1 + \dfrac{R_{\xi_x}(\tau)}{R_x(\tau)}}\,. \tag{18}$$

The bias of a_τ depends on the relative correlation values of the response x and its measurement noise ξ_x. The solution of the least-square method corresponds to the time lag $\tau = 0$, i.e.

$$a_{\mathrm{LS}} = a_0 = \frac{a}{1 + \dfrac{R_{\xi_x}(0)}{R_x(0)}}\,. \tag{19}$$

The variance $R_{\xi_x}(0)$ is greater than zero, therefore the estimation a_{LS} of the least-square method is always biased whether the measurement noises ξ_x and ξ_u are white noises or colored noises. It is known that for time discrete dynamic systems the least-square method gives unbiased estimations if the model error is a white noise, see e.g. (Isermann, 1988). Thus, this statement is not generally true for time continuous dynamic systems as shown for the system (9). According to the authors' opinion, the reason is that the model error

$$\varepsilon(t) = a\xi_x(t) - \xi_u(t) \tag{20}$$

depends on the unknown parameter a considering equations (9) to (11), which results in the noised objective function $J_{\mathrm{LS}}(a)$ for the least-square method.

The question is now for which time lag τ the estimation $\boldsymbol{\theta}_\tau$ is better than $\boldsymbol{\theta}_{\mathrm{LS}}$. From equation (18) it can be seen that the optimal time lag τ should satisfy

$$R_{\xi_x}(\tau) = 0 \qquad \text{and} \qquad R_x(\tau) \neq 0 \tag{21}$$

and the worst time lag τ satisfies $R_x(\tau) = 0$. Since the statistical properties of measurement noises are unknown, the optimal time lag τ is not available. However, we know that contrarily to the signals, measurement noises are more related to white noise processes and, therefore, the correlation time is shorter than for the signals itself. It is reasonable to use estimations for the time lag τ between the correlation time of the noises τ_{noise} and the correlation time of signals τ_{signal}. In other words, the objective function $J_\tau(\boldsymbol{\theta})$ is noised due to measurement noises. The influence of the noises on the function $J_0(\boldsymbol{\theta})$ is relative large. The minimization of the noised objective function does not guarantee to get the optimal estimation of the unknown parameters. Nevertheless, functions $J_\tau(\boldsymbol{\theta})$ with the time lag τ between the correlation time of the noises τ_{noise} and one of signals τ_{signal} may be used as an objective function. Inspired by the idea of the least-square method, one can assume all these reasonable objective functions simultaneously to be stationary, i.e.

$$\frac{\partial J_\tau(\boldsymbol{\theta})}{\partial \boldsymbol{\theta}} \approx \mathbf{0} \quad \text{for all} \quad \tau \in [\tau_{\text{noise}}, \tau_{\text{signal}}] \; . \tag{22}$$

There exist different definitions for the correlation time of a stationary stochastic process. In this paper, a definition given by Stratonovich (1981) is used. For a scalar stochastic process ζ, the correlation time is defined as

$$\tau_\zeta = \frac{\int_0^\infty |R_\zeta(\tau)| d\tau}{R_\zeta(0)} \tag{23}$$

For a stochastic vector process $\boldsymbol{\zeta}$, the correlation time is defined as the maximal correlation time of its all scalar components. Since the correlation time of the noise τ_{noise} is unknown, it is replaced approximately by zero in equation (22). Further, the correlation time of signals τ_{signal} is replaced approximately by one available from measured signals using equation (23). Then, a new objective function

$$J_{\text{CB}}(\boldsymbol{\theta}) = \sum_{\tau=0}^{\tau_{\text{signal}}} \left(\frac{\partial J_\tau(\boldsymbol{\theta})}{\partial \boldsymbol{\theta}}\right)^T \left(\frac{\partial J_\tau(\boldsymbol{\theta})}{\partial \boldsymbol{\theta}}\right) \tag{24}$$

is constituted and correspondingly a parameter estimation can be obtained by its minimization

$$J_{\text{CB}}(\boldsymbol{\theta}) = \min \tag{25}$$

using optimization techniques.

3. Numerical Example

A Duffing oscillator describing nonlinear stiffness has been extensively investigated in nonlinear dynamics, see e.g. (Kreuzer, 1987). It is used here as an example to test the suggested correlation-based method and to compare it with the least-square method. The equation of motion of a Duffing oscillator reads as

$$m\ddot{y} + c\dot{y} + ky + ay^3 = u \tag{26}$$

with given parameters

$$m = 1, \quad c = 0.2, \quad k = 1, \quad a = 0.1. \tag{27}$$

The stochastic excitation u is assumed to be stationary and with zero mean value. It has the following spectral density function and correlation function:

$$S_u(\omega) = \frac{1}{\pi(1+\omega^2)} \qquad \text{and} \qquad R_u(\tau) = e^{-\tau}. \tag{28}$$

The unknown parameters to estimate are

$$\boldsymbol{\theta} = [m \ c \ k \ a]^T. \tag{29}$$

The excitation is simulated with the spectral representation method presented by Shinozuka and Jan (1972) and modified by Hu and Schiehlen (1997). In the numerical simulation, the Duffing oscillator which is initially at rest is integrated from the initial time $t_0 = 0$ to the time $T = 32.6$s with a method given by Shampine and Gordon (1975). The time signals are sampled with sampling frequency 500 Hz. Figure 1 shows the time histories of the excitation and responses without measurement noise.

The measurement noises are usually with zero mean values and assumed to be independent board band white noises which have spectral density function

$$S_\xi(\omega) = \begin{cases} \dfrac{q^2}{2\omega_u} & \text{for} \quad |\omega| < \omega_u \\ 0 & \text{for} \quad |\omega| \geq \omega_u \end{cases} \tag{30}$$

where q means the noise intensity and ω_u the band width or the Nyquist frequency. As an example, the measurement noise with $q = 0.3$ and $\omega_u =$ 100Hz is added extra to the acceleration signal $\ddot{x}$ and other signals remain not noised. For the resulting noise to signal ratio equal to 0.2, Figure 2 shows the estimation values $\boldsymbol{\theta}_\tau$ obtained from equation (7) for every time lag τ. It can be seen that the relative error of estimations $\boldsymbol{\theta}_\tau$ in the neighborhood of $\tau = 0$ are much better than one of the least-square method $\boldsymbol{\theta}_{\text{LS}}$ which corresponds to the time lag $\tau = 0$. Moreover, for some time lags τ the estimations $\boldsymbol{\theta}_\tau$ may have very large errors, for instance, for the parameter m at the time lag $\tau = 1.1$s. The reason is that for these time lags some signals are weakly correlated or not correlated.

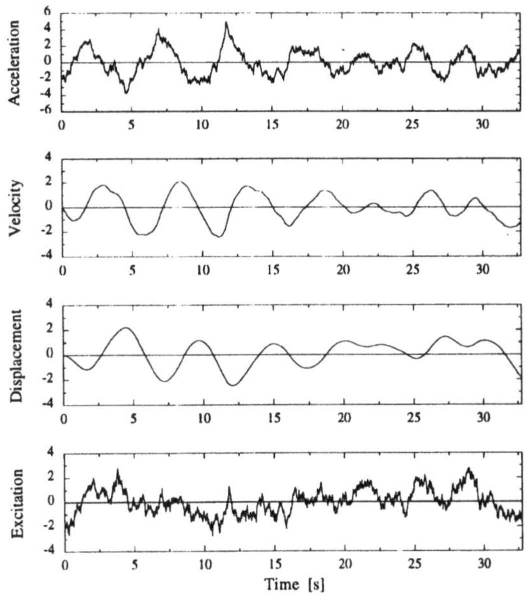

Figure 1. Time histories of the excitation and responses of the Duffing oscillator

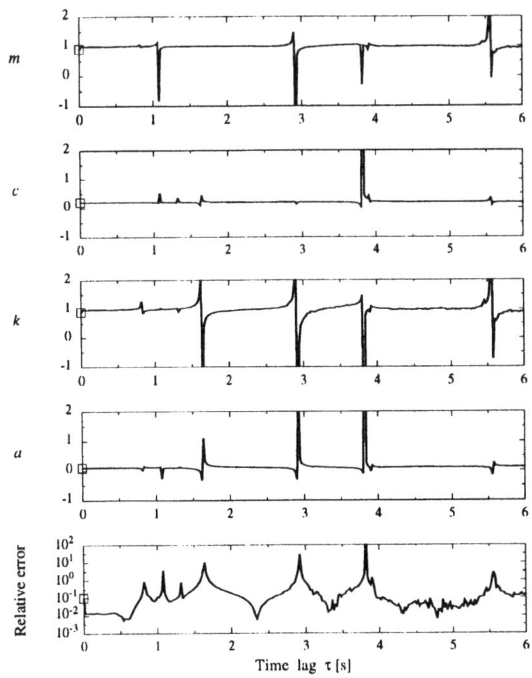

Figure 2. Parameter estimation $\boldsymbol{\theta}_\tau$ for different time lag τ

In the correlation-based method, the correlation functions are considered for discrete time lags τ in the interval $[0, \tau_{\max}]$ where $\tau_{\max}$ is approximated by the correlation time of the measured signals τ_{signal}. The identification results depend on the considered maximal correlation time $\tau_{\max}$. If we regard $\tau_{\max}$ as a variable, Figure 3 shows the dependency of the estimation values on the variable $\tau_{\max}$. It can be seen that the estimation values converge quickly to stable values. Therefore, in the correlation-based method we need only a coarse value for the correlation time of signals.

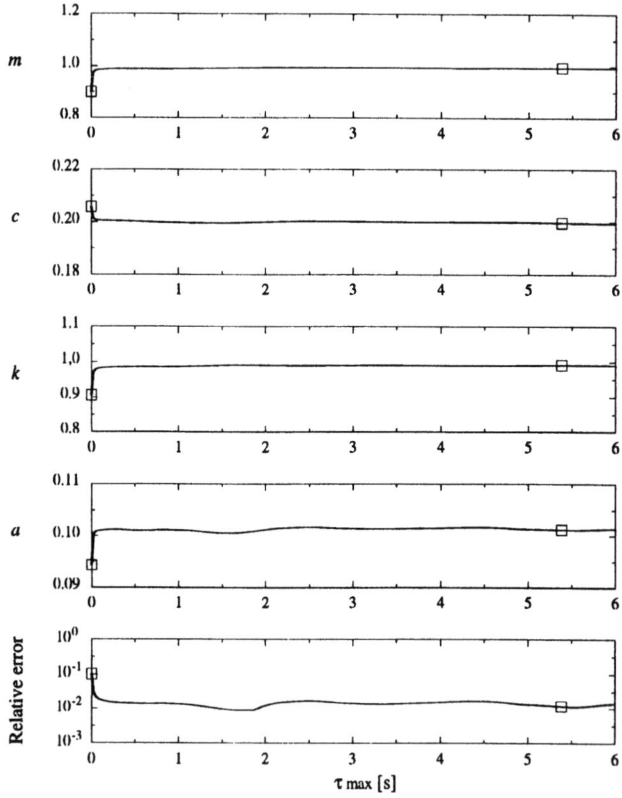

Figure 3. Dependency of correlation-based estimations on the considered maximum time lags $\tau_{\max}$

According to the definition (23), the correlation time $\tau_{\text{signal}} = 5.382$ s. The solution of the correlation-based method is

$$m_{\text{CB}} = 0.9951, \quad c_{\text{CB}} = 0.1998, \quad k_{\text{CB}} = 0.9912, \quad a_{\text{CB}} = 0.1012, \tag{31}$$

while the solution of the least-square method is

$$m_{\text{LS}} = 0.8999, \quad c_{\text{LS}} = 0.2058, \quad k_{\text{LS}} = 0.9058, \quad a_{\text{LS}} = 0.09432. \tag{32}$$

Both estimation values are marked in Figure 3 with a box. The maximal relative error of the estimation values from the correlation-based method is 1.2% and from the least-square method it is 10.0%, respectively. The correlation-based method gives clearly a much better estimation.

Furthermore, for different noised signals, Figures 4 and 5 show the influences of the noise intensity q and the band width ω_u on the identification results.

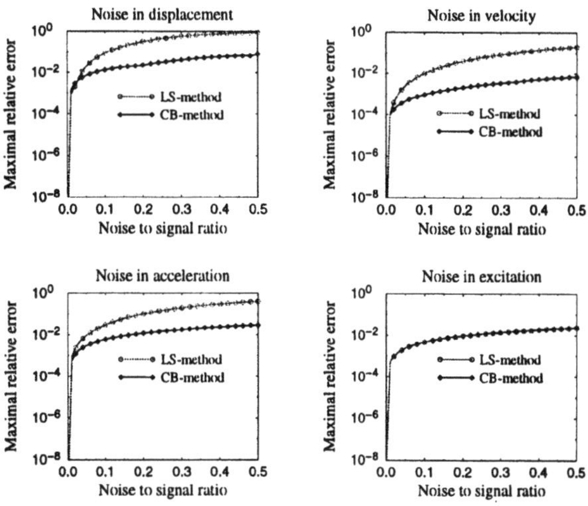

Figure 4. Influence of the noise to signal ratios on the identification results

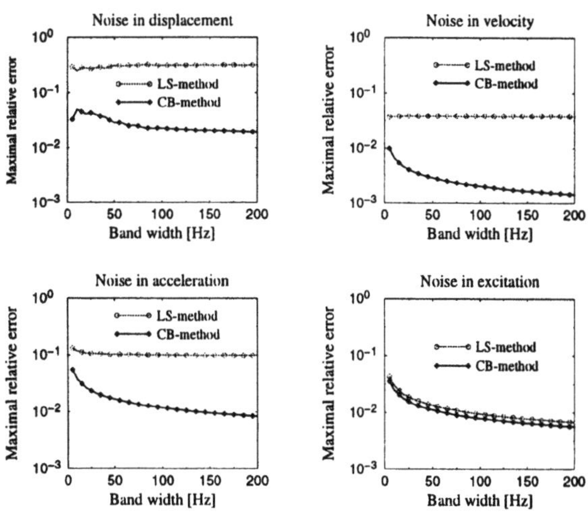

Figure 5. Influence of the band widths of the noise on the identification results

In both figures, CB-method means the correlation-based method and LS-method the least-square method. It is shown that the correlation-based method gives better identification results than the least-square method. For both methods, the stronger the noise intensity, the worse the identification results. The estimation accuracy of the least-square method does not depend on the band width ω_u, while the correlation-based method gives better results for larger ω_u since the correlation time of the noise with larger band width is shorter. More details, for instance, if all signals are noised simultaneously or some signals are not available from measurements, can be found in (Hu, 1998).

4. Conclusions

The correlation-based method presented in this paper takes advantage of the property that the measurement noises have smaller correlation times compared to those of signals. It considers correlation functions of the model errors at a finite number of time lags instead of only variances of the model errors as in the least-square method. To compare the presented method with the least-square method, a stochastic Duffing system is numerically investigated in detail. It turns out that the applicability of the presented method is as wide as the least-square method but it gives much better estimation results.

References

Fisher, R.A. (1950) *Contributions to Mathematics Statistics.* John Wiley, New York.

Gauss, K.F. (1887) *Abhandlungen zur Methode der kleinsten Quadrate*, edited by Börsch, A. and Simon, P., Stankkiewicz' Buchdruckrei, Berlin.

Hu, B. (1998) *Korrealationsbasierte Methode zur Parameteridentifikation nichtlinearer mechanischer Systeme.* VDI-Verlag, Düsseldorf.

Hu, B. and Schiehlen, W. (1997) On the simulation of stochastic process by spectral representation. *Probabilistic Eng. Mech.* **12**, 105–113.

Isermann, R. (1988) *Identification dynamischer Systeme*, Band I. Springer-Verlag, Berlin.

Kreuzer, E. (1987) *Numerische Untersuchung nichtlinearer dynamischer Systeme.* Springer-Verlag, Berlin.

Schiehlen, W. (1997) Multibody System Dynamics: Roots and Perspectives. *Multibody System Dynamics* **1**, 149–188.

Shampine, L.F. and Gordon, M.K. (1975) *Computer Solution of Ordinary Differential Equations.* Freeman, San Francisco.

Shinozuka, M. and Jan, C.M. (1972) Digital simulation of random processes and its application. *J. Sound and Vibration* **25**, 111–128.

Söderström, T. and Stoica, P. (1989) *System Identification.* Prentice–Hall, Cambridge.

Stratonovich, R. (1981) *Topics in the Theory of Random Noise*, Vol. I. Gordon and Breach, New York.

Young, P.C. (1970) An instrumental variable method for real-time identification of a noisy process. *Automatica* **6**, 271–287.

NUMERICAL SOLUTION OF THE FOUR-DIMENSIONAL NONSTATIONARY FOKKER-PLANCK EQUATION

S. F. WOJTKIEWICZ
Department of Aeronautical and Astronautical Engineering, University of Illinois at Urbana-Champaign, Urbana, IL 61801, U.S.A.

L. A. BERGMAN
Department of Aeronautical and Astronautical Engineering, University of Illinois at Urbana-Champaign, Urbana, IL 61801, U.S.A.

B. F. SPENCER, JR.
Department of Civil Engineering and Geological Sciences, University of Notre Dame, Notre Dame, IN 46556, U.S.A.

E.A. JOHNSON
Department of Civil Engineering and Geological Sciences, University of Notre Dame, Notre Dame, IN 46556, U.S.A.

1. Abstract

The response of mechanical and structural systems subjected to environmental loads is often determined through the use of random vibration methods. Many of these loads can be effectively modeled as random processes, and more specifically as Gaussian white noise or filtered Gaussian white noise excitations. This choice of excitation model infers that the response of the system will be Markovian and, thus, expressible in terms of its transition probability density function, which satisfies both the forward (Fokker-Planck) and backward Kolmogorov equations.

Numerical solution of the Kolmogorov equations by finite element and finite difference methods for dynamical systems of engineering interest has been hindered by the problem of dimensionality. Forward and backward Kolmogorov equations corresponding to two and three state systems have been successfully solved by the authors [1-5]. While the extension of these solution algorithms to systems with more than three states is conceptually straightforward, the computational issues involved become daunting. These issues can be separated into two broad classes: those dealing with the spatial and temporal discretization of the initial-boundary value problem, and those that pertain to the actual techniques of linear algebra required to solve the resulting nonsymmetric system of linear, algebraic equations that result.

The authors have successfully implemented high accuracy spatial discretization schemes into their codes thus permitting the use of coarse computational meshes without penalty in solution accuracy [6] and have combined these with iterative solvers to facilitate the extension to four-state dynamical systems [7]. Several four state dynamical systems were recently studied by Wedig *et al.* [8], in which the Fokker-Planck equation was solved using a global weighted residual method. This paper reports on efforts to solve

S. Narayanan and R.N. Iyengar (eds.),
IUTAM Symposium on Nonlinearity and Stochastic Structural Dynamics, 271–287.

the Fokker-Planck equation for a two degree of freedom nonlinear oscillator by a high order finite difference procedure.

2. Introduction

In prior work, the authors successfully solved the Fokker-Planck equation numerically for systems having two and three states (*i.e.*, $n = 2$ and 3) using both finite element and finite difference techniques. The numerical solution of these equations constitutes a formidable computational challenge, and several issues are involved in designing an effective solution algorithm for problems of this type.

These issues can be broadly categorized as those (1) dealing with the spatial discretization of the Fokker-Planck equation, (2) pertaining to the temporal discretization of the resulting system of ordinary differential equations, and (3) involving the actual numerical techniques of linear algebra required to solve the system of linear algebraic equations that result. For four-dimensional problems, the solution of this system becomes a daunting task due to its size, which is on the order of 10^7 equations.

This paper examines how these computational issues were resolved, resulting in the development and implementation of a robust algorithm for the solution of the Fokker-Planck equation for systems up to and including dimension four. After a brief discussion of the algorithm, it will be applied to two, four state dynamical systems.

First, a system of coupled linear oscillators constituting a four-dimensional linear system is studied. This system provides a benchmark to verify the accuracy of the algorithm since exact solutions exist for the probability density function and moments of the system response. The second system consists of a set of coupled oscillators, one of which is strongly nonlinear. Here, the numerical results will be compared to those of Monte Carlo simulation. Results for the density functions and response moments for the two systems are given, and the errors involved in their computation are discussed.

3. Problem Formulation

Consider a time-invariant dynamical system subjected to an additive white noise excitation. The corresponding stochastic differential equation is given by

$$\frac{d\mathbf{X}}{dt} = \mathbf{m}(\mathbf{X}) + \mathbf{G}\mathbf{w}(t) \tag{1}$$

where $\mathbf{X}$ is an R^n-valued stochastic process, $\mathbf{m}(\mathbf{X}) \in R^n$, $\mathbf{G} \in R^n x R^m$, and $\mathbf{w}(t)$ is a m dimensional, zero-mean, vector Gaussian white noise process with autocorrelation matrix given by $E[\mathbf{w}(t+\tau)\mathbf{w}'(t)] = 2\mathbf{D}\delta(\tau)$ where $\mathbf{D}$ is $m \times m$. The aforementioned system forms a Markov process in R^n, the behavior of which is completely determined by the transition probability density function $p(\mathbf{x}, t|\mathbf{x}_0)$. The Fokker-Planck equation for this system is given by

$$\frac{\partial p}{\partial t} = \frac{1}{2}\left(\sum_{i=1}^{n}\sum_{j=1}^{n}\frac{\partial^2}{\partial x_i x_j}\left[H_{ij}p\right]\right) - \sum_{j=1}^{n}\frac{\partial}{\partial x_j}\left[m_j(\mathbf{x})p\right] \tag{2}$$

subject to appropriate initial and boundary conditions, where $\mathbf{H} = 2\mathbf{GDG}'$. The boundary conditions are given by $p \to 0$ as $|x_1|, |x_2|, \ldots |x_n| \to \infty$ and the initial condition by $p(\boldsymbol{x}, 0) = \delta(\boldsymbol{x} - \boldsymbol{x}_0)$ [3].

4. Computational Issues

4.1 SPATIAL DISCRETIZATION

The first set of issues involved in the numerical solution of any initial-boundary value problem is its spatial discretization; *i.e.*, the transformation of the partial differential equation to a system of ordinary differential equations. As stated earlier, the finite element method was applied successfully to the numerical solution of the Fokker Planck equation. However, the solution algorithm was fairly rudimentary, employing shape functions of class C^0 and uniform meshes on the computational domain. Resulting errors for the probability density function of response, computed using the one, two and infinity norms, were on the order of 10^{-4}, which were deemed to be acceptable for the subsequent calculation of quantities such as response moments, but not for analyses involving probabilities in the tails of the computed distributions where errors many orders of magnitude smaller are necessary for reasonable estimates of, for example, system reliability.

With this goal in mind, the authors embarked on a study of higher order discretization schemes, using finite difference methods, to ascertain their limits of accuracy, particular in the tails of the computed distributions. Results have been obtained [6} demonstrating that certain high order schemes are capable of producing very accurate solutions in these regions. These schemes involve the manipulation of Taylor expansions of the density function about uniformly distributed nodal points in a n-dimensional mesh. One can examine [6] for further details regarding the derivation of the difference approximations of the pertinent derivatives appearing in the Fokker-Planck equation.

There are two motivations for using a high order scheme: (1) it gives more accurate solutions on a given mesh, and (2) it allows for the use of fewer mesh points for a given accuracy (*i.e.*, reduces memory requirements). One specific advantage of using high order finite differences rather than finite elements is that the row population of the coefficient matrix is linear in the dimension of the problem rather than exponential, which leads to savings in memory and efficiency for a given mesh. This advantage is exploited fully in the algorithm developed by the authors, thus permitting solution of four-dimensional problems.

After application of the appropriate high order finite difference scheme, one arrives at a system of first order differential equations given by

$$C\dot{f} + Kf = 0\,, \; f \in R^l\,. \tag{3}$$

where l equals the number of nodes in the computational mesh. For the four state systems to be considered here, l is equal to 41^4.

4.2 TEMPORAL DISCRETIZATION

The next set of issues concerns the temporal discretization process; *i.e.*, converting the system of first order ODE's to a set of linear algebraic equations. In the authors' previous efforts [3], the Crank-Nicholson time integration scheme was successfully employed. This method is attractive due to its unconditional stability and second order accuracy.

Two characteristics of the system of ordinary differential equations observed in the course of studying the current four-dimensional problems have necessitated a change in the time discretization process. First, the use of high accuracy spatial discretization schemes produces stiff coefficient matrices, $\boldsymbol{K}$; and second, as the dimensionality of the problem increases so do the variances of the physical time scales present in the system. As a result of the increased stiffness of the coefficient matrix, $\boldsymbol{K}$, a time integration scheme designed specifically for stiff systems known as a Gear or BDF (backward differentiation) method [9] has been implemented to perform the temporal discretization of eq. (3). Using the Gear method, the time derivative of the density function is approximated by a linear combination of the density function evaluated at the current and several previous times. This approximation is given by [9]

$$\dot{f} = \sum_{j=0}^{N_t} \alpha_j f(t_i - j\Delta t)\,, \qquad f = f(t_i) \tag{4}$$

where N_t is the order of the method and the α_j's are the integration weights. Following the temporal discretization, one arrives at the so-called evolution equations for the nodal density values. These equations consist of a system of linear algebraic equations which must be solved at each time step to advance the numerical solution for the density function to stationarity. The evolution equations are given by

$$C^* f(t_i) = \sum_{j=1}^{N_t} \alpha_j f(t_i - j\Delta t) \tag{5}$$

where $C^* = \alpha_0 I + K$.

4.3 ITERATIVE SOLUTION

The final computational issue is that pertaining to the actual solution of eq. (5). Direct methods such as *LU* decomposition, used successfully by the authors for lower dimensional problems, become prohibitively expensive when applied to large systems of equations, such as those obtained from a discretization of the four-dimensional Fokker-

Planck equation. Consequently, iterative techniques were adopted for reasons of storage and parallelizability and, hence, efficiency on high performance computing platforms. Only the action of the coefficient matrix, C^*, upon a vector is computed (*i.e.,* C^*x) rather than the elements of the matrix explicitly. This reduces the memory requirements of the algorithm to those of the iterative solver itself, associated with the generation of the basis vectors in the Krylov subspace used by the solver.

In terms of required resources, the direct solution of a four state problem requires $O(10^4)$ gigabytes of memory (if it were currently available), while the iterative method used in this algorithm requires only $O(1)$ gigabytes, well within the present capabilities of today's computing platforms. Thus, it is clear that an iterative solver is the method of choice for higher dimensional problems. Furthermore, an additional advantage of using the iterative solver is its scalability. Since the bulk of computation time used in an iterative solver involves the computation of matrix-vector products, these methods parallelize extremely well if programmed properly.

To summarize, the robust numerical algorithm implemented herein utilizes a tenth order finite difference scheme for spatial discretization and a sixth order Gear method for temporal discretization. The resulting system of linear algebraic equations is solved at each time step employing the nonsymmetric iterative solver GMRES [10].

Results for two, four state problems will now be presented and compared with the exact solution in the case of the linear system and to a Monte Carlo simulation result in the case of the nonlinear system.

5. Results

5.1 LINEAR SYSTEM

The equations of motion for the four-dimensional linear system are

$$\dot{X} = \begin{bmatrix} 0 & 1 & 0 & 0 \\ -(k_1+k_2) & -c_1 & k_2 & 0 \\ 0 & 0 & 0 & 1 \\ k_2 & 0 & -(k_2+k_3) & -c_2 \end{bmatrix} X + \begin{bmatrix} 0 & 0 \\ 0 & \sqrt{2D_1} \\ 0 & 0 \\ 0 & \sqrt{2D_2} \end{bmatrix} \mathbf{w}(t) \tag{6}$$

with $\mathbf{w}(t)$ a Gaussian vector white noise stochastic process such that

$$E[\mathbf{w}(t)] = 0\,, \qquad E[\mathbf{w}(t_2)\mathbf{w}'(t_1)] = \begin{bmatrix} 2D_1 & 0 \\ 0 & 2D_2 \end{bmatrix} \delta(\tau) \tag{7}$$

with $\tau = t_2 - t_1$.

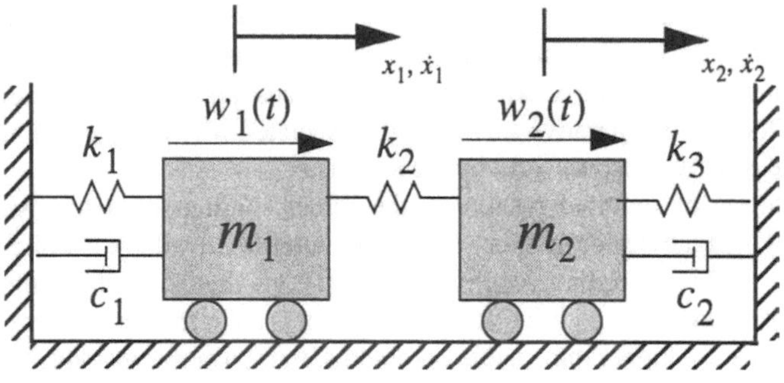

Figure 1. Four-dimensional linear system

The Fokker-Planck equation for this system is given by

$$\frac{\partial p}{\partial t} = D_1\frac{\partial^2 p}{\partial x_2^2} + D_2\frac{\partial^2 p}{\partial x_4^2} - x_2\frac{\partial}{\partial x_1}(p) - x_4\frac{\partial}{\partial x_3}(p) + (k_1 + k_2)x_1\frac{\partial}{\partial x_2}(p) + c_1 x_2\frac{\partial}{\partial x_2}(p) \tag{8}$$

$$- k_2 x_3\frac{\partial}{\partial x_2}(p) - k_2 x_1\frac{\partial}{\partial x_4}(p) + (k_2 + k_3)x_3\frac{\partial}{\partial x_4}(p) + c_2 x_2\frac{\partial}{\partial x_4}(p) + (c_1 + c_2)p$$

System parameters, $k_1 = k_2 = k_3 = 1$, $c_1 = c_2 = 0.4$, and $D_1 = D_2 = 0.2$, were chosen. The problem was discretized using a tenth order finite difference scheme on a computational mesh of 41^4 nodes ($41 \times 41 \times 41 \times 41$) of extent $[-6, 6] \times [-6, 6] \times [-6, 6] \times [-6, 6]$. The initial condition is a zero mean, four-dimensional Gaussian distribution with variance $\sigma_i^2 = 0.5$; a time step of 0.0025 was used in the Gear method.

The numerical and exact values of the centerpoint of the density function through time are in excellent agreement through time, as shown in Fig. 2.

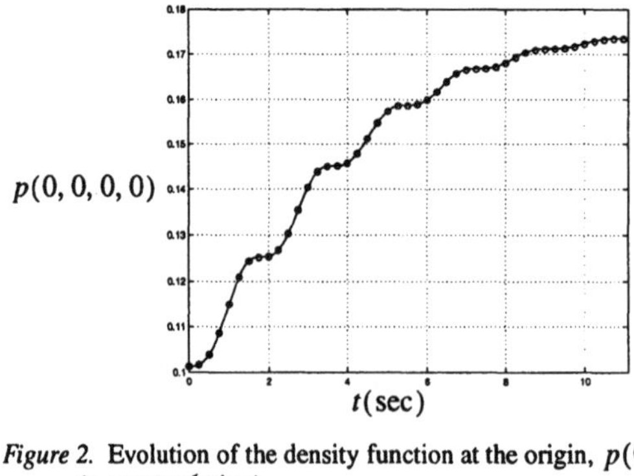

Figure 2. Evolution of the density function at the origin, $p(0, 0, 0, 0)$; (o-exact solution).

At stationarity, an error analysis was performed. Two error measures were computed,

$$\|e\|_\infty = \|p^{exact} - p^{numerical}\|_\infty = 1.226 \times 10^{-3} \tag{9}$$

$$\|e\|_2 = \|p^{exact} - p^{numerical}\|_2 = 1.612\times10^{-2} \tag{10}$$

the first being a measure of the maximum error across the entire mesh and the second corresponding to average nodal error; *i.e.*, $\|e\|_2/41^4 = 5.706\times10^{-9}$. It is clear from these measures that the numerical solution is very accurate throughout the domain. Another way of assessing the accuracy of the numerical solution is by comparing the computed response moments to their exact values. At stationarity, the exact and numerical covariance matrices, $\mathbf{K}_{XX}$, are given by

$$\mathbf{K}_{\mathbf{XX}}^{exact} = \begin{bmatrix} 0.3333 & 0 & 0.16667 & 0 \\ 0 & 0.5 & 0 & 0 \\ 0.1667 & 0 & 0.3333 & 0 \\ 0 & 0 & 0 & 0.5 \end{bmatrix}, \quad \mathbf{K}_{\mathbf{XX}}^{numerical} = \begin{bmatrix} 0.3268 & 0.0007 & 0.1647 & -0.0017 \\ 0.0007 & 0.4985 & 0.001 & 0.0001 \\ 0.1647 & 0.001 & 0.3310 & 0.0001 \\ -0.0017 & 0.0001 & 0.0001 & 0.5017 \end{bmatrix} \tag{11}$$

All computed covariance values are within two percent of their exact values. The numerical and exact response moments can also be compared through their evolution as shown in Figs. 3-8. It should be noted that, due to the symmetry of the problem and initial condition, many of the response moments appear in identical pairs. The computed numerical moments are in close agreement with their exact counterparts throughout the evolution with small errors occurring in some of the second order moments, especially the mean square displacement of the two masses. The source of this error will be discussed shortly.

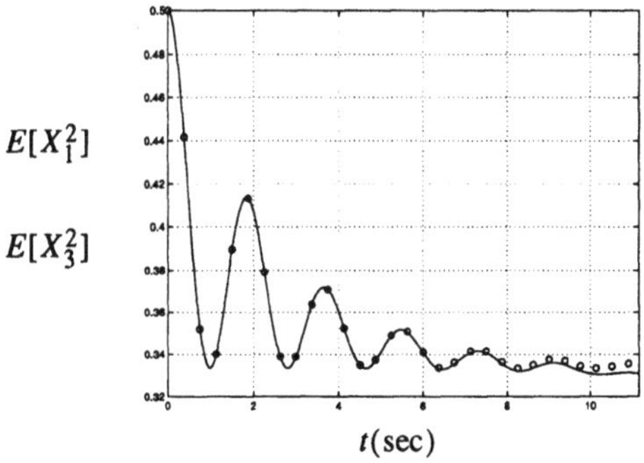

Figure 3. Evolution of the mean square displacement of each mass; (o-exact solution).

The numerical and exact density functions are compared by plotting cross sections through the origin at two time instants as shown in Figs. 9 and 10. Again, excellent agreement is shown between the numerical and exact solutions. The peakedness of the density function in the x_1 and x_3 axes should be noted as it affects the subsequent calculation of the response moments.

$E[X_2^2]$

$E[X_4^2]$

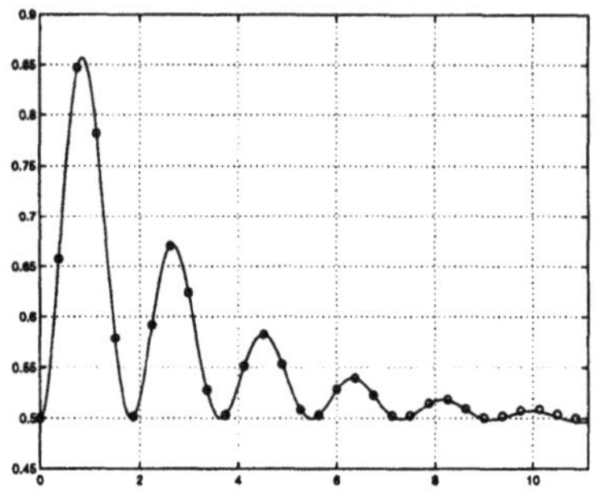

t(sec)

Figure 4. Evolution of the mean square velocity of each mass; (o-exact solution).

$E[X_1X_2]$

$E[X_3X_4]$

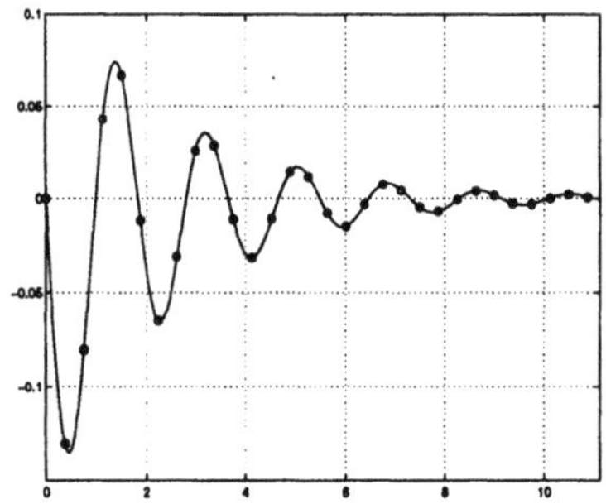

t(sec)

Figure 5. Evolution of the cross-correlation between the displacement and velocity of the same mass; (o-exact solution).

$E[X_1X_4]$

$E[X_2X_3]$

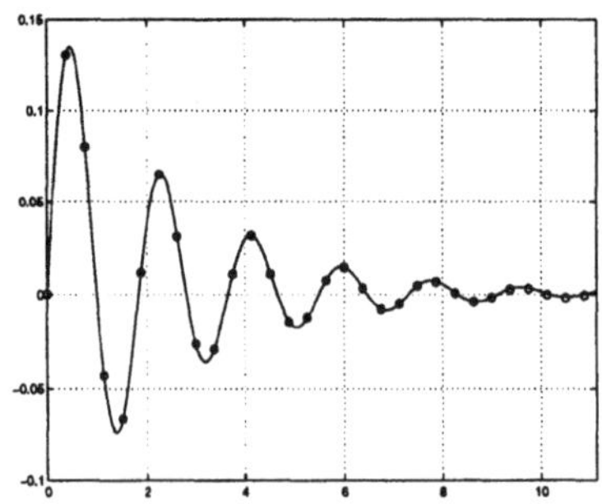

t(sec)

Figure 6. Evolution of the cross-correlation between the displacement and velocity of different masses; (o-exact solution).

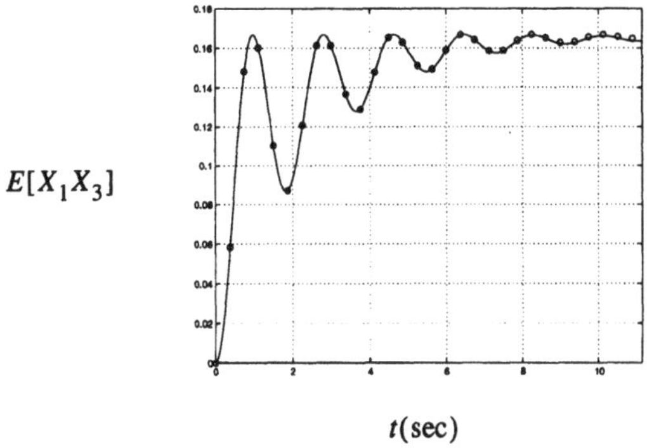

Figure 7. Evolution of the cross-correlation between the displacements of the two masses; (o-exact solution).

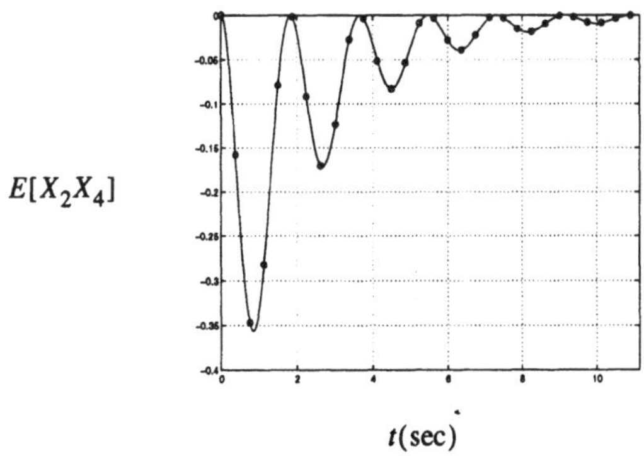

Figure 8. Evolution of the cross-correlation between the velocities of the two masses; (o-exact solution).

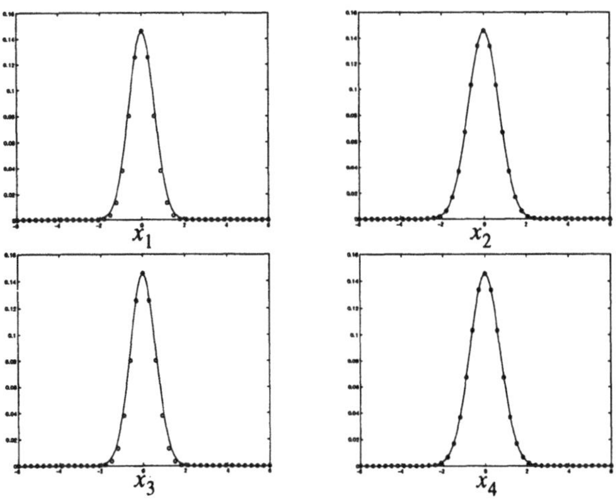

Figure 9. Cross sections of the density function of the linear system through the origin at $t = 4$ sec. (o-exact solution).

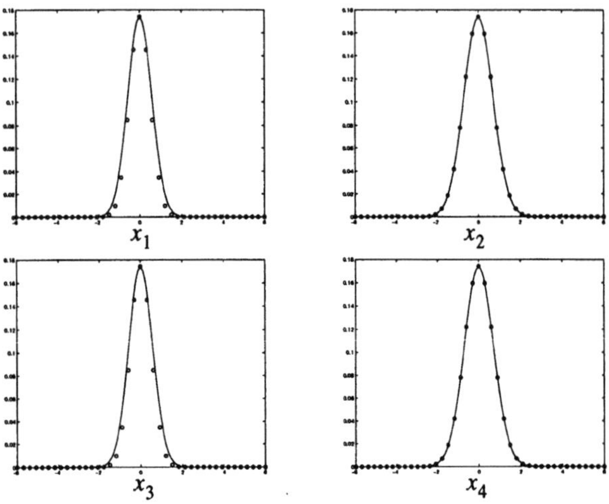

Figure 10. Cross sections of the density function of the linear system through the origin at $t = 12.25$ sec. (o-exact solution).

5.2 NONLINEAR SYSTEM

The equations of motion of the four-dimensional nonlinear system are

$$\dot{X} = \begin{bmatrix} 0 & 1 & 0 & 0 \\ -(k_1+k_2) & -c_1 & k_2 & 0 \\ 0 & 0 & 0 & 1 \\ k_2 & 0 & -(k_2+k_3) & -c_2 \end{bmatrix} X + \begin{bmatrix} 0 \\ -\varepsilon X_1^3 \\ 0 \\ 0 \end{bmatrix} + \begin{bmatrix} 0 & 0 \\ 0 & \sqrt{2D_1} \\ 0 & 0 \\ 0 & \sqrt{2D_2} \end{bmatrix} \mathbf{w}(t) \qquad (12)$$

with $\mathbf{w}(t)$ a Gaussian vector white noise stochastic process such that

$$E[\mathbf{w}(t)] = 0\,, \qquad E[\mathbf{w}(t_2)\mathbf{w}'(t_1)] = \begin{bmatrix} 2D_1 & 0 \\ 0 & 2D_2 \end{bmatrix} \delta(\tau) \qquad (13)$$

and $\tau = t_2 - t_1$. The Fokker-Planck equation for this system is given by

$$\frac{\partial p}{\partial t} = D_1 \frac{\partial^2 p}{\partial x_2^2} + D_2 \frac{\partial^2 p}{\partial x_4^2} - x_2 \frac{\partial}{\partial x_1}(p) - x_4 \frac{\partial}{\partial x_3}(p) + (k_1+k_2)x_1 \frac{\partial}{\partial x_2}(p) + c_1 x_2 \frac{\partial}{\partial x_2}(p) \qquad (14)$$

$$- k_2 x_3 \frac{\partial}{\partial x_2}(p) - k_2 x_1 \frac{\partial}{\partial x_4}(p) + (k_2+k_3)x_3 \frac{\partial}{\partial x_4}(p) + c_2 x_2 \frac{\partial}{\partial x_4}(p) + \varepsilon x_1^3 \frac{\partial p}{\partial x_2} + (c_1+c_2)p$$

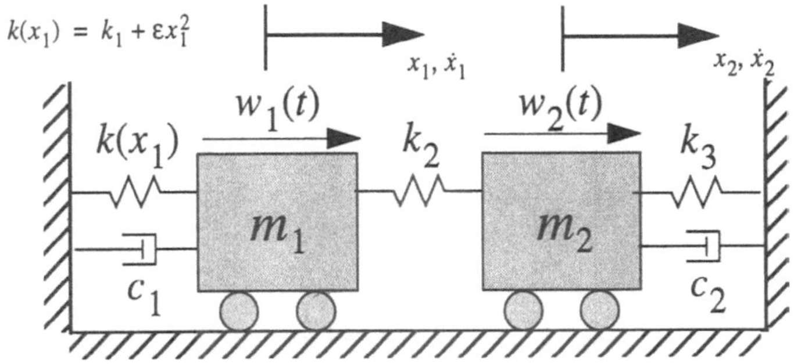

Figure 11. Four-dimensional nonlinear system

System parameters $k_1 = k_2 = k_3 = 1$, $c_1 = c_2 = 0.4$, $\varepsilon = 0.2$, and $D_1 = D_2 = 0.2$ were chosen. The problem was discretized on a computational mesh of 41^4 nodes ($41 \times 41 \times 41 \times 41$) of extent $[-8, 8] \times [-8, 8] \times [-8, 8] \times [-8, 8]$. The initial condition is a zero mean, four-dimensional Gaussian distribution with variance $\sigma_i^2 = 0.5$; a time step of 0.001 was used in the Gear method.

The numerical and estimated values of the centerpoint of the density function are in excellent agreement through time, as shown in Fig. 12; the estimated values were obtained from a Monte Carlo simulation of five million realizations. The numerical and simulated response moments can also be compared through their evolution as shown in Figs.13-22. Most of the computed numerical moments agree qualitatively with the results from Monte Carlo simulation. However, a sizeable difference is noted between the numerical and estimated values of the mean square displacements of the two masses. The source of this error will be discussed in the next section.

5.3 DISCUSSION OF ERROR

In the previous two sections, results for the numerical solution of the four-dimensional Fokker-Planck equation were given for a four state linear and four state nonlinear system. From Figs. 2 and 11, it is clear that the density functions are highly accurate, as expected, with a maximum pointwise error of $O(10^{-3})$ in the case of the linear system. The density function is the only quantity obtained directly in the numerical solution algorithm. Other response measures, such as the moments, are derived from the numerical solution of the density function during the postprocessing phase. Herein lies the source of the discrepancy between the computed numerical moments and those from the exact solution, or alternatively, from the Monte Carlo simulation.

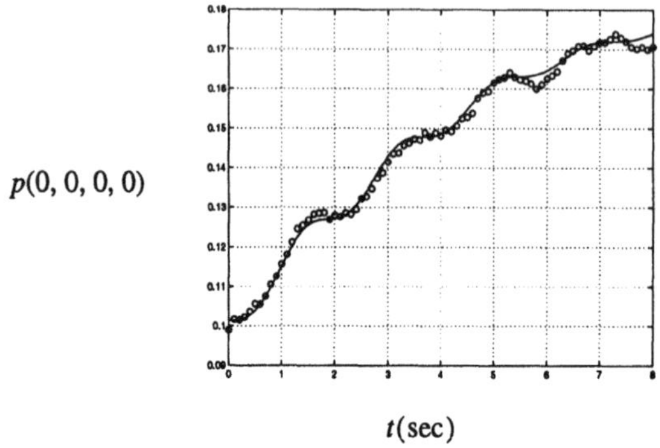

Figure 12. Evolution of the density function at the origin, $p(0, 0, 0, 0)$; (o-simulation).

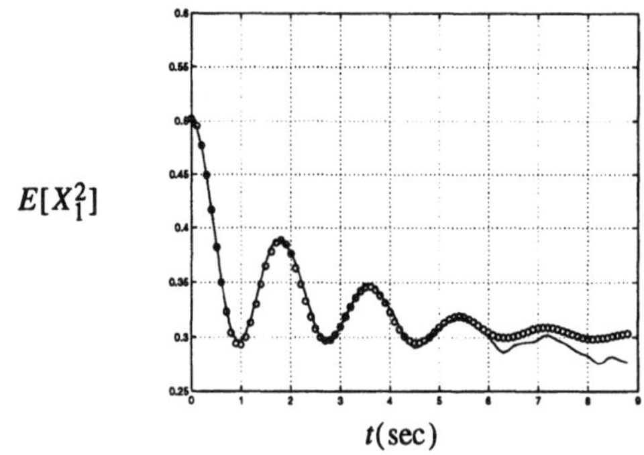

Figure 13. Evolution of the mean square displacement of the first mass; (o-simulation).

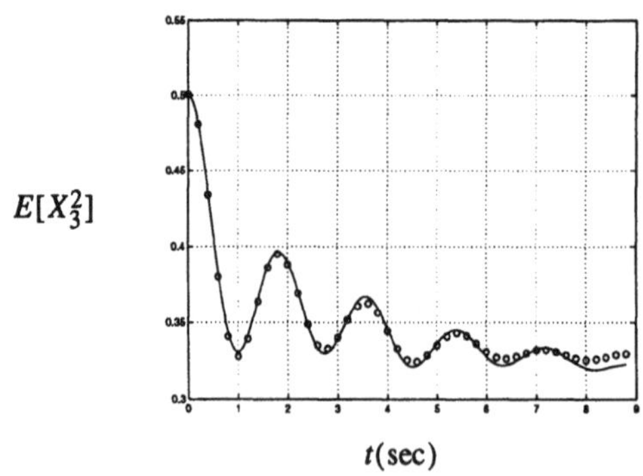

Figure 14. Evolution of the mean square displacement of the second mass; (o-simulation).

$E[X_2^2]$

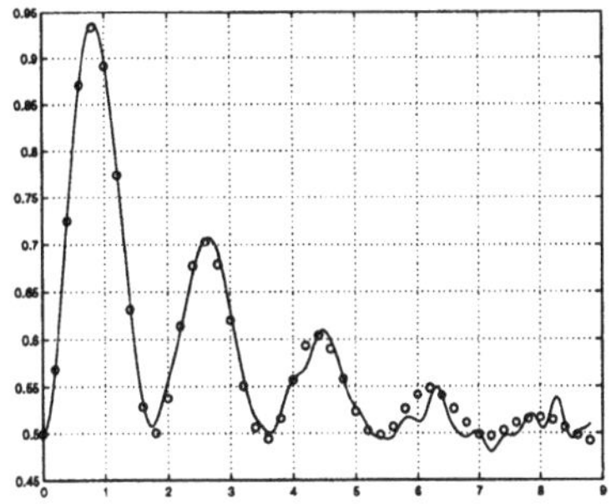

t(sec)

Figure 15. Evolution of the mean square velocity of the first mass; (o-simulation).

$E[X_4^2]$

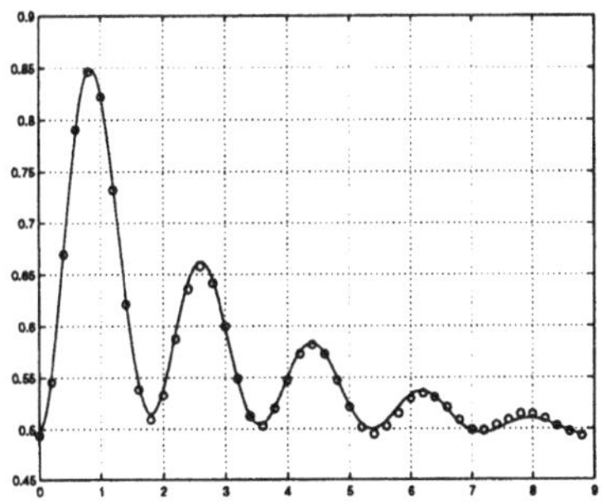

t(sec)

Figure 16. Evolution of the mean square velocity of the second mass; (o-simulation).

$E[X_1X_2]$

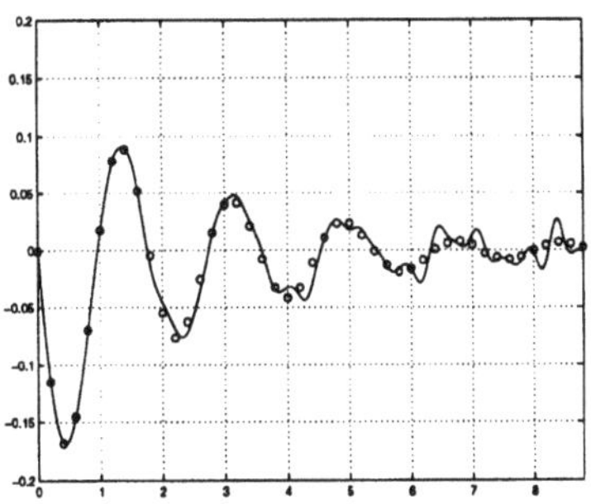

t(sec)

Figure 17. Evolution of the cross-correlation between the displacement and velocity of the first mass; (o-simulation).

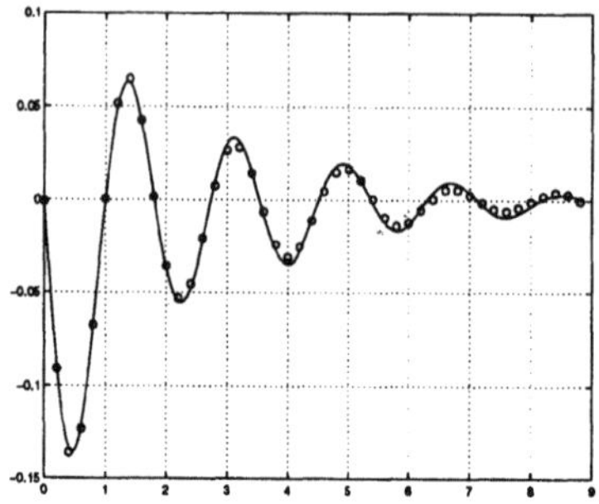

$t(\text{sec})$

Figure 18. Evolution of the cross-correlation between the displacement and velocity of the second mass; (o-simulation).

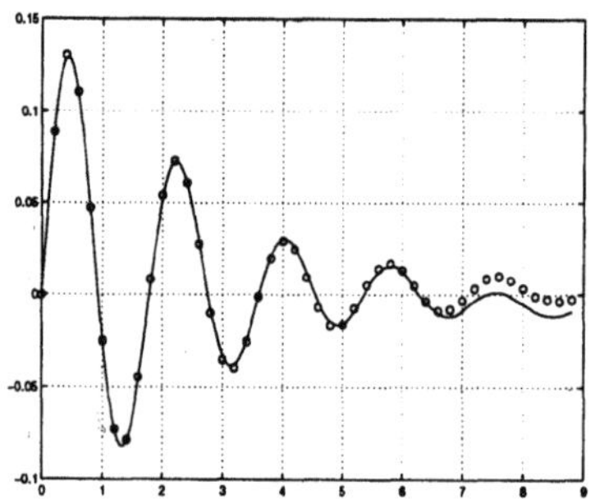

$t(\text{sec})$

Figure 19. Evolution of the cross-correlation between the displacement of the first mass and the velocity of the second mass; (o-simulation).

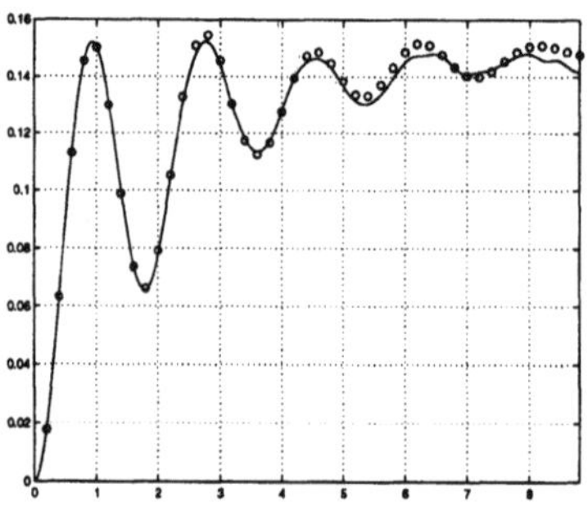

$t(\text{sec})$

Figure 20. Evolution of the cross-correlation between the displacement of the two masses; (o-simulation).

t

$E[X_2X_3]$

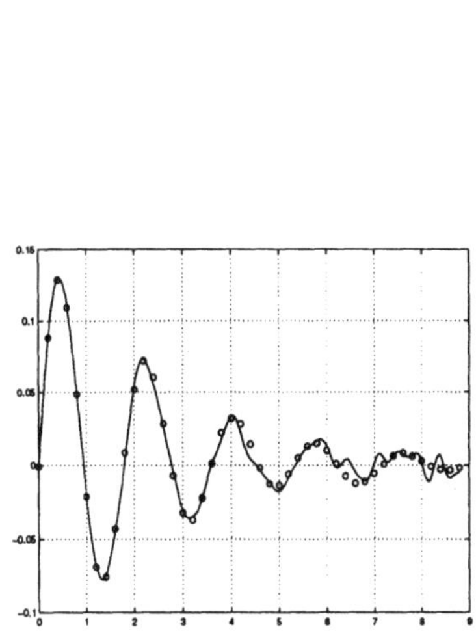

t(sec)

Figure 21. Evolution of the cross-correlation between the velocity of the first mass and displacement of the second mass; (o-simulation).

$E[X_2X_4]$

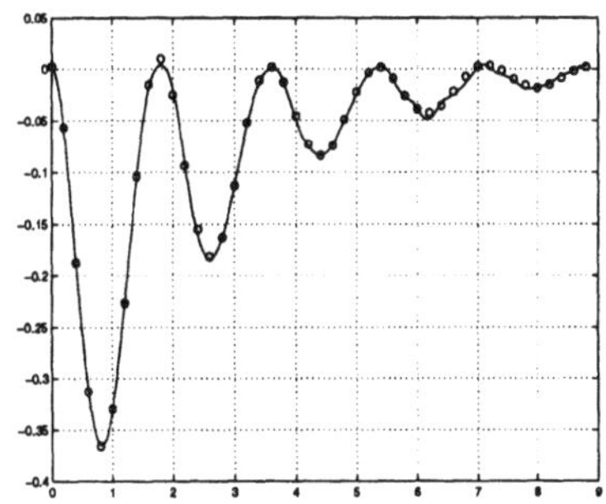

t(sec)

Figure 22. Evolution of the cross-correlation between the velocities of the two masses; (o-simulation).

In previous work [4-5], the authors computed the response moments using simple weighted row and column sums of the density function. These sums are just the Riemann approximations of the appropriate integrals. For instance, the mean square displacement of the first mass using this method is approximated by

$$E[X_1^2] = \int_{R^4} x_1^2 p(\boldsymbol{x}) d\boldsymbol{x} = \sum_{i=1}^{l} x_{1_i}^2 p_i \delta_1 \delta_2 \delta_3 \delta_4 \tag{15}$$

where l is the number of nodes in the computational mesh and δ_j is the nodal spacing in the j-th direction. This simple quadrature rule has worked effectively in the past but here causes errors. This change can be attributed to two factors. First, this approximation is less accurate when the density function is sharply peaked, as it is for both of these systems in the x_1 and x_3 directions. Second, the approximation is of a much lower order than the spatial scheme used to discretize the initial-boundary value problem. Thus, the savings in mesh density afforded by the use of a high order finite difference scheme does not translate into moment computations of equivalent accuracy unless a higher order quadrature rule is utilized.

This lack of consistency between the order of the discretization scheme and the integration rule is the source of the error in the computed response moments. The error is greatest for the smallest variance being predicted (*i.e.*, the mean square displacement of the first mass of the nonlinear system, as shown in Fig. 13).

To further understand this phenomenon, a trapezoidal rule was implemented to integrate the density function in the moment calculations, and an improvement in accuracy was obtained. For example, the stationary mean square displacement of the first mass in the linear system using the exact solution, Riemann sum, and trapezoidal rule are compared in Table 1. The authors believe that higher order quadrature rules will further reduce the errors in the moment computation to a level consistent with the accuracy of the density function, and this effort is currently under way.

TABLE 1. Effect of quadrature rule on moment computation

Method	$E[X_1^2]$
Riemann Sum	.3268
Trapezoidal Rule	.3290
Exact Solution	.3333

6. Conclusions

A high accuracy grid based solution method for high dimensional problems can now be reported. The density function obtained is very accurate as evidenced by the linear result. The accuracy of the computed quantities, such as response moments, decreases when the distribution is very peaked. It is anticipated that a higher order quadrature rule will circumvent this problem and is being currently developed. Problems of up to and including four states with arbitrary nonlinearities can now be studied. The algorithm is scalable to even higher dimensional systems with minimal additional programming effort.

7. Acknowledgments

This work has been supported in part by National Science Foundation contracts ECS-9224828, ECS-9411589, CMS-9818145, CEE-92000-4N, and MSS-94000-1N, the latter two through the National Center for Supercomputing Applications at the University of Illinois at Urbana-Champaign.

8. References

1. Spencer, Jr., B.F., and Bergman, L.A. (1985) On the Reliability of a Simple Hysteretic System, *ASCE Journal of Engineering Mechanics* **111**, 1502-1514.
2. Bergman, L.A. (1989) Numerical Solutions of the First Passage Problem in Stochastic Structural Dynamics, in W.K. Liu and T. Belytschko, (eds.), *Computational Mechanics of Probabilistic and Reliability Analysis*, ElmePress International, Lausanne, pp. 379-408.
3. Spencer, B.F., Jr. and Bergman, L.A. (1993) On the Numerical Solution of the Fokker–Planck Equation for Nonlinear Stochastic Systems, *Nonlinear Dynamics,* **4**, 357-372.
4. Wojtkiewicz, S.F., Bergman, L.A. and Spencer, B.F. Jr., (1995) Numerical Solution of Some Three-State Random Vibration Problems, in S.C. Sinha *et al.*, (eds.), *Vibration of Nonlinear, Random, and Time-Varying Systems*, ASME DE-84-1, pp. 939-947.
5. Bergman, L.A., Spencer, Jr., B.F., Wojtkiewicz, S.F. and Johnson, E.A. (1996) Robust Numerical Solution of the Fokker-Planck Equation for Second Order Dynamical Systems Under Parametric and External White Noise Excitations, in W. Langford, W. Kliemann and N. Sri Namachchivaya, (eds.), *Fields Institute Communications, Nonlinear Dynamics and Stochastic Mechanics*, American Mathematical Society, Providence, pp. 23-37.
6. Wojtkiewicz, S.F., L.A. Bergman, and Spencer, Jr., B.F. (1998) High Fidelity Numerical Solutions of the Fokker-Planck Equation, in N. Shiraishi, M. Shinozuka and Y.K. Wen (eds.), *ICOSSAR '97, Proceedings*, **2**, pp. 933-940.
7. Wojtkiewicz, S.F., Bergman, L.A. and Spencer, Jr., B.F., (1998) Computational Issues Arising in the Solution of the Fokker-Planck Equation in Higher Dimensions: Use of Iterative Solution Methods, in N. Shiraishi, M. Shinozuka, Y.K. Wen (eds.), *ICOSSAR '97, Proceedings*, **2**, pp. 851-858.
8. Wedig, W.V. and von Wagner, U. (1999) Extended Laguerre Polynomials for Nonlinear Stochastic Polynomials, *Proceedings of the Third International Conference on Computational Stochastic Mechanics*, in press.
9. Gear, C.W. (1971) *Numerical Initial Value Problems in Ordinary Differential Equations*, Prentice-Hall, Englewood Cliffs, N.J.
10. Saad, Y. (1996) *Iterative Methods for Sparse Linear Systems*, Boston, PWS.

SYSTEM IDENTIFICATION OF HYSTERETIC STRUCTURES

H. ZHANG
Department of Mechanical Engineering
University of California at Berkeley
USA

P. PAEVERE
Department of Civil Engineering
University of Melbourne
Australia

Y. YANG
Department of Mechatronic Engineering and Instrumentation
Changsha Institute of Technology
People's Republic of China

G. C. FOLIENTE
CSIRO Building, Construction and Engineering
Australia

F. MA
Department of Mechanical Engineering
University of California at Berkeley
USA

Abstract

Three different time-domain system identification algorithms are presented in this paper to estimate the parameters of an extended Bouc-Wen hysteretic model. These algorithms include the simplex, extended Kalman filter, and generalized reduced gradient methods. The effectiveness of the proposed algorithms is demonstrated through simulations of nonlinear systems with pinching and degradation characteristics. Due to very modest computing requirements, the proposed identification algorithms can be acceptable as a basic tool for estimating hysteretic parameters in engineering design.

1. Introduction

S. Narayanan and R.N. Iyengar (eds.),
IUTAM Symposium on Nonlinearity and Stochastic Structural Dynamics, 289–306.

Building structures usually exhibit nonlinear and inelastic behavior under severe dynamic loading such as earthquakes, austere winds and waves. Inelastic behavior often manifests itself in the form of hysteresis loops in the force-displacement relation, when the structure is subjected to a cyclic load. The area enclosed by each loop is a measure of the energy dissipated over a complete cycle as a result of internal friction within the structure. Hysteresis refers to the hereditary and the memory nature of an inelastic system, in which the restoring force depends not only on the instantaneous deformation but also on the past history of deformation. Hysteresis is particularly important in depicting the nonlinear response of wood buildings, reinforced concrete, and structures with a high proportion of composite materials. These structural systems exhibit appreciable energy dissipation and tend to possess large hysteresis loops. Any exposition of the significant damping capacity in these structures must include hysteresis (Sozen 1974). Yet, a comprehensive theory of hysteresis has not been developed.

In the past few decades, various empirical models of restoring forces were proposed to account for the hereditary nature of inelastic systems. One of the widely accepted model is a differential model of hysteresis originally proposed by Bouc (1967) and subsequently generalized by Wen (1980) and other researchers. In this model the applied force and deformation are connected through a nonlinear differential equation containing unspecified loop parameters. By specifying the loop parameters, it is possible to generate a large variety of different shapes of the hysteresis loops. On the other hand, suppose a set of actual, measured hysteresis data from cyclic performance tests is given. A very meaningful question is how the loop parameters can be calibrated from the measured data. Only after the loop parameters have been identified can the hysteretic model be used for prediction of system response when the structure is subjected to other severe load reversals.

System identification offers a class of techniques for estimating the loop parameters of hysteresis from a set of given data. In general, system identification methods fall into two classes: parametric and nonparametric (Imai *et al.*, 1989). Parametric identification involves estimation of system parameters, while nonparametric identification determines the transfer function of the system in terms of analytical representation. Identification methods can also be classified as those in the time domain and the frequency domain. In the time domain method, system parameters are determined from observational data sampled in time. On the other hand, in the frequency domain method, modal quantities such as natural frequencies, damping ratio and modal shapes are identified using measurements in the frequency domain. The prior knowledge of the system and the intended use of the model determine the choice between time domain and frequency domain. Hjelmstad and Banan (1995) indicate that a time-domain model is required when the system is governed by differential or difference equations, or when the model is intended to predict future response or simulate the system, or when a stochastic control is desired. Frequency domain algorithms involve averaging temporal information, thus discarding most of their details. For structural systems, whose parameters are expected to degrade with time, this tradeoff of temporal information for frequency information may not be justifiable.

Only time-domain system identification methods are considered in this paper. Three relatively powerful algorithms are constructed for use with an extended hysteretic model that incorporates both degradation and pinching parameters. Hysteretic models with such generalities have not been considered in earlier studies in system identification. In addition, the algorithms presented herein utilize streamlining schemes such as global search and constraints. Effectiveness of the algorithms is demonstrated through numerical simulations of nonlinear inelastic structures. Although a limited number of simulations are reported in this paper, extensive numerical calculations have been performed to support any qualitative results presented herein. The organization of the paper is as follows. The generalized differential model of hysteresis is described in Section 2. In Section 3, a constrained downhill simplex method is used to identify the parameters of a hysteretic system without pinching. It is shown that this method can achieve the same performance as the modified Gauss-Newton method. The constrained simplex method, together with a constrained extended Kalman Filter algorithm and a generalized reduced gradient method are then used to estimate the parameters of two hysteretic systems with pinching and degradation. A comparison of the effectiveness of the three identification algorithms is given in Section 4. In Section 5, a summary of findings is given.

2. The extended Bouc-Wen hysteretic model

In the past few decades, various models of hysteresis have been developed to account for the hereditary nature of an inelastic system. One of the widely accepted model is the Bouc-Wen differential model (Bouc, 1967 and Wen, 1976). In this study, several parameter identification algorithms are applied to an extended version of the Bouc-Wen hysteretic model. This model is generally applicable to a wide variety of material types and structural configurations (Foliente, 1995). It is capable of modeling strength degradation, stiffness degradation and pinching. Thirteen control parameters are introduced into the model. Since this model has more parameters than other models being proposed before, new parameter identification methods are needed to correctly identify all the parameters.

The equation of motion of the a single-degree-of-freedom hysteretic system may be generally written as

$$m\ddot{u} + c\dot{u} + R_T[u(t), z(t); t] = F(t)\,, \tag{1}$$

where u is the relative displacement of the mass with respect to the ground motion; $F(t)$ is the forcing function; and c is the coefficient of viscous damping. The non-damping restoring force can be written as

$$R_T[u(t), z(t); t] = \alpha k u + (1-\alpha) k z \,, \tag{2}$$

where k is the stiffness of the system, z is the hysteretic displacement, α is the rigidity ratio, which is a weighting constant representing the relative participation's of the linear and nonlinear restoring force terms. Dividing both sides of Eq. (1) by m, the following standard form is obtained

$$\ddot{u} + 2\zeta\omega_0\dot{u} + \alpha\omega_0^2 u + (1-\alpha)\omega_0^2 z = f(t) \,, \tag{3}$$

where ζ is the viscous damping ratio, and ω_0 is the system natural frequency. The constitutive law is given by the following first-order nonlinear differential equation

$$\dot{z} = h(z)\frac{A\dot{u} - \nu\left(\beta|\dot{u}||z|^{n-1} z + \gamma\dot{u}|z|^n\right)}{\eta} \,, \tag{4}$$

with pinching function

$$h(z) = 1.0 - \xi_1 e^{\left[-\left(z\cdot\mathrm{sgn}(\dot{u}) - q\cdot z_u\right)^2/\xi_2^2\right]} \,, \tag{5}$$

where sgn$(\cdot)$ is the signum function, z_u is the ultimate value of z given by

$$z_u = \left[\frac{A}{\nu(\beta+\gamma)}\right]^{1/n} \,, \tag{6}$$

and

$$\xi_1(\varepsilon) = \xi_0\left[1.0 - e^{-p\varepsilon}\right] \,, \tag{7}$$

$$\xi_2(\varepsilon) = (\psi_0 + \delta_\psi \varepsilon)(\lambda + \xi_1) \,. \tag{8}$$

Strength and stiffness degradation are modeled by allowing the parameters ν and η to vary as a function of the response duration and severity. A convenient measure of the combined effect of duration and severity is the total energy dissipated through hysteresis as the system evolves. This total energy, measured from an initial time $t = t_0$ to t_f, is proportional to

$$\varepsilon(t) = \int_{t_0}^{t_f} z\dot{u}\,dt \tag{9}$$

As explained by Baber and Wen (1981), physically realistic models of degradation can be obtained by putting

$$\nu(\varepsilon) = 1.0 + \delta_\nu \varepsilon , \tag{10}$$

$$\eta(\varepsilon) = 1.0 + \delta_\eta \varepsilon . \tag{11}$$

The thirteen system parameters A, β, γ, n, δ_ν, δ_η, ξ_0, q, p, ψ, δ_ψ, λ, and α control the shape and magnitude of the hysteresis loops. They must be specified in order to describe the system response unambiguously.

3. Parameter Identification

3.1. SIMPLEX METHOD

The downhill simplex method due to Nelder and Mead (1965) is a non-gradient search method. This method requires only function evaluations, not derivatives. This method is less prone to divergence, comparing with gradient methods. It is also not sensitive to initial values. However, it is not very efficient in terms of the number of function evaluations that it requires because it makes no use of the information supplied by the gradient and the Hessian matrix.

A simplex is a geometrical figure consisting of, in N dimensions, N + 1 points (or vertices) and all their interconnecting line segments, polygonal faces, etc. (Press *et al.*, 1988). In two dimensions, a simplex is a triangle. In three dimensions it is a tetrahedron. The simplex method starts out by initializing the vertices. One of the vertices is the initial approximation to the solution. Unlike gradient methods the simplex method is insensitive to the initial approximation. The other vertices are set to the initial approximation varied by an amount τ reasonable for the problem at hand,

$$P_i = P_0 + \tau e_i ,\ i = 1, \cdots, N, \tag{12}$$

where P_0 is the initial approximation, P_i the other vertices, and e_i the unit vectors.

The simplex method calculates the objective function at each vertex, rejects the worst, i.e., the one with the highest value, and calculates a new vertex to replace it in such a way that the non-degeneracy of the simplex is maintained. Four mechanisms for the calculation of the new vertex are used: reflection, expansion, contraction along one dimension, and contraction along all dimensions (shrinkage). To find a set of meaningful system parameters, some constraints of the set of parameters based on experimental data should be added to the minimization procedure.

For the extended Bouc-Wen hysteretic model, the objective function is defined as

$$E=\sum_{i=1}^{m}\left[\dot{z}_i-h(z)\left(\frac{A\dot{u}_i-\nu\left(\beta|\dot{u}_i||z_i|^{n-1}z_i+\gamma\dot{u}_i|z_i|^{n}\right)}{\eta}\right)\right]^2. \tag{13}$$

The best estimate of the model parameters is obtained when the objective function achieves global minimum.

A numerical test is performed to test the performance of the simplex method for parameter identification. A hypothetical hysteretic structure with degradation but no pinching is simulated with the following modal parameters: $\zeta=0.1$, $\omega_0=6.28$, $f(t)=(20.0+1.2t)\sin(10.0t)$, $h(z)=1.0$, $A=1.0$, $\beta=1.5$, $\gamma=-0.5$, $\delta_v=0.005$, $\delta_\eta=0.05$, and $n=1.0$. These parameters are used to integrate Eq. (4) to generate a set of data points. The total number of data points is 1000 and the size of the time step is 0.001. The simplex method and a reliable gradient technique called the Levenberg-Marquardt method, are then employed to recover the original set of parameters A, β, γ, δ_v, δ_η, and n from the data points. No constraints are used in the simplex method in this case.

Table 1. shows the two sets of hysteretic parameters estimated by the two methods. The unconstrained simplex works as well as the Levenberg-Marquardt method in identifying the parameters of the hysteretic model. However, when the above hysteretic model is extended to include pinching effect by adding more parameters, the Levenberg-Marquardt method is easily trapped into local minima of the objective function, which are far from the true value of the actual parameters, instead of converging to the global minimum. In this case, the constrained simplex method will be shown to be able to give a better estimation.

Most system parameters are not permitted to vary freely. Proper incorporation of constraints can substantially accelerate the rate of convergence. The appropriate set of constraints can be obtained from experimental data. For typical timber structures, the following set of constraints can be incorporated into the estimation algorithms by variable substitution as follows:

$$\left\{\begin{array}{cc} A\geq 1 & p\geq 0\\ \beta\geq 0 & \psi\geq 0\\ \beta+\gamma>0 & \delta_\psi\geq 0\\ \beta-\gamma>0 & \lambda\geq 0\\ 0.7\leq\xi_0\leq 1.0 & 0\leq\alpha\leq 0.1\\ q\geq 0 & \end{array}\right. \Rightarrow \left\{\begin{array}{cc} A=1.0+x_1^2 & \psi=x_7^2\\ \beta=x_2^2 & \delta_\psi=x_8^2\\ \gamma=x_3^2-x_2^2 & \lambda=x_9^2\\ \xi_0=0.7+0.3\sin^2 x_4 & \alpha=0.1\sin^2 x_{10}\\ q=x_5^2 & \\ p=x_6^2 & \end{array}\right.$$

3.2. CONSTRAINED KALMAN FILTER METHOD

The extended Kalman filter is a recursive algorithm for estimating the optimal state of a nonlinear system from noise-corrupted data. It utilizes an extended state vector that includes, in addition to the response and velocity, all the parameters to be identified. Starting from an initial guess, this extended state space is recursively updated as new observational data are made available. The update is based upon the traditional formulation of the Kalman filter. The extended Kalman filter has been used in many applications. Since it is rather sensitive to the initial guess, a global weighted iteration scheme for the extended filter (Hoshia and Saito, 1984) can be used to improve stability. Such a scheme was also discussed by Maruyama *et al.* (1989), Loh and Chung (1993), and Lin and Zhang (1994). However, constraints were not considered in their investigations.

For the system governed by Eq. (1) to Eq. (11), let the extended state be $\mathbf{x}^T = [\mathbf{a}^T, \mathbf{b}^T]$, where $\mathbf{a} = [u,\ \dot{u},\ z,\ \varepsilon]^T$, and $\mathbf{b} = [A, \beta, \gamma, \cdots \lambda]^T$. The state equation can be written as

$$\dot{\mathbf{x}}(t) = \mathbf{g}(\mathbf{x}(t), t)\,, \tag{14}$$

where $\dot{\mathbf{x}}^T = [\dot{\mathbf{a}}^T, \dot{\mathbf{b}}^T] = [\dot{u}, \ddot{u}, \dot{z}, z\dot{u}, \mathbf{0}^T]$. The observed vector $\mathbf{y} = [u, \dot{u}]^T$ at instant k is related to $\mathbf{x}$ by

$$\mathbf{y}(k) = \begin{bmatrix} 1 & 0 & 0 & \ldots & 0 \\ 0 & 1 & 0 & \ldots & 0 \end{bmatrix} \mathbf{x}(k) + \mathbf{w}(k) \tag{15}$$

where $\mathbf{w}(k)$ is zero-mean Gaussian white noise. The extended Kalman filter seeks to linearize equation (14) in the vicinity of each estimated state. Let $\hat{\mathbf{x}}(k|k)$ be the estimated state at instant k. Upon linearization about $\hat{\mathbf{x}}(k|k)$, the state equation becomes

$$\dot{\mathbf{x}}(t) = \mathbf{H}\mathbf{x}(t) + \mathbf{r}(t)\,, \tag{16}$$

where

$$\mathbf{H} = \frac{\partial \mathbf{g}}{\partial \mathbf{x}}\Big|_{\mathbf{x} = \hat{\mathbf{x}}(k|k)}\,, \tag{17}$$

and $\mathbf{r}(t) = \mathbf{g}(\hat{\mathbf{x}}(k|k), t) - \mathbf{H}\hat{\mathbf{x}}(k|k)$. The traditional Kalman algorithm can be used in the linear system (16) with a state transition matrix

$$\Phi(k+1, k) = \mathbf{I} + T\mathbf{H} \tag{18}$$

with a sufficiently small sampling interval T.

High sensitivity of the extended Kalman filter to initial values often leads to problems of convergence and bias. To improve convergence and stability, a weighted global iteration scheme may be employed. Starting from an arbitrary value $\mathbf{x}_1^T(0) = [\mathbf{a}_1^T(0), \mathbf{b}_1^T(0)]$ and a weighted error variance matrix, a sequence of N recursive calculations using all available measured data is performed. This generates an estimated state $\mathbf{x}_1^T(N) = [\mathbf{a}_1^T(N), \mathbf{b}_1^T(N)]$ and a new error covariance matrix. A second sequence of N recursive calculations are then performed starting with the initial state $\mathbf{x}_2^T(0) = [\mathbf{a}_1^T(0), \mathbf{b}_1^T(N)]$ and an updated error covariance matrix. This process can be repeated until $\mathbf{b}_M(0) \approx \mathbf{b}_M(N)$, at which time convergence is obtained. A detailed description of the weighted global iteration scheme is provided, for example, by Maruyama *et al.* (1989). Other remedial schemes to improve convergence and stability may also be used. To speed up the convergence, the earlier proposed set of constraints is also incorporated here in the Kalman filter algorithm. The extended state and the error functional of the filter are reformulated in terms of variables of the new extended state. As discussed by Tugnait (1985), optimal values for the new extended state may not always correspond to optimal values of the original extended state, and there is a slight probability of model mismatch. For simple inequality constraints of the above type, this probability is small. However, additional investigation of this matter is needed for more complex constraints

3.3. GENERALIZED REDUCED GRADIENT METHOD (GRG)

The Solver tool in Microsoft Excel program is also employed to estimate modal parameters. The core of the Solver tool is the generalized reduced gradient algorithm (Carpentier and Abadie, 1966), which is an extension of the Wolfe algorithm (Charnes and Cooper, 1957) to accommodate both a nonlinear objective function and nonlinear constraints. In essence the method employs linear, or linearized constraints, defines new variables that are normal to some of the constraints, and transform the gradient to this new basis (Himmelblau, 1972). Further details of the generalized reduced gradient method can be found in Lasdon (1978).

Table 1. Parameter identification results from the Levenberg-Marquardt method and the simplex method (1000 data points)

	Real Value	Initial Value	L-M Method	Simplex Method
A	1.0	5.0	0.99999	0.99999
β	1.5	5.0	1.49991	1.49993
γ	-0.5	5.0	-0.49997	-0.49998
δ_v	0.005	5.0	0.00519	0.00516
δ_η	0.05	5.0	0.04991	0.04991
n	1.0	2.0	0.99999	0.99999

4. Comparison of Parameter Identification Methods

4.1. SIMULATION DATA DESCRIPTION

In order to demonstrate the utility of the three parameter estimation methods presented above, identification of two timber structure models will be made. Wood structures are well known for their ability to dissipate energy through hysteresis. They also show strong pinching effect. Extensive pseudo-dynamic tests of shear walls subjected to cyclic loads have been conducted at the Building Research Institute in Japan. A typical test configuration is shown in Figure 1. In the experiment, a magnet-type displacement transducer is used to monitor the displacement at the middle of two walls just lacing an oil-jack actuator. Further details of the experimental configuration are provided by Kawai (1997). If actual test data are used for identification, it would be difficult to determine the accuracy of identification. For this reason, simulated data resembling the test data will be used to validate the three methods presented earlier. Two sets of realistic parameters are chosen and displayed in Table 1 and Table 2. These parameters are chosen in such a way that, together with a cyclic load used in actual experiments, they can numerically generate response data using Eq. (1) to Eq. (11) that resemble the response measured in actual experiments. Figure 2 and 3 show the restoring force diagram of the sets of simulation data, respectively.

Example 1: Japanese Timber House Under Kobe Earthquake

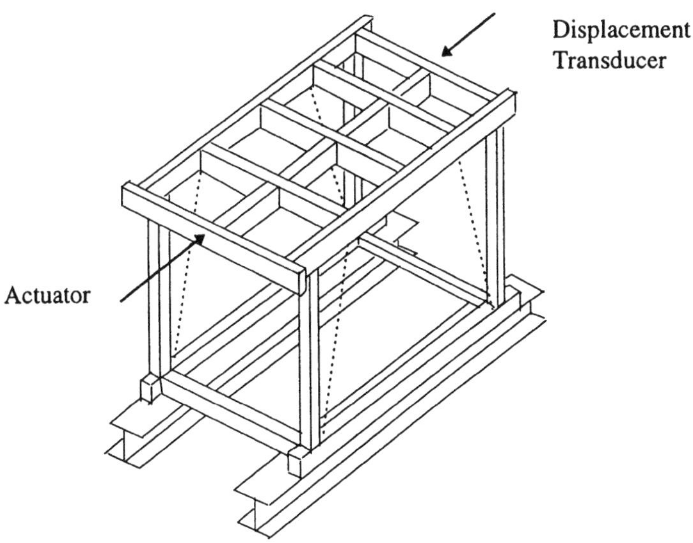

Figure 1. Testing configuration

System Details

- Natural Frequency = 2.63 Hz
- Peak Displacement = 30 cm

Excitation Details

- Kobe Earthquake
- Duration = 11.56 sec
- Peak Ground Acceleration = 0.8g

Example 2: Timber building under severe earthquake

System Details

- Natural Frequency = 3.6 Hz
- Peak Displacement = 6.4 cm

Excitation Details

- EW El Centro Earthquake x 2
- Duration = 20 sec
- Peak Ground Acceleration = 0.7g

4.2. RESULTS:

The three different parameter identification algorithms presented in this paper are used to estimated the actual sets of parameters of the two simulated hysteretic models. In all the three estimation processes, n is set to be 1 in order to simplify the process. This simplification is based on the fact that n generally assumes small integral values. It is then easy to choose n by comparing the overall error with n = 1, 2, ... 7 after the other parameters have been identified. The estimated parameters of the Japanese shear wall model and the timber building model are shown in Table 2 and Table 3, respectively. Figure 4, and 5 show the displacement diagram and energy dissipation diagram of the Japanese shear wall model, respectively. Figure 6, and 7 show the displacement diagram and energy dissipation diagram of the timber building model, respectively. All three methods give good estimation of the modal parameters, although for different model the performance of each of the three algorithms may vary. Some parameters identified in Table 2 and 3 are quite different from their true value. However, the resulting diagrams are not far off the original curves. This may point out that the hysteretic loop is not sensitive to some of the parameters. In later study, these insensitive parameters can be set to constant values over certain range in order to speed up the convergence of estimation.

Table 2. Parameter identification results from simplex method, GRG method and extended Kalman filter method for the Japanese shear wall data

	Real Value	Simplex Method	GRG Method	Extended Kalman Filter
A	1.0	1.0011	1.0000	1.0016
β	0.6	0.4886	0.8436	0.6532
γ	-0.35	-0.2415	-0.5265	-0.4072
δ_ν	0.037	0.0402	0.0213	0.0375
δ_η	0.3	0.2807	0.3098	0.3116
ξ_0	0.972	0.9655	0.9119	0.9733
q	0.003	0.0047	0.0166	0.0050
p	4.6	4.9990	1.9330	2.3765
ψ	0.06	0.1017	0.0738	0.0521
δ_ψ	0.0005	0.0001	0.0009	0.0004
λ	0.0001	0.0049	0.4723	0.0716
α	0.001	0.0009	0.0005	0.0027

Table 3. Parameter identification results from simplex method, GRG method and extended Kalman filter method for the timber building data

	Real Value	Simplex Method	GRG Method	Extended Kalman Filter
A	1.0	1.0127	1.0000	1.0135
β	1.4	1.3283	1.0743	1.4424
γ	0.2	0.2076	0.6096	0.1074
δ_ν	0.002	0.0046	0.0001	0.0042
δ_η	0.001	0.0248	0.0000	0.0013
ξ_0	0.95	0.9147	0.9552	0.9502
q	0.25	0.2513	0.2582	0.2450
p	2.0	2.3829	2.2090	1.8765
ψ	0.1	0.0620	0.0499	0.1133
δ_ψ	0.005	0.0007	0.0042	0.0054
λ	1.0	3.6083	1.8536	0.8429
α	0.0001	0.0092	0.0095	0.0097

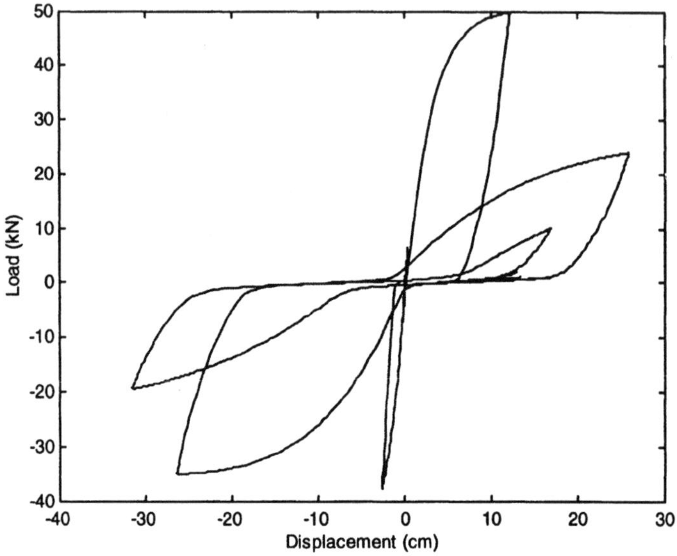

Figure 2. Restoring force diagram of the Japanese shear wall model

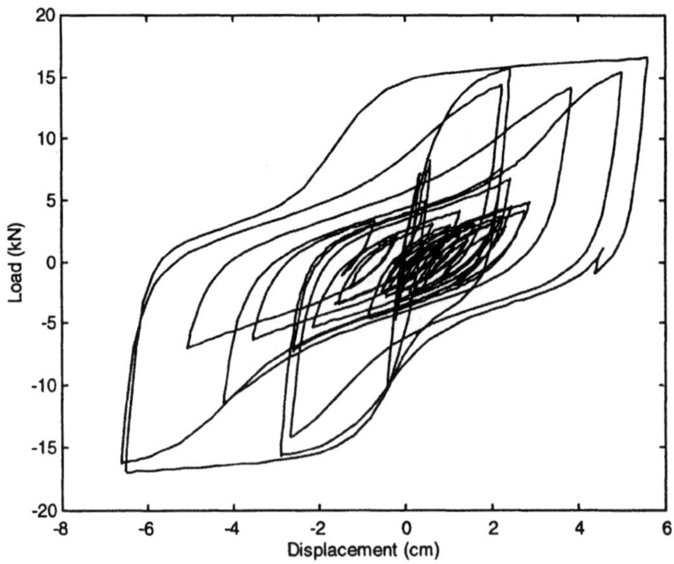

Figure 3. Restoring force diagram of the timber building model

5. Conclusions

Three parameter identification methods, including constrained simplex method, constrained extended Kalman filter method and generalized reduced gradient method, have been used to identify the parameters of an extended version of the Bouc-Wen differential hysteretic model. The simplex method is a non-gradient search method. It is not sensitive to the initial value of iteration. The generalized reduced gradient method used by the Microsoft Excel Solver is an extension of the Wolfe algorithm to accommodate both a nonlinear objective function and nonlinear constraints. The extended Kalman filter method is a recursive algorithm for estimating the optimal state of a nonlinear system. Feasibility of the three algorithms has been demonstrated by applying to two sets of simulated data resembling typical timber structure cyclic testing data. All system parameters of the two hysteretic structures are quickly identified. The resulting hysteretic loops generated by the estimated parameters resemble the actual hysteretic loops closely. These parameter estimation methods should be accepted as a basic tool for estimating hysteretic parameters in engineering design. Among other things, this would permit evaluation and performance of energy-dissipating devices to be made with added confidence. It would also point to new directions along which experimental testing programs may be directed. It is also found from the identification procedure that the hysteretic loop shape is not sensitive to some of the parameters of the extended hysteretic model. These insensitive parameters can be set to constant values over certain ranges in order to speed up the convergence of the estimation procedure.

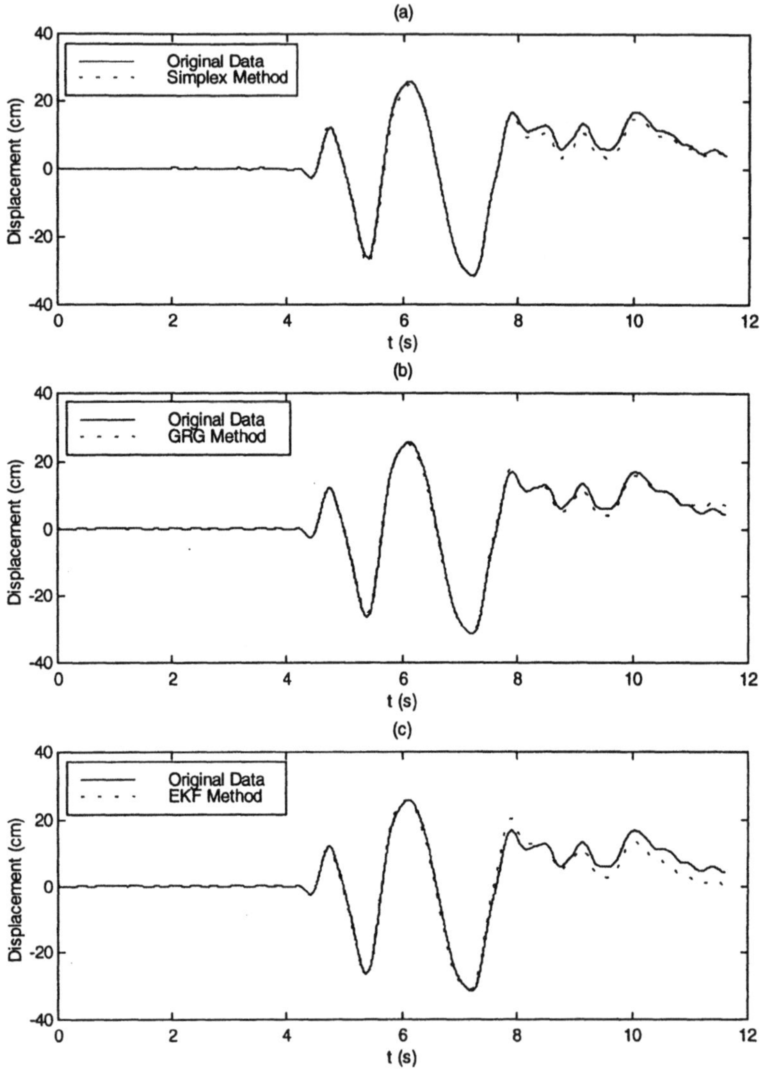

Figure 4. Displacement diagram of the Japanese shear wall model

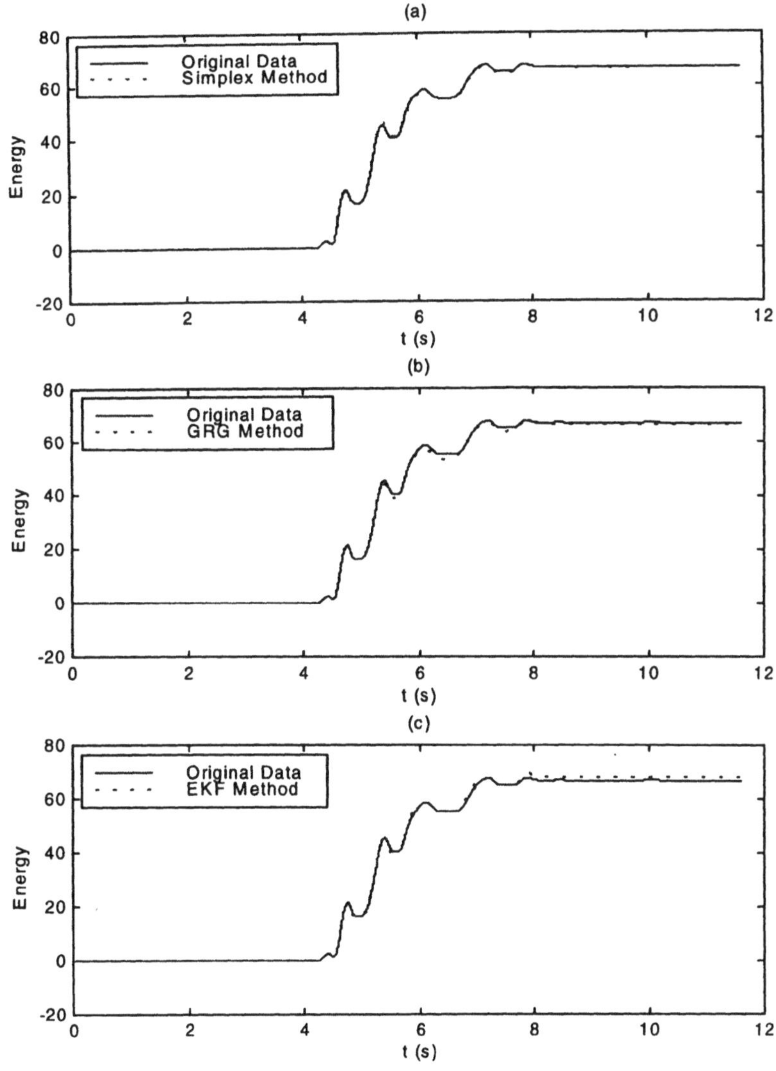

Figure 5. Energy dissipation diagram of the Japanese shear wall model

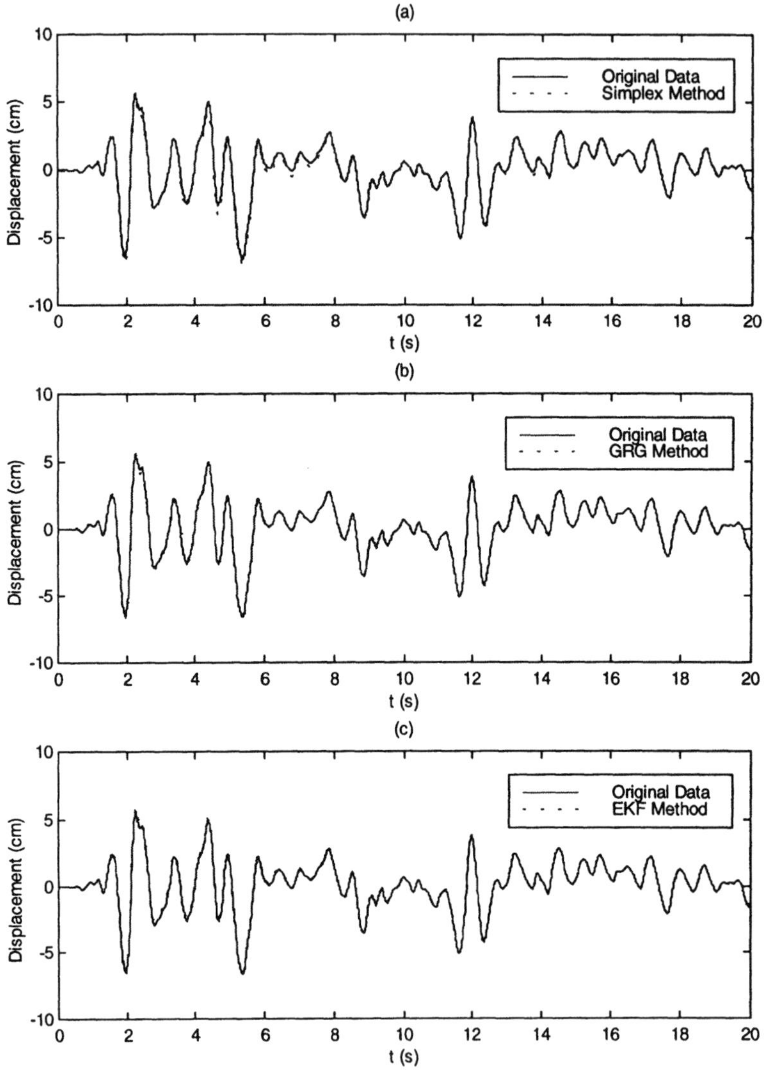

Figure 6. Displacement diagram of the timber building model

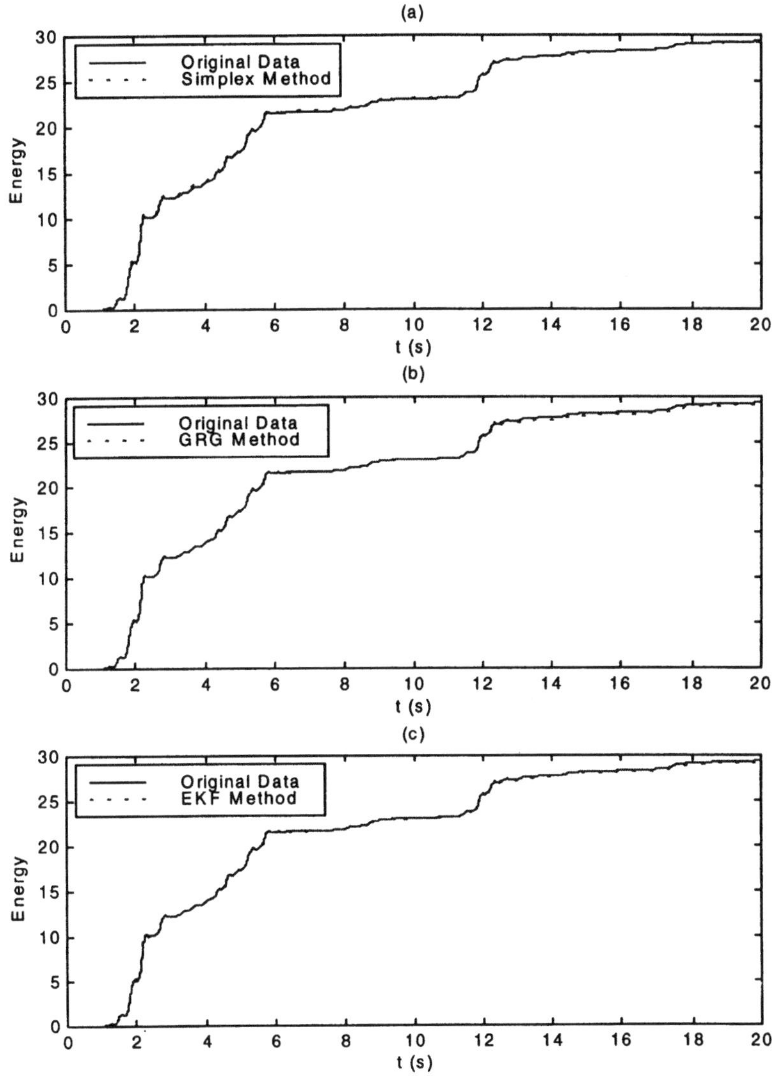

Figure 7. Energy dissipation diagram of the timber building model

Reference

Bouc, R. (1967) Forced Vibration of Mechanical Systems with Hysteresis, Proceedings of the 4-th International Conference. on Nonlinear Oscillations, Prague, Czechoslovakia.

Carpentier, J. and Abadie, J. (1966) Generalisation de la methode du gradient reduit de Wolfe au cas de constraintes nonlineaires, Proceedings of IFORS Congress, Cambridge, Mass., August 29-September 2.

Charnes, A. and Cooper, W.W. (1957) Nonlinear Power of Adjacent Extreme Point Methods in Linear Programming, Econometrica, 25:132.

Foliente, G. C. (1995) Hysteretic Modeling of Wood Joints and Structural Systems, Journal of Structural Engineering, ASCE 121, 1013-1022.

Himmelblau, D. M. (1972) Applied Nonlinear Programming, McGraw-Hill, New York.

Hjelmstad, K.D., Banan, M.R. and Banan, M.R. (1995) Time-Domain Parameter Estimation Algorithm For Structures. I: Computational Aspects, Journal of Engineering Mechanics Vol. 121, No. 3, 424-434.

Hoshia, M., and Saito, E. (1984) Structural Identification by Extended Kalman Filter, ASCE Journal of Engineering Mechanics, Vol. 110, pp. 1757-1770.

Imai, H., Yun, C.-B., Maruyama, O. and Shinozuka, M. (1989) Fundamentals of system identification in structural dynamics, Probabilistic Engineering Mechanics, Vol. 4, No. 4, 162-173.

Lasdon, L.S., Warren, A.D., Jain, A., and Ratner, M. (1978) Design and Testing of a Generalized Reduced Gradient Code for Nonlinear Programming, ACM Trans. Math. Software 4.

Maruyama, O., Yun, C. B., Hoshiya, M., and Shinozuka, M. (1989) Program EXKAL2 for Identification of Structural Dynamic Systems, Technical Report NCEER-89-0014, National Center for Earthquake Engineering Research, Buffalo, New York.

Lin, J-S., and Zhang, Y. (1994) Nonlinear Structural Identification using Extended Kalman Filter, Computers & Structures, Vol. 52, pp. 757-764.

Loh, C.-H., and Chung, S.-T. (1993) A Three-Stage Identification Approach for Hysteretic Systems, Earthquake Engineering and Structural Dynamics, Vol. 22, 129-150.

Nelder, J.A., and Mead, R. (1965) A Simplex Method for Function Minimization, Computer Journal, Vol.7, 308.

Press, W. H., Flannery, B. P., Teukolsky, S. A., and Vetterling, W. T. (1988) Numerical Recipes in C: The Art of Scientific Computing, Cambridge University Press, United Kingdom.

Sozen, M.A. (1974) Hysteresis in structural elements, Applied Mechanics in Earthquake Engineering, W.D. Iwan, ed., ASME, New York, NY, 63-98.

Tugnait, J. K. (1985) Constrained Signal Restoration via Iterated Extended Kalman Filtering, IEEE Transaction on Acoustics, speech, and signal processing, Vol. ASSP-33, pp. 472-475.

Wen, Y. K. (1976) Method for Random Vibration of Hysteretic Systems, ASCE Journal of Engineering Mechanics, Vol. 102, 249-263.

STOCHASTIC STABILITY AND BIFURCATION OF QUASI-HAMILTONIAN SYSTEMS

W.Q.ZHU AND Z.L.HUANG
Department of Mechanics, Zhejiang University,
Hangzhou, 310027,P.R.China

Abstract. The paper consists of two parts. In the first part the stochastic averaging of quasi-integrable-Hamiltonian systems with real noise excitations is introduced. The expression for the largest Lyapunov exponent of the square root of the Hamiltonian is formulated by generalizing the Khasminskii's procedure to the averaged systems, based on which the stochastic stability and bifurcation of the original systems are studied. In the second part, an n-degree-of-freedom quasi-non-integrable-Hamiltonian system is reduced to an *Itô* equation of one-dimensional averaged Hamiltonian by using the stochastic averaging method for quasi-non-integrable-Hamiltonian systems. The necessary and sufficient conditions for the asymptotic stability in probability of the trivial solution and the condition for the Hopf bifurcation of the original systems are obtained approximately by examining the sample behaviors of the one-dimensional diffusion process of the square-root of averaged Hamiltonian and the averaged Hamiltonian , respectively, at the two boundaries.

1. Introduction

For linear systems of *Itô* stochastic differential equations, a necessary and sufficient condition for almost sure asymptotic stability was given by Khasminskii(1967). According to the multiplicative ergodic theorem of Oseledec(1968),vanishing of the largest Lyapunov exponent implies a change in the almost-sure asymptotic stability of the systems. Thus, the Khasminskii's theorem(1967) actually provide a procedure to evaluate the largest Lyapunov exponent for linear systems of *Itô* stochastic differential equations. This procedure has been successfully employed to determine numerically the largest Lyapunov exponent, almost-sure asymptotic stability and bifurcation of certain two-dimensional linear stochastic systems(Kozin and Prodromou,1971; Nishioka, 1976; Pardoux and Wihstutz,1988; Sri Namachchivaya, 1991; Wedig, 1988; Ariaratnam and Xie, 1989, and others) and of coupled linear stochastic systems(Ariaratnam and Xie, 1992). The procedure has also been applied to a special type of nonlinear stochastic systems, for which the drift and diffusion coefficients of *Itô* equations are homogeneous of order one (Kozin and Zhang, 1990).

S. Narayanan and R.N. Iyengar (eds.),
IUTAM Symposium on Nonlinearity and Stochastic Structural Dynamics, 307–318.

For a dynamic systems governed by an one-dimensional *Itô* stochastic differential equation, Kozin and Sunahara(1987), Sri Namachchivaya(1989, and Zhang(1991) showed that the necessary and sufficient conditions for the asymptotic stability in probability of the trivial solution can be obtained by examining the sample behavior of the system at the trivial solution. Later, Lin and Cai (1994) showed that the sample behaviors at both of the two boundaries, not just the one at the trivial solution, should be examined in order that the necessary and sufficient conditions for the asymptotic stability in probability can be obtained. They concluded that the trivial solution was asymptotic stable in probability if and only if it is an exit or attractively natural, and the other boundary was an entrance or repulsively natural.

The present paper is a summary of some recent results on the stochastic stability and bifurcation of quasi-Hamiltonian systems obtained by the present authors. In the first part of the paper, the stochastic averaging of quasi-integrable-Hamiltonian systems with real noise excitations is introduced. The expression for the largest Lyapunov exponent of the square-root of the Hamiltonian is formulated by generalizing the Khasminskii's procedure to the averaged *Itô* systems, based on which the stochastic stability and bifurcation of the original systems are studied. In the second part of the paper, an n-degree-of-freedom quasi-non-integrable-Hamiltonian system is reduced to an *Itô* equation of one-dimensional averaged Hamiltonian by using the stochastic averaging method for quasi-non-integrable-Hamiltonian systems. The necessary and sufficient conditions for the asymptotic stability in probability of the trivial solution of the original system are obtained approximately by examining the sample behaviors of the one-dimensional diffusion process of the square root of averaged Hamiltonian at the two boundaries. The relationship between the qualitative behavior of the stationary probability density of the averaged Hamiltonian and the sample behaviors of the one-dimensional diffusion process of the averaged Hamiltonian at the two boundaries is established. The Hopf bifurcation of the original system is determined approximately by examining the sample behaviors of the averaged Hamiltonian at the two boundaries.

2. Stochastic Averaging of Quasi-Integrable-Hamiltonian Systems With Real Noise Excitations

Consider a system governed by the following equations of motion

$$\dot{Q}_i = \frac{\partial H}{\partial P_i} + \varepsilon G_i + \varepsilon^{\frac{1}{2}} F_{ik}\xi_k(t)$$

$$\dot{P}_i = -\left(\frac{\partial H}{\partial Q_i} + \varepsilon g_i\right) + \varepsilon^{\frac{1}{2}} f_{ik}\xi_k(t) \tag{1}$$

$$i = 1,2,\cdots,n; k = 1,2,\cdots,m$$

where Q_i and P_i are generalized displacements and momenta, respectively; $H(\mathbf{Q},\mathbf{P})$ is a Hamiltonian with continuous second-order derivatives; G_i and g_i are differentiable functions of Q_i and P_i; F_{ik} and f_{ik} are twice-differentiable functions of Q_i and P_i; ε is a small parameter; $\xi_k(t)$ are jointly stationary and ergodic processes with correlation functions $R_{kl}(\tau)$ or with spectral densities $S_{kl}(\omega)$. Assume that the Hamiltonian system governed by Eq.(1) with $\varepsilon = 0$ is integrable and nonresonant, and has n independent integrals of motion, $\mathbf{H}=[H_1,H_2,\cdots,H_n]^T$, which are in involution. Hence, Eq.(1) describes a quasi-integrable-nonresonant-Hamiltonian system.

It has been shown (Huang and Zhu, 1998) that if the correlation times of $\xi_k(t)$ are much shorter than the relaxation times of H_r (of order of ε^{-1}), then H_r converge in probability to an n-dimensional vector of diffusion processes as $\varepsilon \to 0$ in a time interval $0 \le t \le T$, where $T \sim O(\varepsilon^{-1})$. The evolution of this vector of diffusion processes is governed by the following *Itô* stochastic differential equations

$$dH_r = \varepsilon m_r(\mathbf{H})dt + \varepsilon^{\frac{1}{2}}\sigma_{ru}(\mathbf{H})dB_u(t) \qquad r = 1,2,\cdots,n; u = 1,2,\cdots,2m \tag{2}$$

where $B_u(t)$ are independent unit Wiener processes,

$$\begin{aligned} m_r =& <(\frac{\partial H_r}{\partial Q_i}G_i - \frac{\partial H_r}{\partial P_i}g_i) + \int_{-\infty}^{0}[(\frac{\partial H_s}{\partial Q_j}F_{jl} + \frac{\partial H_s}{\partial P_j}f_{jl})\Big|_{t+\tau} \times \\ & \frac{\partial}{\partial H_s}(\frac{\partial H_r}{\partial Q_i}F_{ik} + \frac{\partial H_r}{\partial P_i}f_{ik})\Big|_t + (\frac{\partial \theta_s}{\partial Q_j}F_{jl} + \frac{\partial \theta_s}{\partial P_j}f_{jl})\Big|_{t+\tau} \times \\ & \frac{\partial}{\partial \theta_s}\left(\frac{\partial H_r}{\partial Q_i}F_{ik} + \frac{\partial H_r}{\partial P_i}f_{ik}\right)\Big|_t R_{kl}(\tau)d\tau >_t \\ \sigma_{ru}\sigma_{su} =& <\int_{-\infty}^{\infty}\left(\frac{\partial H_r}{\partial Q_i}F_{ik} + \frac{\partial H_r}{\partial P_i}f_{ik}\right)\Big|_t (\frac{\partial H_s}{\partial Q_j}F_{jl} + \frac{\partial H_s}{\partial P_j}f_{jl})\Big|_{t+\tau} R_{kl}(\tau)d\tau >_t \end{aligned} \tag{3}$$

the symbol $<[.]>_t$ represents the time-averaging operation. Note that any dynamical quantity $f(\mathbf{H},t)$ on the right sides of Eq.(3) is quasi-periodic function with incommensurable periods $T_r = 2\pi/\omega_r(\mathbf{H})$ and can be expanded as n-fold Fourier series.

3. The Largest Lyapunov Exponent of Averaged Equations

Consider the averaging equations(2). Assume that the Hamiltonian is a sum of n independent integrals of motion

$$H = \sum_{r=1}^{n} H_r(\mathbf{Q},\mathbf{P}) \tag{4}$$

and the drift and diffusion coefficients satisfy the following conditions

$$\lim_{|\mathbf{H}|\to 0} F_r(\mathbf{H}) = 0; \lim_{|\mathbf{H}|\to 0} G_{ru}(\mathbf{H}) = 0$$
$$(\mathbf{G}(\mathbf{H})\mathbf{G}^T(\mathbf{H})\mathbf{c},\mathbf{c}) \geq d|\mathbf{H}|^2|\mathbf{c}|^2 \tag{5}$$
$$km_r(\mathbf{H}) = m_r(k\mathbf{H}); k\sigma_{ru}(\mathbf{H}) = \sigma_{ru}(k\mathbf{H})$$

where $\mathbf{c}$ is an arbitrary vector and $d > 0$ is a scalar. Introduce new variables

$$\rho = \frac{1}{2}\log \mathrm{H}$$
$$\alpha_r = H_r/\mathrm{H} \tag{6}$$

The *Itô* equations for ρ and α_i can be obtained from Eq.(2) by using *Itô* differential rule as follows

$$d\rho = \varepsilon Q(\mathbf{a})dt + \varepsilon^{\frac{1}{2}}\sum_u(\mathbf{a})dB_u(t)$$
$$d\alpha_r = \varepsilon b_r(\mathbf{a})dt + \varepsilon^{\frac{1}{2}}\Phi_{ru}(\mathbf{a})dB_u(t) \tag{7}$$
$$r = 1,2,\cdots,n; u = 1,2,\cdots,2m$$

where

$$\mathbf{a} = [\alpha_1, \alpha_2, \cdots, \alpha_n]^T$$
$$Q = \sum_{r=1}^{n} m_r(\mathbf{a}) - \frac{1}{2}\sum_{r,s=1}^{n}\sum_{u=1}^{2m}\sigma_{ru}(\mathbf{a})\sigma_{su}(\mathbf{a})$$
$$b_r = -\alpha_r\sum_{j=1}^{n} m_j(\mathbf{a}) + m_r(\mathbf{a}) + \frac{\alpha_r}{2}\sum_{k_1,k_2=1}^{n}\sum_{u=1}^{2m}\sigma_{k_1u}(\mathbf{a})\sigma_{k_2u}(\mathbf{a}) - \frac{1}{2}\sum_{k=1}^{n}\sum_{u=1}^{2m}\sigma_{ku}(\mathbf{a})\sigma_{ru}(\mathbf{a}) \tag{8}$$
$$\Phi_{ru} = \sigma_{ru}(\mathbf{a}) - \alpha_r\sum_{k=1}^{n}\sigma_{ku}(\mathbf{a})$$

Note that only n-1 equations, e.g., the first n-1 equations, for α_r in Eq.(7) are independent since $\sum_{r=1}^{n}\alpha_r = 1$.

Define the Lyapunov exponents of *Itô* equation (2) as the asymptotic rate of exponential growth of the square-root of the Hamiltonian, i.e.,

$$\lambda = \lim_{t\to\infty}\frac{1}{2t}\log \mathrm{H} \tag{9}$$

which is consistent with that usually defined in terms of Euclidean norm for linear Hamiltonian systems but different for nonlinear Hamiltonian systems.

Following Khasminskii's procedure(1967), the following expression for the largest Lyapunov exponent of the averaged system(2) has been derived(Zhu and Huang, 1998a)

$$\lambda_{\max} = E[Q(\mathbf{a}')] = \int Q(\mathbf{a}')p(\mathbf{a}')d\mathbf{a}' \tag{10}$$

where $p(\mathbf{a}')$ is the invariant measure of diffusion process $\mathbf{a}' = [\alpha_1, \alpha_2, \ldots, \alpha_{n-1}]^T$, which can be obtained from solving the reduced FPK equation associated the *Itô* equations for $\mathbf{a}'$ in (7).

4. Stochastic Stability and Bifurcation of Quasi-Integrable-Hamiltonian Systems, Example

The boundary for stochastic stability and the parameter value for bifurcation of a quasi-integrable-Hamiltonian system can be determined approximately by $\lambda_{max} = 0$. Note that here the stochastic stability and asymptotic stability of the trivial solution of a quasi-integrable-Hamiltonian system are defined in terms of the square-root of the Hamiltonian rather than Euclidean norm. In the following one example is shown to demonstrate the application of Eq.(10). The application to coupled linear non-gyroscopic and gyroscopic systems can be found in Zhu and Huang(1998a) and Huang and Zhu (1998).

Consider a linear oscillator coupled with a nonlinear oscillator by dampings and stochastically parametric excitations. The equations of motion of the system are of the form

$$\begin{aligned} &\ddot{X}_1 + \beta_{11}\dot{X}_1 + \beta_{12}\dot{X}_2 + \omega_1^2 X_1 = k_{11} X_1 W_1(t) + k_{12} X_2^2 W_2(t) \\ &\ddot{X}_2 + \beta_{21}\dot{X}_1 + \beta_{22}\dot{X}_2 + \gamma X_2^3 = k_{21} X_1 W_1(t) + k_{22} X_2^2 W_2(t) \end{aligned} \tag{11}$$

where β_{ij}、 k_{ij} 、 ω_1 、 $\lambda > 0$ are constants; $W_i(t)$ are independent Gaussian white noises in the sense of Stratonovich with small intensities $2D_{ii}$. The Wong-Zakai correction terms vanish and the associated Hamiltonian is separable

$$H = H_1 + H_2 \tag{12}$$

where

$$\begin{aligned} &H_1 = \frac{1}{2}(\dot{X}_1^2 + \omega_1^2 X_1^2) \\ &H_2 = \frac{1}{2}(\dot{X}_2^2 + \frac{1}{2}\gamma X_2^4) \end{aligned} \tag{13}$$

Following the procedure in section 2 we obtained the averaged equations for H_1 and H_2 which satisfy the conditions in Eq.(5). Then, following the procedure in section 3 we obtained the largest Lyapunov exponent

$$\lambda_{max} = \int_0^1 Q(\alpha_1) p(\alpha_1) d\alpha_1 \tag{14}$$

where

$$
\begin{aligned}
&Q(\alpha_1) = (\lambda_1 + \frac{k_{21}^2 D_{11}}{\omega_1^2})\alpha_1 + (\lambda_2 + k_{12}^2\delta_1 D_{22})(1-\alpha_1) + \frac{1}{4}A\alpha_1(1-\alpha_1) \\
&p(\alpha_1) = \frac{C}{\alpha_1(1-\alpha_1)}(\frac{\alpha_1}{1-\alpha_1})^{\frac{4(\lambda_1-\lambda_2)}{A}} \exp[-\frac{2}{A}(\frac{k_{12}^2\delta_1 D_{22}}{\alpha_1} + \frac{k_{21}^2 D_{11}}{\omega_1^2(1-\alpha_1)})] \\
&\lambda_1 = -\frac{1}{2}\beta_{11} + \frac{1}{4}\frac{k_{11}^2 D_{11}}{\omega_1^2}; \\
&\lambda_2 = -\frac{1}{2}\delta_2\beta_{22} + \frac{1}{2}(\delta_1 - \delta_3)k_{22}^2 D_{22}; \\
&A = \frac{k_{11}^2 D_{11}}{\omega_1^2} + 2\delta_3 k_{22}^2 D_{22} \\
&\delta_1 = \frac{2}{3}\sqrt{\frac{2}{\gamma}} / \int_0^1 \frac{dt}{\sqrt{1-t^4}}; \delta_2 = 2\int_0^1 \sqrt{1-t^4}\,dt / \int_0^1 \frac{dt}{\sqrt{1-t^4}}; \delta_3 = \frac{4}{\gamma}\delta_2
\end{aligned}
\tag{15}
$$

and C is a normalization constant. Some numerical result for the stochastic stability boundary on plane (β_1, β_2) obtained from $\lambda_{\max} = 0$ is shown by solid lines in Fig.1, where the result from digit simulation denoted by dot is also shown for comparison.

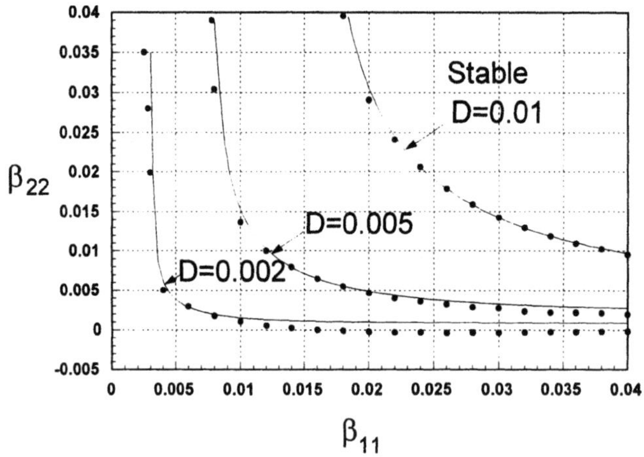

Fig 1 Stability boundaries for different values of parameter $D = D_{11} = D_{22}$; $\omega = \gamma = k_{ij} = 1; (i, j = 1,2)$

5. Stochastic Averaging of Quasi-Non-Integrable-Hamiltonian Systems

Consider a quasi-Hamiltonian system governed by the following equations of motion

$$\dot{Q}_i = \frac{\partial H'}{\partial P_i}$$
$$\dot{P}_i = -\frac{\partial H'}{\partial Q_i} - \varepsilon c_{ij}\frac{\partial H'}{\partial P_j} + \varepsilon^{1/2} f_{ik} w_k(t) \tag{16}$$
$$i, j = 1,2,\cdots,n; k = 1,2,\cdots,m$$

where Q_i and P_i are generalized displacements and momenta, respectively; $H' = H'(\mathbf{Q},\mathbf{P})$ is a Hamiltonian with continuous second-order derivatives; $c_{ij} = c_{ij}(\mathbf{Q},\mathbf{P})$ are differentiable functions; $f_{ik} = f_{ik}(\mathbf{Q},\mathbf{P})$ are twice-differentiable functions; ε is a small parameter; $W_k(t)$ are Gaussian white noises in the sense of Stratonovich with intensities $2D_{kl}$.

After adding the possible Wong-Zakai correction terms Eq.(16) is equivalent to the following *Itô* equations (Zhu and Yang, 1997)

$$dQ_i = \frac{\partial H}{\partial P_i} dt$$
$$dP_i = -(\frac{\partial H}{\partial Q_i} + \varepsilon m_{ij}\frac{\partial H}{\partial P_j})dt + \varepsilon^{1/2} f_{ik} dB_k(t) \tag{17}$$
$$i, j = 1,2,\cdots,n; k = 1,2,\cdots,m$$

Assume that the associated Hamiltonian system with Hamiltonian H governed by Eq.(17) with $\varepsilon = 0$ is nonintegrable, i.e., H is the only independent integral of motion of the Hamiltonian system.Then Eq.(17) describes a quasi-non-integrable-Hamiltonian system.

It has been shown (Zhu and Yang,1997) that the response of the system (17) converges in probability to an one-dimensional diffusion process of averaged Hamiltonian as $\varepsilon \to 0$ in a time interval $0 \le t \le T$, where $T \sim O(\varepsilon^{-1})$. The *Itô* equation or averaged Hamiltonian is of the form

$$dH = \varepsilon U(H)dt + \varepsilon^{1/2} V(H)dB(t) \tag{18}$$

where

$$U(H) = \frac{1}{T(H)V_{\Omega_1}} \int_\Omega \left[(-m_{ij}\frac{\partial H}{\partial p_i}\frac{\partial H}{\partial p_j} + D_{kl} f_{ik} f_{jl} \frac{\partial^2 H}{\partial p_i \partial p_j}) \Big/ \frac{\partial H}{\partial p_1} \right] dq_1 \cdots dq_n dp_2 \cdots dp_n$$
$$V^2(H) = \frac{1}{T(H)V_{\Omega_1}} \int_\Omega \left[2D_{kl} f_{ik} f_{jl} \frac{\partial H}{\partial p_i}\frac{\partial H}{\partial p_j} \Big/ \frac{\partial H}{\partial p_1} \right] dq_1 \cdots dq_n dp_2 \cdots dp_n$$
$$T(H) = \frac{1}{V_{\Omega_1}} \int_\Omega \left(1 \Big/ \frac{\partial H}{\partial p_1} \right) dq_1 \cdots dq_n dp_2 \cdots dp_n \tag{19}$$
$$V_{\Omega_1} = \int_{\Omega_1} dq_2 \cdots dq_n dp_2 \cdots dp_n$$
$$\Omega = \{(q_1,\cdots,q_n,p_2,\cdots,p_n) | H(q_1,\cdots,q_n,0,p_2,\cdots,p_n) \le H\}$$
$$\Omega_1 = \{(q_2,\cdots,q_n,p_2,\cdots,p_n) | H(0,q_2\cdots,q_n,0,p_2,\cdots,p_n) \le H\}$$

6. Stability in Probability of Quasi-Non-Integrable-Hamiltonian Systems

To be consistent with the definition of stochastic stability for quasi-integrable-Hamiltonian systems, the stability in probability of the trivial solution of a quasi-non-integrable-Hamiltonian systems is defined also in terms of the square-root of the Hamiltonian. Let the square-root of the averaged Hamiltonian be

$$Y(t) = H^{\frac{1}{2}}(t) \tag{20}$$

$Y(t)$ is also a one-dimensional diffusion process and governed by the following *Itô* equation

$$dY = \varepsilon a(Y)dt + \varepsilon^{1/2}\sigma(Y)dB(t) \tag{21}$$

where $a(Y)$ and $\sigma(Y)$ can be obtained from $U(H)$ and $V(H)$ in Eq.(19) by using *Itô* differential rule as follows

$$a(Y) = \frac{1}{2}Y^{-1}U(Y) - \frac{1}{8}Y^{-3}V^2(Y), b(Y) = \sigma^2(Y) = \frac{1}{4}Y^{-2}V^2(Y) \tag{22}$$

For $Y(t)$, the left boundary is at the trivial solution $Y = 0$ while the right boundary is usually at infinity, $H = \infty$, if H varies monotonously from 0 to ∞ and if there is no constraint imposed on the response of the system. For quasi-non-integrable-Hamiltonian system with parametric excitations of Gaussian white noises, the two boundaries are often singular. Stability in probability implies that most sample functions of response process remains near the trivial solution on the entire semi-infinite time domain $0 \le t < \infty$ provided they are near the trivial solution initially. Asymptotic stability in probability means that except the boundedness of most sample functions on $0 \le t < \infty$, almost all sample functions of the response process converge to the trivial solution as time goes to infinity. For one-dimensional diffusion process $Y(t)$, it is obvious that the requirements of asymptotic stability in probability can be satisfied only if the trivial boundary is an exit or attractively natural while the infinite boundary is an entrance or repulsively natural (Lin and Cai, 1995). The classification of singular boundary is determined based on the diffusion exponent , drift exponent ,and character value, which in turn are calculated using the drift and diffusion coefficients of $Y(t)$ near the singular boundary. Since the expressions for the drift and diffusion coefficients of $Y(t)$ are complicated multi-fold integrals, asymptotic analysis is necessary for determining the classification of the boundaries.

For example, consider a system of linearly and nonlinearly coupled two nonlinearly damped oscillators subject to parametric excitations of Gaussian white noises. The equations of motion are of the form

$$\begin{aligned} &\ddot{X}_1 + \varsigma_1\dot{X}_1 + k_1X_1^2\dot{X}_1 + \omega_1^2X_1 + aX_2 + b|X_1 - X_2|^\delta sign(X_1 - X_2) = C_1X_1W_1(t) \\ &\ddot{X}_2 + \varsigma_2\dot{X}_2 + k_2X_2^2\dot{X}_2 + \omega_2^2X_2 + aX_1 + b|X_2 - X_1|^\delta sign(X_2 - X_1) = C_2X_2W_2(t) \end{aligned} \tag{23}$$

where ω_i are the natural frequencies of the two uncoupled oscillators; ς_i and k_i are coefficients of linear and nonlinear dampings; a and b are the constants of linear and nonlinear couplings; δ is the power of nonlinear coupling; C_i represent the amplitudes of excitations; k_i, ς_i and C_i^2 are assumed of order ε; $W_i(t)$ are independent Gaussian white noises with intensities $2D_{ii}$.

It has been shown (Zhu and Huang, 1998b) that in the case of $\delta > 1$ and $k_i > 0$, the diffusion exponent, drift exponent and character value for the left boundary of $Y(t)$ are

$$\alpha_l = 2, \beta_l = 1$$

$$c_l = \frac{\left[-(\varsigma_1+\varsigma_2)+\frac{5}{6}\left(\frac{C_1^2 D_1}{\omega_1^2}+\frac{C_2^2 D_2}{\omega_2^2}\right)\eta\right]}{\left[\frac{1}{6}\left(\frac{C_1^2 D_1}{\omega_1^2}+\frac{C_2^2 D_2}{\omega_2^2}\right)\eta\right]} \tag{24}$$

where

$$\eta = \int_0^\pi \frac{d\theta}{\left(1+\frac{a}{\omega_1\omega_2}\sin 2\theta\right)^2} \Bigg/ \int_0^\pi \frac{d\theta}{1+\frac{a}{\omega_1\omega_2}\sin 2\theta} \tag{25}$$

According to the table for classification of singular boundary (Lin and Cai, 1995), the left boundary $Y(t)=0$ is attractively natural, strictly natural or repulsively natural depending on whether $c_l < 1, c_l = 1$ *or* $c_l > 1$ respectively. The diffusion exponent, drift exponent and character value for right boundary of $Y(t)$ are

$$\alpha_r = 2, \beta_r = 3$$

$$c_r = \frac{2(k_1+k_2)}{(C_1^2 D_1 + C_2^2 D_2)} \tag{26}$$

and $a(\infty) < 0$. The right boundary is an entrance. Thus, the necessary and sufficient condition for the asymptotic stability in probability of the trivial solution of system (23) is approximately $c_l < 1$, i.e.,

$$\varsigma_1 + \varsigma_2 > \frac{2}{3}\eta\left(\frac{C_1^2 D_1}{\omega_1^2}+\frac{C_2^2 D_2}{\omega_2^2}\right) \tag{27}$$

The stability condition of system (23) for other cases of parameter values can be obtained in a similar way(Zhu and Huang, 1998b).

7. Stochastic Hopf Bifurcation of Quasi-Non-Integrable-Hamiltonian Systems

Stochastic Hopf bifurcation has been studied for two-dimensional dynamical systems, i. g., van der Pol-Duffing oscillator with parametric excitation of Gaussian white noise (Arnold, et al., 1996).The joint stationary probability density of displacement and velocity undergoes two times changes during stochastic Hopf bifurcation. It transits from a delta function to a normalizable function with the peak at the center at a lower value of bifurcation parameter and then to a crater-like function at a higher value of bifurcation parameter. Correspondingly, the stationary probability density of the amplitude or energy envelope transits from a delta function to a normalizable function with the peak at the origin at the first bifurcation and then to a normalizable function with the peak away from the origin at the second bifurcation. Before the first bifurcation the trivial solution of the system is asymptotic stable in probability and after that it is unstable in probability. Thus, the parameter value of the first bifurcation can be determined from the necessary and sufficient conditions for the asymptotic stability in probability of the trivial solution of the system. For one-dimensional diffusion process, the necessary and sufficient condition for the asymptotic stability in probability is that the trivial solution is either an exit or attractively natural while the other boundary is either an entrance or repulsively natural as stated in the last section.

After the first bifurcation the response of the system is described by the stationary probability density. In order that a stationary probability density exists, the right boundary $H=\infty$ must still be an entrance or repulsively natural. In other words, the behavior of response near $H=\infty$ is unchanged during the second bifurcation. Thus, we can consider only the qualitative change in sample behavior of $H(t)$ near the boundary $H=0$ during the second bifurcation. It has been shown (Zhu and Huang, 1998c) that the second bifurcation can occur only if $\beta_l-\alpha_l=-1$. Under this condition

$$p(H)=O(H^{\lambda}), as H\to 0 \tag{28}$$

where

$$\lambda = c_l - \alpha_l \tag{29}$$

$p(H)$ exists with a peak at origin when $-1<\lambda<0$ and with a peak away from origin when $\lambda>0$. Thus, $\beta_l-\alpha_l=-1$ and $\lambda = c_l - \alpha_l = 0$ are the conditions for the second bifurcation.

For example, consider two linearly and nonlinearly coupled van der Pol oscillators with both stiffness and damping subject to parametric excitations of Gaussian white noises. The equations of motion are

$$\ddot{X}_1-(\varsigma_1-k_1X_1^2)\dot{X}_1+\omega_1{}^2X_1+aX_2+b|X_1-X_2|^{\delta}sign(X_1-X_2)=C_1X_1W_1(t)+C_3\dot{X}_1W_3(t)$$
$$\ddot{X}_2-(\varsigma_2-k_2X_2^2)\dot{X}_2+\omega_2{}^2X_2+aX_1+b|X_2-X_1|^{\delta}sign(X_2-X_1)=C_2X_2W_2(t)+C_4\dot{X}_2W_4(t) \tag{30}$$

System (30) is the same as (23) except parametric excitations in damping terms. ς_i are bifurcation parameters. It has been shown (Zhu and Huang, 1998c) that in the case of $\delta>1$ and $k_i>0$, the diffusion exponent, drift exponent and character value for the left boundary of $H(t)$ are

$$\alpha_l=2, \beta_l=1, c_l=\frac{\varsigma_1+\varsigma_2+C_3^2D_3+C_4^2D_4+(\frac{C_1^2D_1}{\omega_1^2}+\frac{C_2^2D_2}{\omega_2^2})\eta}{C_3^2D_3+C_4^2D_4+\frac{1}{3}(\frac{C_1^2D_1}{\omega_1^2}+\frac{C_2^2D_2}{\omega_2^2})\eta} \tag{31}$$

Thus, the left boundary is attractively natural if $c_l<1$, strictly natural if $c_l=1$, and repulsively natural if $c_l>1$. It has also been shown that the right boundary is an entrance if $k_1 + k_2 > 0$. Therefore, the first bifurcation occurs at $c_l = 1$, i.e.,

$$\varsigma_1 + \varsigma_2 = -\frac{2}{3}\eta\left(\frac{C_1^2 D_1}{\omega_1^2} + \frac{C_2^2 D_2}{\omega_2^2}\right) \tag{32}$$

which is consistent with stability condition (27). Since $\beta_l - \alpha_l = -1$ the second bifurcation occurs at $c_l = \alpha_l = 2$, i.e.,

$$\varsigma_1 + \varsigma_2 = -\frac{1}{3}\eta\left(\frac{C_1^2 D_1}{\omega_1^2} + \frac{C_2^2 D_2}{\omega_2^2}\right) + C_3^2 D_3 + C_4^2 D_4 \tag{33}$$

which is larger than that in Eq.(32)

8. Concluding Remarks

In the present paper some recent results on the stochastic stability and bifurcation of quasi-Hamiltonian systems obtained by the present authors have bean reviewed. For quasi-integrable-Hamiltonian systems, a procedure for evaluating the largest Lyapunov exponent for the averaged equations of the systems based on the Khasminskii theorem has bean proposed. The largest Lyapunov exponent can be used to approximately study the stochastic stability and bifurcation of quasi-integrable-Hamiltonian systems, linear or nonlinear, gyroscopic or non-gyroscopic. The stochastic stability and Lyapunov exponent have been defined in terms of the square-root of Hamiltonian rather than Euclidean norm. These definitions are the same as those for the almost sure stability and the Lyapunov exponent usually defined in terms of Euclidean norm when the associated Hamiltonian systems are linear,. and they are different when the associated Hamiltonian systems are nonlinear. In the later case the new definitions are physically meanful although mathematically the square-root of Hamiltonian does not satisfy the homogeneous condition of a norm. Furthermore, the new definitions make the problem of determining the stochastic stability and bifurcation of quasi-integrable-Hamiltonian systems more tractable. For quasi-non-integrable-Hamiltonian systems, a procedure for obtaining the necessary and sufficient conditions for the asymptotic stability in probability by examining the sample behaviors of one-dimensional diffusion process of the square-root of the averaged Hamiltonian at the two boundaries and a procedure for obtaining the values of parameters at which the stochastic Hopf bifurcation of the response of the systems occurs by examing the sample behaviors of one-dimensional diffusion process of the averaged Hamiltonian at the two boundaries have been proposed. The procedures are simple but asymptotic analysis is necessary for determining the classification of a boundary for the diffusion process of the averaged Hamiltonian or the square-root of averaged Hamiltonian since the drift and diffusion coefficients are expressed in terms of multi-fold integrals.

Acknowledgment

The work reported in this paper is supported by the National Natural Science Foundation of China under grant No. 19672054 and Special Fund for Doctor Programs in Institutions of Higher Learning of China and the Cai Guang Biao science foundation of Zhejiang university.

References

Arnold, L., Sri.Namachchivaya, N., and Schenk-Hoppe, K.R. (1996) Toward an understanding of stochastic Hopf bifurcation: a case study, *Int. J. Bifurcation and Chaos*, **6**(11), 1947-1975.

Ariaratnam, S.T.,and Xie, W.C.(1989) Stochastic perturbation of Pitchfork bifurcation, *Structure Safety*, **6**, 205-210.

Ariaratnam, S.T., and Xie, W.C. (1992) Lyapunov exponents and stochastic stability of coupled linear systems under real noise excitation, *ASME Journal of Applied Mechanics*, **59**, 664-673.

Huang, Z.L. and Zhu, W.Q. (1998) Lyapunov exponent and almost sure asymptotic stability of quasi-linear gyroscopic systems, submitted to *International Journal of Nonlinear Mechanics*

Khasminskii, R.Z. (1967) Sufficient and necessary conditions of almost sure asymptotic stability of a linear stochastic system, *Theory of Probability and Applications*, **11**, 395-405.

Kozin, F. and Prodromou, S. (1971) Necessary and sufficient conditions for almost sure sample stability of linear $It\hat{o}$ equations, *SIAM Journal of Applied Mathematics*. **21**, 413-424.

Kozin, F. and Zhang, Z.Y. (1990) On almost sure sample stability of nonlinear $It\hat{o}$ differential equations, In Lin, Y.K., and Elishakoff, I.(Eds.),*Stochastic Structural Dynamics, Vol.1: New Theoretical Developments*, Springer-Verlag,Berlin, 147-154.

Lin, Y.K. and Cai, G.Q. (1994) Stochastic stability of nonlinear systems, *International Journal of Non-Linear Mechanics* **29**, 539-553.

Lin, Y.K. and Cai, G.Q. (1995) *Probabilistic Structural Dynamics, Advanced Theory and Applications*,Mcgraw-Hill Inc.,New York.

Nishioka, K. (1976) On the stability of two-dimensional linear stochastic systems, *Kodai Mathematics Seminar Report*, **27**, 211-230.

Oseledec, V.I.,(1968) A multiplicative ergodic theorem: Lyapunov characteristic number for dynamic systems, *Transaction of the Moscow Mathematical Society*, **19**, 197-231.

Pardoux, E. and Wihstutz, V. (1988) Lyapunov exponent and rotation number of two-dimensional linear stochastic systems with small diffusion, *SIAM Journal of Applied Mathematics*, **48**, 442-457.

Sri Namachchivaya, N. (1989) Instability theorem based on the nature of the boundary behavior for one-dimensional diffusion, *Solid Mechanics Archives* **14**(3/4), 131-142.

Sri Namachchivaya, N. (1991) Co-dimensional two bifurcation in presence of noise, *ASME Journal of Applied Mechanics*, **58**, 259-265.

Wedig, W.V. (1988) Lyapunov exponents of stochastic systems and related bifurcation problems, In Ariaratnam,S.T.,et al.(Eds.), *Stochastic Structural Dynamics, Progress in Theory and Application*, Elsevier Applied Science, London, 315-327.

Zhang, Z.Y. (1991) New developments in almost sure sample stability of nonlinear stochastic dynamical systems, *ph.D dissertation*, Polytechnic University, New York.

Zhu, W.Q. and Huang, Z.L. (1998a) Lyapunov exponent and stochastic stability of quasi-integrable-Hamilotnian systems, to appear in *ASME Journal of Applied Mechanics*.

Zhu, W.Q. and Huang, Z.L. (1998b) Stochastic stability of quasi-non-integrable-Hamilotnian systems, to appear in *Journal of Sound and Vibration*.

Zhu, W.Q. and Huang, Z.L. (1998c) Stochastic Hopf bifurcation of quasi-non-integrable-Hamilotnian systems, to appear in *International Journal of Non-Linear Mechanics*.

Zhu,W.Q.,Huang,Z.L.,and Yang,Y.Q.,(1997) Stochastic averaging of quasi-integrable-Hamiltonian systems, *ASME Journal of Applied Mechanics*, **64**, 975-984.

Zhu,W.Q., and Yang,Y.Q.,(1997) Stochastic Averaging of quasi-non-integrable-Hamiltonian systems, *ASME Journal of Applied Mechanics*, **64**, 157-164.

List of Participants

Anh, N.D., Institute of Mechanics, Vien Co Hoc, 224, Doi Can, Hanoi, Vietnam.

Arnold, L., Institut für dynamische Systeme, Universität Bremen, Postfach 330440 D-28334 Bremen Germany.

Bajaj, A.K., School of Mechanical Engineering, Purdue University, West Lafayette IN 47907-1288, USA.

Balaji Rao, K., Concrete Composites Laboratory SERC, CSIR Campus, Taramani, Chennai 600 113, India

Bhattacharya, S.K., Ocean Engineering Centre, Indian Institute of Technology, Chennai 600 036, India

Casciati, F., Department of Structural Mechanics, University of Pavia, Via Abbiategrasso, 211-27 1 00, Pavia, Italy.

Chellapandi, P., Structural Mechanics Section IGCAR, Kalpakkam 603 102, India.

Davies, H.G., Department of Mechanical Engineering, University of New Brunswick, P.O. Box 4400, Fredericton NB E3B SA3, Canada.

Devdas Menon, Department of Civil Engineering, Indian Institute of Technology, Chennai 600 036, India

Faravelli, L., Department of Structural Mechanics, University of Pavia, Via Abbiategrasso, 211-27 1 00, Pavia, Italy.

Gupta, S., Department of Civil Engineering, Indian Institute of Science, Bangalore 560 012, India

Haldar, A., Department of Civil Engineering and Engineering Mechanics, University of Arizona Tucson, AZ 85721 USA.

Iyengar, N.C.N., Department of Mathematics Vellore Engineering College Vellore 632 014, India.

Iyengar, R.N., Central Building Research Institute, Roorkee 247667, India.

Jayatheertha Rao, K., ENTEST, IRARAT DRDL, Vignayana Kancha PO, Hyderabad 500 069, India.

Lin, Y.K., Center for Applied Stochastics Research, Florida Atlantic University, Boca Raton, FL 33431 USA.

Lutes, L.D., Department of Civil Engineering, Texas A&M University, College Station, TX 77843-3136, USA.

Ma, F., Department of Mechanical Engineering, University of California 6135 Etcheverry Hall, Berkeley, CA 94720, USA.

Manohar, C.S., Department of Civil Engineering, Indian institute of science, Bangalore, 560 012, India.

McWilliam, S., Dynamics Research Group, Department of Mechanical Engineering University of Nottingham, University Park, Nottingham, NG7 2RD, UK.

Meher Prasad, A., Department of Civil Engineering, Indian Institute Of Technology, Madras Chennai 600 036, India.

Naess, A., Department of Civil Engineering, The Norwegian Institute of Technology Rith Birkelands V. 1 a, N - 7034, Trondheim, Norway.

Nambudiripad, K.B.M., Department of Mechanical Engineering, Calicut Regional Engineering College Calicut 6, India.

Narayanan, S., Machine Dynamics Laboratory, Department of Applied Mechanics, Indian Institute of Technology, Madras Chennai 600 036, India

Nigam, N.C., Roorkee University Roorkee 247 667, India.

Pandey, M.D., Solid Mechanics Division, Faculty of Engineering, University of Waterloo, Ontario N2L3G1, Canada.

Pathak, A.G., National Institute of Ocean Technology Chennai 600 036, India.

Popp, K., Institut für Mechanik Universität Hamover, Appelstraβe 11, 30167 Hamover, Germany.

Pratap, G., Structures Division, National Aerospace Laboratory Bangalore 560 017, India

Pratap, R., Department of Mechanical Engineering Indian Institute of Science Bangalore, 560 012, India.

Rajagopal, P., Structures Division, National Aerospace Laboratory Bangalore 560 017, India.

Ramesh, M., Machine Dynamics Laboratory Department of Applied Mechanics Indian Institute of Technology, Madras Chennai 600 036, India.

Reddy, C.K., Department of Mechanical Engineering Indian Institute of Science Bangalore 560 012, India

Roberts, J.B., School of Engineering University of Sussex, Falmer, Brighton, BNI9QT, UK.

Savkoor, A.R., Vehicle Research Lab., Department of Mechanical Engineering and Marine Technology, Mckelweg 2, 2628 CD DELFT, Netherlands.

Schenk-Hoppé, K.R., Wirtschaftswissenschaften, Universität Biciefeld, 33501 Bielefeld, Germany.

Schiehlen, W., Institute B of Mechanics, University of Stuttgad, Pfaffenwaldring 9, 70550, Stuttgart, Germany.

Sharma, B.S., Defence Research and Development Laboratory Kanchanbagh, Hyderabad 500 058, India

Srinivasa, Y.G., Department of Mechanical Engineering Indian Institute of Technology, Madras Chennai 600 036, India

Wedig, W.V., Institute of Technical Mechanics, University of Karisruhe, D-76128, Karlsnffie, Germany.

Wojtkiewicz, S.F., Department of Aeronautical and Astronautical Engineering, University of Illinois at Urbana-Champaign, Urbana, Illinois 61801 USA.

Zhu, W.Q., Department of Mechanics, Zhejiang University, Hangzhou 310027, Peoples Republic of China.

Mechanics

SOLID MECHANICS AND ITS APPLICATIONS

Series Editor: G.M.L. Gladwell

Aims and Scope of the Series

The fundamental questions arising in mechanics are: *Why?*, *How?*, and *How much?* The aim of this series is to provide lucid accounts written by authoritative researchers giving vision and insight in answering these questions on the subject of mechanics as it relates to solids. The scope of the series covers the entire spectrum of solid mechanics. Thus it includes the foundation of mechanics; variational formulations; computational mechanics; statics, kinematics and dynamics of rigid and elastic bodies; vibrations of solids and structures; dynamical systems and chaos; the theories of elasticity, plasticity and viscoelasticity; composite materials; rods, beams, shells and membranes; structural control and stability; soils, rocks and geomechanics; fracture; tribology; experimental mechanics; biomechanics and machine design.

1. R.T. Haftka, Z. Gürdal and M.P. Kamat: *Elements of Structural Optimization*. 2nd rev.ed., 1990 ISBN 0-7923-0608-2
2. J.J. Kalker: *Three-Dimensional Elastic Bodies in Rolling Contact*. 1990 ISBN 0-7923-0712-7
3. P. Karasudhi: *Foundations of Solid Mechanics*. 1991 ISBN 0-7923-0772-0
4. *Not published*
5. *Not published.*
6. J.F. Doyle: *Static and Dynamic Analysis of Structures*. With an Emphasis on Mechanics and Computer Matrix Methods. 1991 ISBN 0-7923-1124-8; Pb 0-7923-1208-2
7. O.O. Ochoa and J.N. Reddy: *Finite Element Analysis of Composite Laminates*. ISBN 0-7923-1125-6
8. M.H. Aliabadi and D.P. Rooke: *Numerical Fracture Mechanics*. ISBN 0-7923-1175-2
9. J. Angeles and C.S. López-Cajún: *Optimization of Cam Mechanisms*. 1991 ISBN 0-7923-1355-0
10. D.E. Grierson, A. Franchi and P. Riva (eds.): *Progress in Structural Engineering*. 1991 ISBN 0-7923-1396-8
11. R.T. Haftka and Z. Gürdal: *Elements of Structural Optimization*. 3rd rev. and exp. ed. 1992 ISBN 0-7923-1504-9; Pb 0-7923-1505-7
12. J.R. Barber: *Elasticity*. 1992 ISBN 0-7923-1609-6; Pb 0-7923-1610-X
13. H.S. Tzou and G.L. Anderson (eds.): *Intelligent Structural Systems*. 1992 ISBN 0-7923-1920-6
14. E.E. Gdoutos: *Fracture Mechanics*. An Introduction. 1993 ISBN 0-7923-1932-X
15. J.P. Ward: *Solid Mechanics*. An Introduction. 1992 ISBN 0-7923-1949-4
16. M. Farshad: *Design and Analysis of Shell Structures*. 1992 ISBN 0-7923-1950-8
17. H.S. Tzou and T. Fukuda (eds.): *Precision Sensors, Actuators and Systems*. 1992 ISBN 0-7923-2015-8
18. J.R. Vinson: *The Behavior of Shells Composed of Isotropic and Composite Materials*. 1993 ISBN 0-7923-2113-8
19. H.S. Tzou: *Piezoelectric Shells*. Distributed Sensing and Control of Continua. 1993 ISBN 0-7923-2186-3
20. W. Schiehlen (ed.): *Advanced Multibody System Dynamics*. Simulation and Software Tools. 1993 ISBN 0-7923-2192-8
21. C.-W. Lee: *Vibration Analysis of Rotors*. 1993 ISBN 0-7923-2300-9
22. D.R. Smith: *An Introduction to Continuum Mechanics*. 1993 ISBN 0-7923-2454-4
23. G.M.L. Gladwell: *Inverse Problems in Scattering*. An Introduction. 1993 ISBN 0-7923-2478-1

Mechanics

SOLID MECHANICS AND ITS APPLICATIONS

Series Editor: G.M.L. Gladwell

24. G. Prathap: *The Finite Element Method in Structural Mechanics*. 1993 ISBN 0-7923-2492-7
25. J. Herskovits (ed.): *Advances in Structural Optimization*. 1995 ISBN 0-7923-2510-9
26. M.A. González-Palacios and J. Angeles: *Cam Synthesis*. 1993 ISBN 0-7923-2536-2
27. W.S. Hall: *The Boundary Element Method*. 1993 ISBN 0-7923-2580-X
28. J. Angeles, G. Hommel and P. Kovács (eds.): *Computational Kinematics*. 1993 ISBN 0-7923-2585-0
29. A. Curnier: *Computational Methods in Solid Mechanics*. 1994 ISBN 0-7923-2761-6
30. D.A. Hills and D. Nowell: *Mechanics of Fretting Fatigue*. 1994 ISBN 0-7923-2866-3
31. B. Tabarrok and F.P.J. Rimrott: *Variational Methods and Complementary Formulations in Dynamics*. 1994 ISBN 0-7923-2923-6
32. E.H. Dowell (ed.), E.F. Crawley, H.C. Curtiss Jr., D.A. Peters, R. H. Scanlan and F. Sisto: *A Modern Course in Aeroelasticity*. Third Revised and Enlarged Edition. 1995 ISBN 0-7923-2788-8; Pb: 0-7923-2789-6
33. A. Preumont: *Random Vibration and Spectral Analysis*. 1994 ISBN 0-7923-3036-6
34. J.N. Reddy (ed.): *Mechanics of Composite Materials*. Selected works of Nicholas J. Pagano. 1994 ISBN 0-7923-3041-2
35. A.P.S. Selvadurai (ed.): *Mechanics of Poroelastic Media*. 1996 ISBN 0-7923-3329-2
36. Z. Mróz, D. Weichert, S. Dorosz (eds.): *Inelastic Behaviour of Structures under Variable Loads*. 1995 ISBN 0-7923-3397-7
37. R. Pyrz (ed.): *IUTAM Symposium on Microstructure-Property Interactions in Composite Materials*. Proceedings of the IUTAM Symposium held in Aalborg, Denmark. 1995 ISBN 0-7923-3427-2
38. M.I. Friswell and J.E. Mottershead: *Finite Element Model Updating in Structural Dynamics*. 1995 ISBN 0-7923-3431-0
39. D.F. Parker and A.H. England (eds.): *IUTAM Symposium on Anisotropy, Inhomogeneity and Nonlinearity in Solid Mechanics*. Proceedings of the IUTAM Symposium held in Nottingham, U.K. 1995 ISBN 0-7923-3594-5
40. J.-P. Merlet and B. Ravani (eds.): *Computational Kinematics '95*. 1995 ISBN 0-7923-3673-9
41. L.P. Lebedev, I.I. Vorovich and G.M.L. Gladwell: *Functional Analysis*. Applications in Mechanics and Inverse Problems. 1996 ISBN 0-7923-3849-9
42. J. Menčik: *Mechanics of Components with Treated or Coated Surfaces*. 1996 ISBN 0-7923-3700-X
43. D. Bestle and W. Schiehlen (eds.): *IUTAM Symposium on Optimization of Mechanical Systems*. Proceedings of the IUTAM Symposium held in Stuttgart, Germany. 1996 ISBN 0-7923-3830-8
44. D.A. Hills, P.A. Kelly, D.N. Dai and A.M. Korsunsky: *Solution of Crack Problems*. The Distributed Dislocation Technique. 1996 ISBN 0-7923-3848-0
45. V.A. Squire, R.J. Hosking, A.D. Kerr and P.J. Langhorne: *Moving Loads on Ice Plates*. 1996 ISBN 0-7923-3953-3
46. A. Pineau and A. Zaoui (eds.): *IUTAM Symposium on Micromechanics of Plasticity and Damage of Multiphase Materials*. Proceedings of the IUTAM Symposium held in Sèvres, Paris, France. 1996 ISBN 0-7923-4188-0
47. A. Naess and S. Krenk (eds.): *IUTAM Symposium on Advances in Nonlinear Stochastic Mechanics*. Proceedings of the IUTAM Symposium held in Trondheim, Norway. 1996 ISBN 0-7923-4193-7
48. D. Ieşan and A. Scalia: *Thermoelastic Deformations*. 1996 ISBN 0-7923-4230-5

Mechanics

SOLID MECHANICS AND ITS APPLICATIONS

Series Editor: G.M.L. Gladwell

49. J.R. Willis (ed.): *IUTAM Symposium on Nonlinear Analysis of Fracture*. Proceedings of the IUTAM Symposium held in Cambridge, U.K. 1997 ISBN 0-7923-4378-6
50. A. Preumont: *Vibration Control of Active Structures*. An Introduction. 1997 ISBN 0-7923-4392-1
51. G.P. Cherepanov: *Methods of Fracture Mechanics: Solid Matter Physics*. 1997 ISBN 0-7923-4408-1
52. D.H. van Campen (ed.): *IUTAM Symposium on Interaction between Dynamics and Control in Advanced Mechanical Systems*. Proceedings of the IUTAM Symposium held in Eindhoven, The Netherlands. 1997 ISBN 0-7923-4429-4
53. N.A. Fleck and A.C.F. Cocks (eds.): *IUTAM Symposium on Mechanics of Granular and Porous Materials*. Proceedings of the IUTAM Symposium held in Cambridge, U.K. 1997 ISBN 0-7923-4553-3
54. J. Roorda and N.K. Srivastava (eds.): *Trends in Structural Mechanics*. Theory, Practice, Education. 1997 ISBN 0-7923-4603-3
55. Yu.A. Mitropolskii and N. Van Dao: *Applied Asymptotic Methods in Nonlinear Oscillations*. 1997 ISBN 0-7923-4605-X
56. C. Guedes Soares (ed.): *Probabilistic Methods for Structural Design*. 1997 ISBN 0-7923-4670-X
57. D. François, A. Pineau and A. Zaoui: *Mechanical Behaviour of Materials*. Volume I: Elasticity and Plasticity. 1998 ISBN 0-7923-4894-X
58. D. François, A. Pineau and A. Zaoui: *Mechanical Behaviour of Materials*. Volume II: Viscoplasticity, Damage, Fracture and Contact Mechanics. 1998 ISBN 0-7923-4895-8
59. L.T. Tenek and J. Argyris: *Finite Element Analysis for Composite Structures*. 1998 ISBN 0-7923-4899-0
60. Y.A. Bahei-El-Din and G.J. Dvorak (eds.): *IUTAM Symposium on Transformation Problems in Composite and Active Materials*. Proceedings of the IUTAM Symposium held in Cairo, Egypt. 1998 ISBN 0-7923-5122-3
61. I.G. Goryacheva: *Contact Mechanics in Tribology*. 1998 ISBN 0-7923-5257-2
62. O.T. Bruhns and E. Stein (eds.): *IUTAM Symposium on Micro- and Macrostructural Aspects of Thermoplasticity*. Proceedings of the IUTAM Symposium held in Bochum, Germany. 1999 ISBN 0-7923-5265-3
63. F.C. Moon: *IUTAM Symposium on New Applications of Nonlinear and Chaotic Dynamics in Mechanics*. Proceedings of the IUTAM Symposium held in Ithaca, NY, USA. 1998 ISBN 0-7923-5276-9
64. R. Wang: *IUTAM Symposium on Rheology of Bodies with Defects*. Proceedings of the IUTAM Symposium held in Beijing, China. 1999 ISBN 0-7923-5297-1
65. Yu.I. Dimitrienko: *Thermomechanics of Composites under High Temperatures*. 1999 ISBN 0-7923-4899-0
66. P. Argoul, M. Frémond and Q.S. Nguyen (eds.): *IUTAM Symposium on Variations of Domains and Free-Boundary Problems in Solid Mechanics*. Proceedings of the IUTAM Symposium held in Paris, France. 1999 ISBN 0-7923-5450-8
67. F.J. Fahy and W.G. Price (eds.): *IUTAM Symposium on Statistical Energy Analysis*. Proceedings of the IUTAM Symposium held in Southampton, U.K. 1999 ISBN 0-7923-5457-5
68. H.A. Mang and F.G. Rammerstorfer (eds.): *IUTAM Symposium on Discretization Methods in Structural Mechanics*. Proceedings of the IUTAM Symposium held in Vienna, Austria. 1999 ISBN 0-7923-5591-1

Mechanics

SOLID MECHANICS AND ITS APPLICATIONS

Series Editor: G.M.L. Gladwell

69. P. Pedersen and M.P. Bendsøe (eds.): *IUTAM Symposium on Synthesis in Bio Solid Mechanics*. Proceedings of the IUTAM Symposium held in Copenhagen, Denmark. 1999 ISBN 0-7923-5615-2
70. S.K. Agrawal and B.C. Fabien: *Optimization of Dynamic Systems*. 1999 ISBN 0-7923-5681-0
71. A. Carpinteri: *Nonlinear Crack Models for Nonmetallic Materials*. 1999 ISBN 0-7923-5750-7
72. F. Pfeifer (ed.): *IUTAM Symposium on Unilateral Multibody Contacts*. Proceedings of the IUTAM Symposium held in Munich, Germany. 1999 ISBN 0-7923-6030-3
73. E. Lavendelis and M. Zakrzhevsky (eds.): *IUTAM/IFToMM Symposium on Synthesis of Nonlinear Dynamical Systems*. Proceedings of the IUTAM/IFToMM Symposium held in Riga, Latvia. 2000 ISBN 0-7923-6106-7
74. J.-P. Merlet: *Parallel Robots*. 2000 ISBN 0-7923-6308-6
75. J.T. Pindera: *Techniques of Tomographic Isodyne Stress Analysis*. 2000 ISBN 0-7923-6388-4
76. G.A. Maugin, R. Drouot and F. Sidoroff (eds.): *Continuum Thermomechanics*. The Art and Science of Modelling Material Behaviour. 2000 ISBN 0-7923-6407-4
77. N. Van Dao and E.J. Kreuzer (eds.): *IUTAM Symposium on Recent Developments in Non-linear Oscillations of Mechanical Systems*. 2000 ISBN 0-7923-6470-8
78. S.D. Akbarov and A.N. Guz: *Mechanics of Curved Composites*. 2000 ISBN 0-7923-6477-5
79. M.B. Rubin: *Cosserat Theories: Shells, Rods and Points*. 2000 ISBN 0-7923-6489-9
80. S. Pellegrino and S.D. Guest (eds.): *IUTAM-IASS Symposium on Deployable Structures: Theory and Applications*. Proceedings of the IUTAM-IASS Symposium held in Cambridge, U.K., 6–9 September 1998. 2000 ISBN 0-7923-6516-X
81. A.D. Rosato and D.L. Blackmore (eds.): *IUTAM Symposium on Segregation in Granular Flows*. Proceedings of the IUTAM Symposium held in Cape May, NJ, U.S.A., June 5–10, 1999. 2000 ISBN 0-7923-6547-X
82. A. Lagarde (ed.): *IUTAM Symposium on Advanced Optical Methods and Applications in Solid Mechanics*. Proceedings of the IUTAM Symposium held in Futuroscope, Poitiers, France, August 31–September 4, 1998. 2000 ISBN 0-7923-6604-2
83. D. Weichert and G. Maier (eds.): *Inelastic Analysis of Structures under Variable Loads*. Theory and Engineering Applications. 2000 ISBN 0-7923-6645-X
84. T.-J. Chuang and J.W. Rudnicki (eds.): *Multiscale Deformation and Fracture in Materials and Structures*. The James R. Rice 60th Anniversary Volume. 2001 ISBN 0-7923-6718-9
85. S. Narayanan and R.N. Iyengar (eds.): *IUTAM Symposium on Nonlinearity and Stochastic Structural Dynamics*. Proceedings of the IUTAM Symposium held in Madras, Chennai, India, 4–8 January 1999 ISBN 0-7923-6733-2
86. S. Murakami and N. Ohno (eds.): *IUTAM Symposium on Creep in Structures*. Proceedings of the IUTAM Symposium held in Nagoya, Japan, 3-7 April 2000. 2001 ISBN 0-7923-6737-5
87. W. Ehlers (ed.): *IUTAM Symposium on Theoretical and Numerical Methods in Continuum Mechanics of Porous Materials*. Proceedings of the IUTAM Symposium held at the University of Stuttgart, Germany, September 5-10, 1999. 2001 ISBN 0-7923-6766-9

Kluwer Academic Publishers – Dordrecht / Boston / London

Mechanics

FLUID MECHANICS AND ITS APPLICATIONS

Series Editor: R. Moreau

Aims and Scope of the Series

The purpose of this series is to focus on subjects in which fluid mechanics plays a fundamental role. As well as the more traditional applications of aeronautics, hydraulics, heat and mass transfer etc., books will be published dealing with topics which are currently in a state of rapid development, such as turbulence, suspensions and multiphase fluids, super and hypersonic flows and numerical modelling techniques. It is a widely held view that it is the interdisciplinary subjects that will receive intense scientific attention, bringing them to the forefront of technological advancement. Fluids have the ability to transport matter and its properties as well as transmit force, therefore fluid mechanics is a subject that is particularly open to cross fertilisation with other sciences and disciplines of engineering. The subject of fluid mechanics will be highly relevant in domains such as chemical, metallurgical, biological and ecological engineering. This series is particularly open to such new multidisciplinary domains.

1. M. Lesieur: *Turbulence in Fluids*. 2nd rev. ed., 1990 ISBN 0-7923-0645-7
2. O. Métais and M. Lesieur (eds.): *Turbulence and Coherent Structures*. 1991 ISBN 0-7923-0646-5
3. R. Moreau: *Magnetohydrodynamics*. 1990 ISBN 0-7923-0937-5
4. E. Coustols (ed.): *Turbulence Control by Passive Means*. 1990 ISBN 0-7923-1020-9
5. A.A. Borissov (ed.): *Dynamic Structure of Detonation in Gaseous and Dispersed Media*. 1991 ISBN 0-7923-1340-2
6. K.-S. Choi (ed.): *Recent Developments in Turbulence Management*. 1991 ISBN 0-7923-1477-8
7. E.P. Evans and B. Coulbeck (eds.): *Pipeline Systems*. 1992 ISBN 0-7923-1668-1
8. B. Nau (ed.): *Fluid Sealing*. 1992 ISBN 0-7923-1669-X
9. T.K.S. Murthy (ed.): *Computational Methods in Hypersonic Aerodynamics*. 1992 ISBN 0-7923-1673-8
10. R. King (ed.): *Fluid Mechanics of Mixing*. Modelling, Operations and Experimental Techniques. 1992 ISBN 0-7923-1720-3
11. Z. Han and X. Yin: *Shock Dynamics*. 1993 ISBN 0-7923-1746-7
12. L. Svarovsky and M.T. Thew (eds.): *Hydroclones*. Analysis and Applications. 1992 ISBN 0-7923-1876-5
13. A. Lichtarowicz (ed.): *Jet Cutting Technology*. 1992 ISBN 0-7923-1979-6
14. F.T.M. Nieuwstadt (ed.): *Flow Visualization and Image Analysis*. 1993 ISBN 0-7923-1994-X
15. A.J. Saul (ed.): *Floods and Flood Management*. 1992 ISBN 0-7923-2078-6
16. D.E. Ashpis, T.B. Gatski and R. Hirsh (eds.): *Instabilities and Turbulence in Engineering Flows*. 1993 ISBN 0-7923-2161-8
17. R.S. Azad: *The Atmospheric Boundary Layer for Engineers*. 1993 ISBN 0-7923-2187-1
18. F.T.M. Nieuwstadt (ed.): *Advances in Turbulence IV*. 1993 ISBN 0-7923-2282-7
19. K.K. Prasad (ed.): *Further Developments in Turbulence Management*. 1993 ISBN 0-7923-2291-6
20. Y.A. Tatarchenko: *Shaped Crystal Growth*. 1993 ISBN 0-7923-2419-6
21. J.P. Bonnet and M.N. Glauser (eds.): *Eddy Structure Identification in Free Turbulent Shear Flows*. 1993 ISBN 0-7923-2449-8
22. R.S. Srivastava: *Interaction of Shock Waves*. 1994 ISBN 0-7923-2920-1
23. J.R. Blake, J.M. Boulton-Stone and N.H. Thomas (eds.): *Bubble Dynamics and Interface Phenomena*. 1994 ISBN 0-7923-3008-0

Mechanics

FLUID MECHANICS AND ITS APPLICATIONS

Series Editor: R. Moreau

24. R. Benzi (ed.): *Advances in Turbulence V.* 1995 ISBN 0-7923-3032-3
25. B.I. Rabinovich, V.G. Lebedev and A.I. Mytarev: *Vortex Processes and Solid Body Dynamics.* The Dynamic Problems of Spacecrafts and Magnetic Levitation Systems. 1994 ISBN 0-7923-3092-7
26. P.R. Voke, L. Kleiser and J.-P. Chollet (eds.): *Direct and Large-Eddy Simulation I.* Selected papers from the First ERCOFTAC Workshop on Direct and Large-Eddy Simulation. 1994 ISBN 0-7923-3106-0
27. J.A. Sparenberg: *Hydrodynamic Propulsion and its Optimization.* Analytic Theory. 1995 ISBN 0-7923-3201-6
28. J.F. Dijksman and G.D.C. Kuiken (eds.): *IUTAM Symposium on Numerical Simulation of Non-Isothermal Flow of Viscoelastic Liquids.* Proceedings of an IUTAM Symposium held in Kerkrade, The Netherlands. 1995 ISBN 0-7923-3262-8
29. B.M. Boubnov and G.S. Golitsyn: *Convection in Rotating Fluids.* 1995 ISBN 0-7923-3371-3
30. S.I. Green (ed.): *Fluid Vortices.* 1995 ISBN 0-7923-3376-4
31. S. Morioka and L. van Wijngaarden (eds.): *IUTAM Symposium on Waves in Liquid/Gas and Liquid/Vapour Two-Phase Systems.* 1995 ISBN 0-7923-3424-8
32. A. Gyr and H.-W. Bewersdorff: *Drag Reduction of Turbulent Flows by Additives.* 1995 ISBN 0-7923-3485-X
33. Y.P. Golovachov: *Numerical Simulation of Viscous Shock Layer Flows.* 1995 ISBN 0-7923-3626-7
34. J. Grue, B. Gjevik and J.E. Weber (eds.): *Waves and Nonlinear Processes in Hydrodynamics.* 1996 ISBN 0-7923-4031-0
35. P.W. Duck and P. Hall (eds.): *IUTAM Symposium on Nonlinear Instability and Transition in Three-Dimensional Boundary Layers.* 1996 ISBN 0-7923-4079-5
36. S. Gavrilakis, L. Machiels and P.A. Monkewitz (eds.): *Advances in Turbulence VI.* Proceedings of the 6th European Turbulence Conference. 1996 ISBN 0-7923-4132-5
37. K. Gersten (ed.): *IUTAM Symposium on Asymptotic Methods for Turbulent Shear Flows at High Reynolds Numbers.* Proceedings of the IUTAM Symposium held in Bochum, Germany. 1996 ISBN 0-7923-4138-4
38. J. Verhás: *Thermodynamics and Rheology.* 1997 ISBN 0-7923-4251-8
39. M. Champion and B. Deshaies (eds.): *IUTAM Symposium on Combustion in Supersonic Flows.* Proceedings of the IUTAM Symposium held in Poitiers, France. 1997 ISBN 0-7923-4313-1
40. M. Lesieur: *Turbulence in Fluids.* Third Revised and Enlarged Edition. 1997 ISBN 0-7923-4415-4; Pb: 0-7923-4416-2
41. L. Fulachier, J.L. Lumley and F. Anselmet (eds.): *IUTAM Symposium on Variable Density Low-Speed Turbulent Flows.* Proceedings of the IUTAM Symposium held in Marseille, France. 1997 ISBN 0-7923-4602-5
42. B.K. Shivamoggi: *Nonlinear Dynamics and Chaotic Phenomena.* An Introduction. 1997 ISBN 0-7923-4772-2
43. H. Ramkissoon, *IUTAM Symposium on Lubricated Transport of Viscous Materials.* Proceedings of the IUTAM Symposium held in Tobago, West Indies. 1998 ISBN 0-7923-4897-4
44. E. Krause and K. Gersten, *IUTAM Symposium on Dynamics of Slender Vortices.* Proceedings of the IUTAM Symposium held in Aachen, Germany. 1998 ISBN 0-7923-5041-3
45. A. Biesheuvel and G.J.F. van Heyst (eds.): *In Fascination of Fluid Dynamics.* A Symposium in honour of Leen van Wijngaarden. 1998 ISBN 0-7923-5078-2

Mechanics

FLUID MECHANICS AND ITS APPLICATIONS

Series Editor: R. Moreau

46. U. Frisch (ed.): *Advances in Turbulence VII*. Proceedings of the Seventh European Turbulence Conference, held in Saint-Jean Cap Ferrat, 30 June–3 July 1998. 1998 ISBN 0-7923-5115-0
47. E.F. Toro and J.F. Clarke: *Numerical Methods for Wave Propagation*. Selected Contributions from the Workshop held in Manchester, UK. 1998 ISBN 0-7923-5125-8
48. A. Yoshizawa: *Hydrodynamic and Magnetohydrodynamic Turbulent Flows*. Modelling and Statistical Theory. 1998 ISBN 0-7923-5225-4
49. T.L. Geers (ed.): *IUTAM Symposium on Computational Methods for Unbounded Domains*. 1998 ISBN 0-7923-5266-1
50. Z. Zapryanov and S. Tabakova: *Dynamics of Bubbles, Drops and Rigid Particles*. 1999 ISBN 0-7923-5347-1
51. A. Alemany, Ph. Marty and J.P. Thibault (eds.): *Transfer Phenomena in Magnetohydrodynamic and Electroconducting Flows*. 1999 ISBN 0-7923-5532-6
52. J.N. Sørensen, E.J. Hopfinger and N. Aubry (eds.): *IUTAM Symposium on Simulation and Identification of Organized Structures in Flows*. 1999 ISBN 0-7923-5603-9
53. G.E.A. Meier and P.R. Viswanath (eds.): *IUTAM Symposium on Mechanics of Passive and Active Flow Control*. 1999 ISBN 0-7923-5928-3
54. D. Knight and L. Sakell (eds.): *Recent Advances in DNS and LES*. 1999 ISBN 0-7923-6004-4
55. P. Orlandi: *Fluid Flow Phenomena*. A Numerical Toolkit. 2000 ISBN 0-7923-6095-8
56. M. Stanislas, J. Kompenhans and J. Westerveel (eds.): *Particle Image Velocimetry*. Progress towards Industrial Application. 2000 ISBN 0-7923-6160-1
57. H.-C. Chang (ed.): *IUTAM Symposium on Nonlinear Waves in Multi-Phase Flow*. 2000 ISBN 0-7923-6454-6

Kluwer Academic Publishers – Dordrecht / Boston / London

ERCOFTAC SERIES

1. A. Gyr and F.-S. Rys (eds.): *Diffusion and Transport of Pollutants in Atmospheric Mesoscale Flow Fields*. 1995 ISBN 0-7923-3260-1
2. M. Hallbäck, D.S. Henningson, A.V. Johansson and P.H. Alfredsson (eds.): *Turbulence and Transition Modelling*. Lecture Notes from the ERCOFTAC/IUTAM Summerschool held in Stockholm. 1996 ISBN 0-7923-4060-4
3. P. Wesseling (ed.): *High Performance Computing in Fluid Dynamics*. Proceedings of the Summerschool held in Delft, The Netherlands. 1996 ISBN 0-7923-4063-9
4. Th. Dracos (ed.): *Three-Dimensional Velocity and Vorticity Measuring and Image Analysis Techniques*. Lecture Notes from the Short Course held in Zürich, Switzerland. 1996 ISBN 0-7923-4256-9
5. J.-P. Chollet, P.R. Voke and L. Kleiser (eds.): *Direct and Large-Eddy Simulation II*. Proceedings of the ERCOFTAC Workshop held in Grenoble, France. 1997 ISBN 0-7923-4687-4

KLUWER ACADEMIC PUBLISHERS – DORDRECHT / BOSTON / LONDON

ICASE/LaRC Interdisciplinary Series in Science and Engineering

1. J. Buckmaster, T.L. Jackson and A. Kumar (eds.): *Combustion in High-Speed Flows.* 1994 ISBN 0-7923-2086-X
2. M.Y. Hussaini, T.B. Gatski and T.L. Jackson (eds.): *Transition, Turbulence and Combustion.* Volume I: Transition. 1994 ISBN 0-7923-3084-6; set 0-7923-3086-2
3. M.Y. Hussaini, T.B. Gatski and T.L. Jackson (eds.): *Transition, Turbulence and Combustion.* Volume II: Turbulence and Combustion. 1994 ISBN 0-7923-3085-4; set 0-7923-3086-2
4. D.E. Keyes, A. Sameh and V. Venkatakrishnan (eds): *Parallel Numerical Algorithms.* 1997 ISBN 0-7923-4282-8
5. T.G. Campbell, R.A. Nicolaides and M.D. Salas (eds.): *Computational Electromagnetics and Its Applications.* 1997 ISBN 0-7923-4733-1
6. V. Venkatakrishnan, M.D. Salas and S.R. Chakravarthy (eds.): *Barriers and Challenges in Computational Fluid Dynamics.* 1998 ISBN 0-7923-4855-9
7. M.D. Salas, J.N. Hefner and L. Sakell (eds.): *Modeling Complex Turbulent Flows.* 1999 ISBN 0-7923-5590-3

KLUWER ACADEMIC PUBLISHERS – DORDRECHT / BOSTON / LONDON